VOLUME FIFTY THREE

ADVANCES IN ECOLOGICAL RESEARCH

Ecosystem Services: From Biodiversity to Society, Part 1

ADVANCES IN ECOLOGICAL RESEARCH

Series Editor

GUY WOODWARD
Professor of Ecology
Imperial College London
Silwood Park Campus
Buckhurst Road
Ascot Berkshire
SL5 7PY United Kingdom

VOLUME FIFTY THREE

Advances in ECOLOGICAL RESEARCH

Ecosystem Services: From Biodiversity to Society, Part 1

Edited by

GUY WOODWARD
Department of Life Sciences, Imperial College London, Ascot, United Kingdom

DAVID A. BOHAN
UMR 1347 Agroécologie, AgroSup/UB/INRA, Pôle Ecologie des Communautés et Durabilité de Systèmes Agricoles, Dijon Cedex, France

AMSTERDAM • BOSTON • HEIDELBERG • LONDON
NEW YORK • OXFORD • PARIS • SAN DIEGO
SAN FRANCISCO • SINGAPORE • SYDNEY • TOKYO
Academic Press is an imprint of Elsevier

Academic Press is an imprint of Elsevier
125 London Wall, London, EC2Y 5AS, UK
The Boulevard, Langford Lane, Kidlington, Oxford OX5 1GB, UK
525 B Street, Suite 1800, San Diego, CA 92101-4495, USA
225 Wyman Street, Waltham, MA 02451, USA

First edition 2015

Notices
Knowledge and best practice in this field are constantly changing. As new research and experience broaden our understanding, changes in research methods, professional practices, or medical treatment may become necessary.

Practitioners and researchers must always rely on their own experience and knowledge in evaluating and using any information, methods, compounds, or experiments described herein. In using such information or methods they should be mindful of their own safety and the safety of others, including parties for whom they have a professional responsibility.

To the fullest extent of the law, neither the Publisher nor the authors, contributors, or editors, assume any liability for any injury and/or damage to persons or property as a matter of products liability, negligence or otherwise, or from any use or operation of any methods, products, instructions, or ideas contained in the material herein.

ISBN: 978-0-12-803885-7
ISSN: 0065-2504

For information on all Academic Press publications visit our website at http://store.elsevier.com/

CONTENTS

Contributors *ix*
Preface *xv*

1. 10 Years Later: Revisiting Priorities for Science and Society a Decade After the Millennium Ecosystem Assessment **1**

Christian Mulder, Elena M. Bennett, David A. Bohan, Michael Bonkowski, Stephen R. Carpenter, Rachel Chalmers, Wolfgang Cramer, Isabelle Durance, Nico Eisenhauer, Colin Fontaine, Alison J. Haughton, Jean-Paul Hettelingh, Jes Hines, Sébastien Ibanez, Erik Jeppesen, Jennifer Adams Krumins, Athen Ma, Giorgio Mancinelli, François Massol, Órla McLaughlin, Shahid Naeem, Unai Pascual, Josep Peñuelas, Nathalie Pettorelli, Michael J.O. Pocock, Dave Raffaelli, Jes J. Rasmussen, Graciela M. Rusch, Christoph Scherber, Heikki Setälä, William J. Sutherland, Corinne Vacher, Winfried Voigt, J. Arie Vonk, Stephen A. Wood, and Guy Woodward

1. Introduction 3
2. Impact of the MEA 7
3. Functional Attributes and Networks as Frames for Ecosystems and Societies 11
4. Network Approaches to ESs as a Means of Implementing the MEA 15
5. Research Priorities One Decade After the MEA 22
6. Preliminary Conclusions 40

Acknowledgements 41
References 41

2. Linking Biodiversity, Ecosystem Functioning and Services, and Ecological Resilience: Towards an Integrative Framework for Improved Management **55**

Amélie Truchy, David G. Angeler, Ryan A. Sponseller, Richard K. Johnson, and Brendan G. McKie

1. Introduction 56
2. Drivers of Ecosystem Functioning 60
3. Adding Spatiotemporal Dimensions 70
4. Extending and Parameterising a Trait-Based Framework for Predicting Functional Redundancy and Outcomes for Ecosystem Functioning and Services 77

5. Concluding Remarks 82
Acknowledgements 85
References 86

3. Detrital Dynamics and Cascading Effects on Supporting Ecosystem Services 97

Giorgio Mancinelli and Christian Mulder

1. Introduction 98
2. Data Analysis 105
3. Discussion 121
4. Future Ecosystem Services Research 124
Acknowledgements 127
Appendix 127
References 148

4. Towards an Integration of Biodiversity–Ecosystem Functioning and Food Web Theory to Evaluate Relationships between Multiple Ecosystem Services 161

Jes Hines, Wim H. van der Putten, Gerlinde B. De Deyn, Cameron Wagg, Winfried Voigt, Christian Mulder, Wolfgang W. Weisser, Jan Engel, Carlos Melian, Stefan Scheu, Klaus Birkhofer, Anne Ebeling, Christoph Scherber, and Nico Eisenhauer

1. Introduction 162
2. Contributions and Limitations of BEF and FWT 163
3. Principles for Integrating BEF and FWT 181
4. Considering Trends in BEF–FWT Research for Better Management of Multiple ESs 184
5. Conclusions 187
Acknowledgements 187
References 187

5. Persistence of Plants and Pollinators in the Face of Habitat Loss: Insights from Trait-Based Metacommunity Models 201

Julia Astegiano, Paulo R. Guimarães Jr., Pierre-Olivier Cheptou, Mariana Morais Vidal, Camila Yumi Mandai, Lorena Ashworth, and François Massol

1. Introduction 202
2. A Trait-Based Metacommunity Model to Understand Plant and Pollinator Persistence in the Face of Habitat Loss 212
3. Results 220

4. Discussion 231
5. Future Directions: Pollination Services in Human-Dominated Landscapes 241
Acknowledgements 242
Appendix A. Generating Bipartite Incidence Matrices with Determined Degree Sequences 243
Appendix B. Nestedness Depends on the Distribution of Degrees 247
References 248

6. A Network-Based Method to Detect Patterns of Local Crop Biodiversity: Validation at the Species and Infra-Species Levels 259

Mathieu Thomas, Nicolas Verzelen, Pierre Barbillon, Oliver T. Coomes, Sophie Caillon, Doyle McKey, Marianne Elias, Eric Garine, Christine Raimond, Edmond Dounias, Devra Jarvis, Jean Wencélius, Christian Leclerc, Vanesse Labeyrie, Pham Hung Cuong, Nguyen Thi Ngoc Hue, Bhuwon Sthapit, Ram Bahadur Rana, Adeline Barnaud, Chloé Violon, Luis Manuel Arias Reyes, Luis Latournerie Moreno, Paola De Santis, and François Massol

1. Introduction 261
2. Description of the Datasets Used in the Meta-Analysis 266
3. Description of the Methodological Framework 274
4. Patterns of Local Crop Diversity: Results of the Meta-Analysis 292
5. Discussion 300
6. Conclusion 306
Acknowledgements 307
Appendix 308
Glossary 316
References 316

Index *321*
Cumulative List of Titles *327*

CONTRIBUTORS

David G. Angeler
Department of Aquatic Sciences and Assessment, Swedish University of Agricultural Sciences, Uppsala, Sweden

Lorena Ashworth
Instituto Multidisciplinario de Biología Vegetal, Universidad Nacional de Córdoba–CONICET, Córdoba, Argentina

Julia Astegiano
Departamento de Ecologia, Instituto de Biociências, Universidade de São Paulo, São Paulo, Brazil; CEFE UMR 5175, CNRS—Université de Montpellier—Université Paul-Valéry Montpellier—EPHE campus CNRS, Montpellier, France, and Instituto Multidisciplinario de Biología Vegetal, Universidad Nacional de Córdoba–CONICET, Córdoba, Argentina

Pierre Barbillon
AgroParisTech/UMR INRA MIA, Paris, France

Adeline Barnaud
IRD, UMR DIADE, Montpellier, France; LMI LAPSE, and ISRA, LNRPV, Centre de Bel Air, Dakar, Senegal

Elena M. Bennett
Department of Natural Resource Sciences and McGill School of Environment, McGill University, Montreal, Canada

Klaus Birkhofer
Department of Biology, Lund University, Lund, Sweden

David A. Bohan
UMR 1347 Agroécologie, AgroSup/UB/INRA, Pôle Ecologie des Communautés et Durabilité de Systèmes Agricoles, Dijon Cedex, France

Michael Bonkowski
Zoologisches Institut, Terrestrische Ökologie, Universität zu Köln, Köln, Germany

Sophie Caillon
CEFE UMR 5175, CNRS-Université de Montpellier, Université Paul-Valéry Montpellier, EPHE, Montpellier, France

Stephen R. Carpenter
Center for Limnology, University of Wisconsin, Madison, Wisconsin, USA

Rachel Chalmers
National *Cryptosporidium* Reference Unit, Public Health Wales Microbiology, Singleton Hospital, Swansea, United Kingdom

Pierre-Olivier Cheptou
CEFE UMR 5175, CNRS—Université de Montpellier—Université Paul-Valéry Montpellier—EPHE campus CNRS, Montpellier, France

Oliver T. Coomes
Department of Geography, McGill University, Montreal, Quebec Canada

Wolfgang Cramer
Institut Méditerranéen de Biodiversité et d'Ecologie marine et continentale (IMBE), Aix Marseille Université, CNRS, IRD, Avignon Université, Aix-en-Provence Cedex, France

Pham Hung Cuong
Plant Resources Center, VAAS, MARD, Hanoi, Viet Nam

Gerlinde B. De Deyn
Department of Soil Quality, Wageningen University, Wageningen, The Netherlands

Paola De Santis
Bioversity International, Rome, Italy

Edmond Dounias
CEFE UMR 5175, CNRS-Université de Montpellier, Université Paul-Valéry Montpellier, EPHE, Montpellier, France

Isabelle Durance
Cardiff School of Biosciences, Cardiff University, Cardiff, United Kingdom

Anne Ebeling
Institute of Ecology, University of Jena, Jena, Germany

Nico Eisenhauer
German Centre for Integrative Biodiversity Research (iDiv), and Institute of Biology, Leipzig University, Leipzig, Germany

Marianne Elias
Institut de Systématique, Évolution, Biodiversité ISYEB-UMR 7205-CNRS, MNHN, UPMC, EPHE Muséum national d'Histoire naturelle, Sorbonne Universités, Paris, France

Jan Engel
Institute of Ecology, University of Jena, Jena, and Terrestrial Ecology Research Group, Department of Ecology and Ecosystem Management, School of Life Sciences, Technische Universität München, Freising, Germany

Colin Fontaine
Centre d'Ecologie et des Sciences de la Conservation (CESCO UMR7204), Sorbonne Universités, Muséum National d'Histoire Naturelle, Paris, France

Eric Garine
Université Paris Ouest/CNRS, UMR 7186 LESC, Nanterre, France

Paulo R. Guimarães Jr.
Departamento de Ecologia, Instituto de Biociências, Universidade de São Paulo, São Paulo, Brazil

Alison J. Haughton
Rothamsted Research, Harpenden, United Kingdom

Jean-Paul Hettelingh
National Institute for Public Health and the Environment (RIVM), Utrecht, The Netherlands

Jes Hines
German Centre for Integrative Biodiversity Research (iDiv), Halle-Jena-Leipzig, and Institute of Biology, University Leipzig, Leipzig, Germany

Nguyen Thi Ngoc Hue
Plant Resources Center, VAAS, MARD, Hanoi, Viet Nam

Sébastien Ibanez
Laboratoire d'Ecologie Alpine (LECA), UMR 5553, Université de Savoie, Le Bourget-du-Lac Cedex, France

Devra Jarvis
Bioversity International, Rome, Italy, and Department of Crop and Soil Sciences, Washington State University, Pullman, Washington, USA

Erik Jeppesen
Department of Bioscience, Aarhus University, Silkeborg, Denmark, and The Sino-Danish Center for Education and Research, Beijing, China

Richard K. Johnson
Department of Aquatic Sciences and Assessment, Swedish University of Agricultural Sciences, Uppsala, Sweden

Jennifer Adams Krumins
Department of Biology, Montclair State University, Montclair, New Jersey, USA

Vanesse Labeyrie
CIRAD, UMR AGAP, Montpellier, France

Christian Leclerc
CIRAD, UMR AGAP, Montpellier, France

Athen Ma
School of Electronic Engineering and Computer Science, Queen Mary University, London, United Kingdom

Giorgio Mancinelli
Department of Biological and Environmental Sciences and Technologies, Centro Ecotekne, University of Salento, Lecce, Italy

Camila Yumi Mandai
Departamento de Ecologia, Instituto de Biociências, Universidade de São Paulo, São Paulo, Brazil, and Imperial College London, Silwood Park Campus, Ascot, London, United Kingdom

François Massol
Unité Evolution, Ecologie et Paléontologie (EEP), CNRS UMR 8198, Université Lille 1, Villeneuve d'Ascq Cedex, Lille, and CEFE UMR 5175, CNRS—Université de Montpellier—Université Paul-Valéry Montpellier—EPHE campus CNRS, Montpellier, France

Doyle McKey
CEFE UMR 5175, CNRS-Université de Montpellier, Université Paul-Valéry Montpellier, EPHE, Montpellier, and Institut Universitaire de France, Paris, France

Brendan G. McKie
Department of Aquatic Sciences and Assessment, Swedish University of Agricultural Sciences, Uppsala, Sweden

Órla McLaughlin
UMR 1347 Agroécologie, AgroSup/UB/INRA, Pôle Ecologie des Communautés et Durabilité de Systèmes Agricoles, Dijon Cedex, France

Carlos Melian
Swiss Federal Institute of Aquatic Science and Technology (Eawag), Kastanienbaum, Switzerland

Luis Latournerie Moreno
Instituto Tecnológico de Conkal, Division de Estudios de Posgrdo e Investigacion Conkal, Conkal, Yucatan, Mexico

Christian Mulder
Department of Environmental Effects and Ecosystems, Centre for Sustainability, Environment and Health, National Institute for Public Health and the Environment (RIVM), Bilthoven, The Netherlands

Shahid Naeem
Department of Ecology, Evolution, and Environmental Biology (E3B), Columbia University, New York, USA

Unai Pascual
Basque Centre for Climate Change (BC3), IKERBASQUE, Basque Foundation for Science, Bilbao, Spain, and Department of Land Economy, University of Cambridge, Cambridge, United Kingdom

Josep Peñuelas
CSIC, Global Ecology Unit CREAF-CSIC-UAB, Universitat Autonoma de Barcelona, and CREAF, Center for Ecological Research and Forestry Applications, Cerdanyola del Vallès, Barcelona, Spain

Nathalie Pettorelli
Institute of Zoology, Zoological Society of London, Regent's Park, London, United Kingdom

Michael J.O. Pocock
Centre for Ecology & Hydrology, Wallingford, United Kingdom

Dave Raffaelli
Environment Department, University of York, York, United Kingdom

Christine Raimond
CNRS-UMR 8586 PRODIG, Paris, France

Ram Bahadur Rana
Local Initiatives for Biodiversity, Research and Development (LI-BIRD), Pokhara, Kaski, Nepal

Jes J. Rasmussen
Department of Bioscience, Aarhus University, Silkeborg, Denmark, and The Sino-Danish Center for Education and Research, Beijing, China

Luis Manuel Arias Reyes
CINVESTAV-IPN Unidad Mérida, Merida, Yucatan, Mexico

Graciela M. Rusch
Norwegian Institute for Nature Research (NINA), Trondheim, Norway

Christoph Scherber
Agroecology, University of Göttingen, Göttingen, and Institute of Landscape Ecology, University of Münster, Münster, Germany

Stefan Scheu
J. F. Blumenbach Institute of Zoology and Anthropology, University of Göttingen, Göttingen, Germany

Heikki Setälä
Department of Environmental Sciences, University of Helsinki, Lahti, Finland

Ryan A. Sponseller
Department of Ecology and Environmental Sciences, Umeå University, Umeå, Sweden

Bhuwon Sthapit
Bioversity International, Office of Nepal, Pokhara City, Nepal

William J. Sutherland
Conservation Science Group, Department of Zoology, University of Cambridge, Cambridge, United Kingdom

Mathieu Thomas
INRA, UMR 0320/UMR 8120 Génétique Quantitative et Évolution-Le Moulon, Gif-sur-Yvette; CEFE UMR 5175, CNRS-Université de Montpellier, Université Paul-Valéry Montpellier, EPHE, Montpellier, and FRB, CESAB (Centre de synthèse et d'analyse de la biodiversité), Technopôle de l'Environnement Arbois-Méditerranée, Aix-en-Provence, France

Amélie Truchy
Department of Aquatic Sciences and Assessment, Swedish University of Agricultural Sciences, Uppsala, Sweden

Corinne Vacher
UMR BIOGECO, Université de Bordeaux, Bordeaux, France

Wim H. van der Putten
Netherlands Institute of Ecology (NIOO), and Laboratory of Nematology, Wageningen University, Wageningen, The Netherlands

Nicolas Verzelen
UMR 729 Mathématiques, Informatique et STatistique pour l'Environnement et l'Agronomie, INRA SUPAGRO, Montpellier, France

Mariana Morais Vidal
Departamento de Ecologia, Instituto de Biociências, Universidade de São Paulo, São Paulo, Brazil

Chloé Violon
Université Paris Ouest/CNRS, UMR 7186 LESC, Nanterre, France

Winfried Voigt
Institute of Ecology, University of Jena, Jena, Germany

J. Arie Vonk
Institute for Biodiversity and Ecosystem Dynamics (IBED), University of Amsterdam, Amsterdam, The Netherlands

Cameron Wagg
Institute of Evolutionary Biology and Environmental Studies, University of Zürich, Zürich, Switzerland

Wolfgang W. Weisser
Terrestrial Ecology Research Group, Department of Ecology and Ecosystem Management, School of Life Sciences, Technische Universität München, Freising, Germany

Jean Wencélius
Université Paris Ouest/CNRS, UMR 7186 LESC, Nanterre, France

Stephen A. Wood
Department of Ecology, Evolution, and Environmental Biology (E3B), Columbia University, New York, and Yale School of Forestry and Environmental Studies, New Haven, CT, USA

Guy Woodward
Department of Life Sciences, Imperial College London, Ascot, United Kingdom

PREFACE

Ecosystem Services: From Biodiversity to Society, Part 1

Ecosystem services (ES) are the natural functions and processes of ecosystems which are of value to humans. By definition, therefore, ES are an anthropocentric concept: humans are the focus of ES (Fig. 1). This means that it is essential to acknowledge the social, economic and ecological systems within which individuals and human societies are embedded, in order to fully apply the concept of ES. Given the ubiquity these socioeconomic–ecological interrelationships across the globe, the ES framework has almost universal potential and its importance in policymaking is growing. Nonetheless, ES and the way the concept is sometimes applied (e.g. the commodification or monetarisation of nature) are still viewed with caution by many, especially those who see it as a threat to the traditional conservation goals of maximising biodiversity. Even now, a full decade after the publication of the Millennium Ecosystem Assessment (MEA 2005), which catalysed the field, there is surprisingly little empirical data that bring together social, economic and ecology thinking about ecosystems, and much of the theory is similarly embryonic.

Part 1 of this two-part volume of *Advances in Ecological Research* opens with an overview of the major trends in the field and the remaining challenges that need to be addressed since the publication of the MEA in 2005. Although ES had been studied before then, under a variety of different names and from somewhat different perspectives, it was with the MEA that the field really took off. The number of papers on ES has been growing since then, with ES accounting for an ever-larger slice of the total number of papers published in ecological journals (Fig. 2). As attested by the papers published, including those assembled here, the term ES is often used loosely, rather than being strictly limited to studies that explicitly consider humans. This might partly explain why the growth of ES has outstripped the related and more established fields related to ecosystem processes (EP) or functioning (EF), through a rebadging of more traditional EP and EF research under the ES moniker, as well as "genuine" new ES research. The idea that ecosystems provide things of value to humans is hardly new, but the formalisation of these concepts into a (more) unified framework and the strong links to emerging environmental legislation represent a fundamental

Figure 1 Examples of human-modified ecosystems and the services they provide. *Images courtesy of I. Palomo.*

shift in how humans are now recognised as being integral to nature, rather than somehow set apart.

The papers in this volume are arranged in a sequence of increasingly broader scope and scale, from those focused in Part 1 on understanding how human activities can alter local biodiversity and ecosystem processes that ultimately support, deliver and modulate services, through to those in Part 2 that move deeper into the more complex territory where the natural and social sciences overlap. Our aim was to show how these studies lie on a continuum and that ES can permeate all levels of biological organisation, influencing socioeconomic–ecological systems both directly and indirectly.

One theme that emerges throughout the volume is the need to move towards a more unified framework, to develop a clear, unambiguous shared

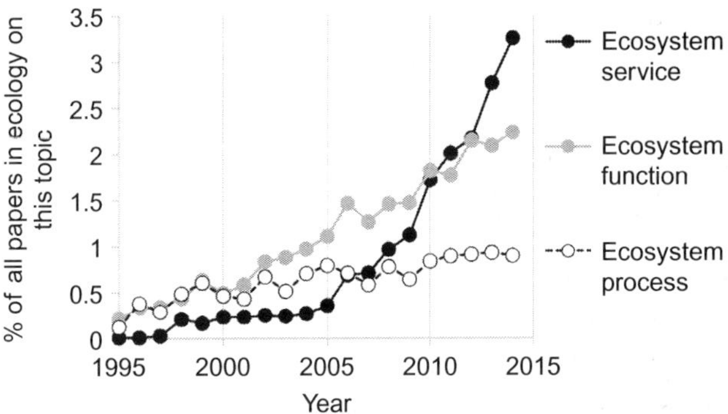

Figure 2 Trends in ES publishing in peer-reviewed journals over the past two decades spanning the publication of the MEA in 2005. Searches were undertaken using Web of Science for the topics "ecosystem servic*", "ecosystem functio*" and "ecosystem proces*" within the category "ecology" and the results were compared to the total number of papers in that category published between 1995 and 2014 inclusive.

lexicon and also, where possible, analytical approaches. This may be done by applying network-based approaches, which are powerful tools for coping with complex systems having multiple drivers, responses and entities that interact with one another—whether these are species in a food web, humans in a farming cooperative or banks within an economic system. This is of course just one of many potential ways of studying ES, but given the multivariate and multidisciplinary nature of the field—and the fact that network theory has already developed in parallel, but largely independently, within each of these disciplines—it represents a promising extension and integration of existing tools that could help provide the more coherent approach we need.

Part 1 opens with a paper by Mulder et al., which sets the tone by assessing the current state of the field and future prospects, in the context of the MEA, its precursors and the revolutionary changes that have occurred since its publication. They identify recurrent themes that have yet to be addressed and suggest how this might be done. This is followed by a series of papers that grapple with some of the more fundamental issues that underpin service provision, but which have yet to be resolved—in particular, these reflect the growing realisation that a synthesis of EF, resilience theory and food web ecology has much to contribute to the development of ES research (Truchy et al.; Mancinelli and Mulder; Hines et al.). These papers highlight the need to be able to understand both the direct cause-and-effect relationships and the subtler, often counterintuitive, indirect effects that can arise when perturbations and drivers "act at a distance". This cluster of papers is followed by an exercise that considers an ecological challenge that is touched upon by the other authors; that the interconnectedness of meta-communities, here of plants and pollinators, forms a spatially explicit set of networks that could confer resilience on service delivery in the face of habitat loss (Astegiano et al.). Network-based approaches are visited again by Thomas et al., but in terms of developing empirical and analytical methods, from theory, to feed into ES studies that continue to be hampered by a shortage of good quality data and transferable methods with which to test and validate hypotheses objectively. Thomas et al. demonstrate how networks that contain both ecological and social elements can be used for managing ES.

The 12 chapters in this two-part volume provide a snapshot of ES research: illustrating the current state of the art and spanning a full spectrum from developing a mechanistic understanding of the biological processes that ultimately deliver services, through to the implementation of policies designed to optimise service delivery. There is clearly much work to be done, but this volume offers an important step towards developing the next generation of approaches that we will need to ensure humanity remains within a "safe operating space" in a more sustainable future.

ACKNOWLEDGEMENTS

This series of papers came, in part, out of the Atelier Reseaux Tophiques (ART) International Workshop, held in Paris on the 11th and 12th of February 2015. The workshop was kindly supported by the ECOSERV Métaprogramme of INRA, as a way of promoting ES approaches in agriculture internationally, and we would like to gratefully acknowledge that support.

David Bohan
Michael J.O. Pocock
Guy Woodward

CHAPTER ONE

10 Years Later: Revisiting Priorities for Science and Society a Decade After the Millennium Ecosystem Assessment

Christian Mulder[*,1], Elena M. Bennett[†], David A. Bohan[‡], Michael Bonkowski[§], Stephen R. Carpenter[¶], Rachel Chalmers[||], Wolfgang Cramer[#], Isabelle Durance[**], Nico Eisenhauer[††,‡‡], Colin Fontaine[§§], Alison J. Haughton[¶¶], Jean-Paul Hettelingh[*], Jes Hines[††,‡‡], Sébastien Ibanez[||||], Erik Jeppesen[##,*], Jennifer Adams Krumins[†††], Athen Ma[‡‡‡], Giorgio Mancinelli[§§§], François Massol[¶¶¶], Órla McLaughlin[‡], Shahid Naeem[||||||], Unai Pascual[###,****], Josep Peñuelas[††††,‡‡‡‡], Nathalie Pettorelli[§§§§], Michael J.O. Pocock[¶¶¶¶], Dave Raffaelli[||||||||], Jes J. Rasmussen[##,***], Graciela M. Rusch[####], Christoph Scherber[*****,†††††], Heikki Setälä[‡‡‡‡‡], William J. Sutherland[§§§§§], Corinne Vacher[¶¶¶¶¶], Winfried Voigt[||||||||||], J. Arie Vonk[#####], Stephen A. Wood[||||||,******], Guy Woodward[††††††,1]**

*Centre for Sustainability, Environment and Health (DMG), National Institute for Public Health and the Environment (RIVM), Utrecht, The Netherlands

†Department of Natural Resource Sciences and McGill School of Environment, McGill University, Montreal, Canada

‡UMR 1347 Agroécologie, AgroSup/UB/INRA, Pôle Ecologie des Communautés et Durabilité de Systèmes Agricoles, Dijon Cedex, France

§Zoologisches Institut, Terrestrische Ökologie, Universität zu Köln, Köln, Germany

¶Center for Limnology, University of Wisconsin, Madison, Wisconsin, USA

||National *Cryptosporidium* Reference Unit, Public Health Wales Microbiology, Singleton Hospital, Swansea, United Kingdom

#Institut Méditerranéen de Biodiversité et d'Ecologie marine et continentale (IMBE), Aix Marseille Université, CNRS, IRD, Avignon Université, Aix-en-Provence Cedex, France

**Cardiff School of Biosciences, Cardiff University, Cardiff, United Kingdom

††German Centre for Integrative Biodiversity Research (iDiv), Leipzig, Germany

‡‡Institute of Biology, Leipzig University, Leipzig, Germany

§§Centre d'Ecologie et des Sciences de la Conservation (CESCO UMR7204), Sorbonne Universités, Muséum National d'Histoire Naturelle, Paris, France

¶¶Rothamsted Research, Harpenden, United Kingdom

||||Laboratoire d'Ecologie Alpine (LECA), UMR 5553, Université de Savoie, Le Bourget-du-Lac Cedex, France

##Department of Bioscience, Aarhus University, Silkeborg, Denmark

***The Sino-Danish Center for Education and Research, Beijing, China

†††Department of Biology, Montclair State University, Montclair, New Jersey, USA

‡‡‡School of Electronic Engineering and Computer Science, Queen Mary University, London, United Kingdom

Advances in Ecological Research, Volume 53
ISSN 0065-2504
http://dx.doi.org/10.1016/bs.aecr.2015.10.005

[§§§]Department of Biological and Environmental Sciences and Technologies, University of Salento, Lecce, Italy
[¶¶¶]Unité Evolution, Ecologie & Paléontologie (EEP), UMR 8198, Université Lille, Villeneuve d'Ascq Cedex, France
[|||||]Department of Ecology, Evolution, and Environmental Biology (E3B), Columbia University, New York, USA
[###]Basque Centre for Climate Change (BC3), IKERBASQUE, Basque Foundation for Science, Bilbao, Spain
[****]Department of Land Economy, University of Cambridge, Cambridge, United Kingdom
[††††]CSIC, Global Ecology Unit CREAF-CSIC-UAB, Universitat Autonoma de Barcelona, Barcelona, Spain
[‡‡‡‡]CREAF, Center for Ecological Research and Forestry Applications, Cerdanyola del Vallès, Barcelona, Spain
[§§§§]Institute of Zoology, Zoological Society of London, Regent's Park, London, United Kingdom
[¶¶¶¶]Centre for Ecology & Hydrology, Wallingford, United Kingdom
[||||||]Environment Department, University of York, York, United Kingdom
[####]Norwegian Institute for Nature Research (NINA), Trondheim, Norway
[*****]Agroecology, Georg-August-Universität, Göttingen, Germany
[†††††]Institute of Landscape Ecology, Westfälische Wilhelms-Universität Münster, Münster, Germany
[‡‡‡‡‡]Department of Environmental Sciences, University of Helsinki, Lahti, Finland
[§§§§§]Conservation Science Group, Department of Zoology, University of Cambridge, Cambridge, United Kingdom
[¶¶¶¶¶]UMR BIOGECO, Université de Bordeaux, Bordeaux, France
[||||||||]Institute of Ecology, Friedrich-Schiller-Universität, Jena, Germany
[#####]Institute for Biodiversity and Ecosystem Dynamics (IBED), University of Amsterdam, Amsterdam, The Netherlands
[******]Yale School of Forestry and Environmental Studies, New Haven, CT, USA
[††††††]Department of Life Sciences, Imperial College London, Ascot, United Kingdom
[1]Corresponding authors: e-mail address: christian.mulder@rivm.nl; guy.woodward@imperial.ac.uk

Contents

1. Introduction 3
2. Impact of the MEA 7
3. Functional Attributes and Networks as Frames for Ecosystems and Societies 11
4. Network Approaches to ESs as a Means of Implementing the MEA 15
5. Research Priorities One Decade After the MEA 22
   5.1 Underpinning Knowledge: From Functioning to Services 23
   5.2 Regulating Services 26
   5.3 Provisioning Services 28
   5.4 Supporting Services 29
   5.5 Cultural and Aesthetic Services 32
   5.6 Synergies Among Services and Multiple Drivers: How Can We Quantify Main Effects and Interactions Among ESs and Their Drivers in the Real World? 33
   5.7 How Are Services Linked in Different Realms? 34
   5.8 How Do We Prioritize the 'Value' of Services? 35
   5.9 Coupling Models to Data: How Do We Develop a Better Predictive Understanding? 37
   5.10 How Can We Manage Systems for Sustainable Delivery of ESs? 38
   5.11 What is the Role of Global Connections in ESs Delivery, and How Should This Impact Our Management and Understanding/Prediction of Future Provision? 39
6. Preliminary Conclusions 40
Acknowledgements 41
References 41

Abstract

The study of ecological services (ESs) is fast becoming a cornerstone of mainstream ecology, largely because they provide a useful means of linking functioning to societal benefits in complex systems by connecting different organizational levels. In order to identify the main challenges facing current and future ES research, we analyzed the effects of the publication of the Millennium Ecosystem Assessment (MEA, 2005) on different disciplines. Within a set of topics framed around concepts embedded within the MEA, each co-author identified five key research challenges and, where feasible, suggested possible solutions. Concepts included those related to specific service types (i.e. provisioning, supporting, regulating, cultural, aesthetic services) as well as more synthetic issues spanning the natural and social sciences, which often linked a wide range of disciplines, as was the case for the application of network theory. By merging similar responses, and removing some of the narrower suggestions from our sample pool, we distilled the key challenges into a smaller subset. We review some of the historical context to the MEA and identify some of the broader scientific and philosophical issues that still permeate discourse in this field. Finally, we consider where the greatest advances are most likely to be made in the next decade and beyond.

1. INTRODUCTION

The concept of ecosystem service (ES) is increasingly coming to the fore across a range of disciplines that span both the natural and social sciences (e.g. Bennett et al., 2015; Bohan et al., 2013; Carpenter et al., 2009; Díaz et al., 2006; Naeem et al., 2012; Naidoo et al., 2010; Pocock et al., 2016). Although many of the underlying tenets are not necessarily novel *per se* and analogous phenomena have been described in various guises over several decades, a unified language has emerged only relatively recently, following the rise of a suite of multidisciplinary approaches. Much of the current predominance of ESs can be traced back to the crystallization of these ideas in the Millennium Ecosystem Assessment (MEA), published a decade ago (MEA, 2005). With the benefit of hindsight, it is clear now that this was a seminal moment in ecological research, assembling a large international community for work that produced repercussions for policy and research during the following decade. It is timely to reflect on the major advances made during these years and to identify the future challenges. Rather than a comprehensive coverage of what is now a vast and varied field of research that is becoming a recognizable discipline in its own right, this chapter presents a collation and distillation of the views of a sample of experts, some of whom helped shape the thinking behind the MEA, and others who represent the new generation of researchers who have emerged within the increasingly multidisciplinary world forged by the MEA. In particular, we sought to explore how new frameworks might be adopted to advance the field, with an emphasis on the potential of network-based approaches, given that we are dealing with complex systems comprised of many interacting parts.

Since the publication of the MEA, a considerable amount of research has centred on strengthening its conceptual framework by providing theoretical and empirical tests of core ideas. Often, the objective of this research was to enable two activities: monetary valuation of ESs and linking ESs to socio-economic systems. Part of that process has inevitably led to a search for indicators of the status of ES and whether human interventions have negative or positive consequences. Many environmental factors that could potentially affect ESs are now being measured to gauge their utility as predictors and indicators of change, some of which are relatively closely linked to biodiversity or ecosystem functioning (e.g. fish production), whereas others are more abstract and challenging to measure rigorously (e.g. cultural significance of riverine bird species), although scenario-building approaches and new visualization tools are helping to bridge these gaps (Pocock et al., 2016; Sutherland et al., 2013). In some cases, a range of indicators of ESs change are currently being employed in management practices associated with ES delivery (cf. Liss et al., 2013). These approaches are still relatively narrow in scope, with the number and type of services restricted to the few that are easiest to measure (Daw et al., 2015; Perrings et al., 2011). This scope needs to be broadened if ES indicators are to be widely applicable, but to do so is difficult given the enormous range of ESs and the many variables that determine their magnitude, dynamics, interactions, and trade-offs at all levels, including whom the beneficiaries are.

Complex (living and non-living) systems comprise relationships among their components, and the number, pattern, and dynamics of such relationships being regarded as measures of system behaviour (Mesarovic, 1984). Complex system theory could provide a valuable means for developing a more comprehensive and integrated understanding of ES dynamics, as it deals explicitly with the mix of direct and indirect actors and consequences that are a defining characteristic of ES research. Can the behaviour of a system capture the value of ESs? According to Holling (1987) and Gunderson and Holling (2002), the behaviour of a living system results from an interaction among four basic functions: (1) exploitation (e.g. via rapid colonization), (2) conservation (e.g. resource accumulation), (3) release (stored resources suddenly released after external disturbances), and (4) reorganization (making the released resources easily accessible for a novel colonization). Holling's classification can be applied to both living and non-living systems (Costanza et al., 1997; Walker and Salt, 2006) and can be adapted to help integrate ecological economics and ESs in a more coherent manner than is currently the case.

General concepts of ESs have been in use for more than three decades (Ehrlich and Mooney, 1983), reflecting longstanding and widespread concerns that global changes have potentially strongly and adversely influenced terrestrial and aquatic communities. The MEA, which grew out of these earlier ideas, is arguably the most successful and enduring framing of scientific questions concerning biodiversity, ecosystem functions, ESs, and human well-being in complex socio-ecological systems. It was established to help develop the knowledge base for improved decision-makings in recognition that 'it is impossible to devise effective environmental policy unless it is based on sound scientific information' (Millennium Report to the United Nations General Assembly: Annan, 2000). This text continues 'While major advances in data collection have been made in many areas, large gaps in our knowledge remain' in how to use the MEA framework for the ever increasing wealth of data on environmental factors and human activities. 'In particular, there has never been a comprehensive global assessment of the world's major ecosystems'. The MEA viewed ecosystems through the complex science-policy lens of society, how ESs provide benefits to people, and how human actions alter ecosystems and the ESs they provide to humanity (Carpenter et al., 2009). Among multiple science-policy frameworks, the ES concept is undoubtedly now by far the most popular (Fig. 1).

Concepts like ESs, which integrate natural and social factors that link ecosystems with human societies, have triggered new waves of scientific research. Any consideration of ESs should centre on linking ecological, socio-economic, and related disciplines and will benefit from the approaches and insights gleaned from MEA, with its broad frameworks that linked nature (i.e. biodiversity and ecosystem functions) with ESs and human well-being (Fig. 1), although some papers have attempted to deal with the difficulties of connecting ESs to human well-being (Carpenter et al., 2009; Fisher et al., 2008). These and similar works clarified the need to separate benefits to people from ecosystem functions (Fisher and Turner, 2008; Fisher et al., 2008). However, governmental bodies have *de facto* a long history of bridging the gap between human well-being and ecosystem functioning.

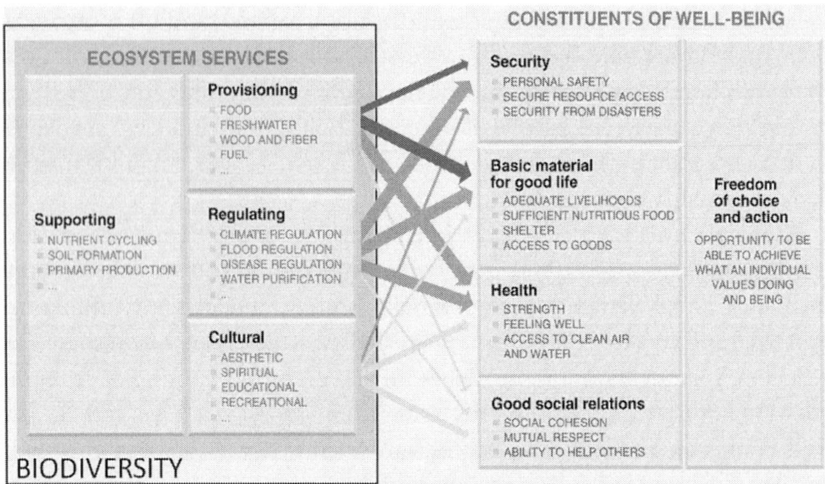

Figure 1 The conceptual framework of ecosystem services (ESs) as presented by the Millennium Ecosystem Assessment (MEA, 2005). The arrows' widths and colours depict the supposed interaction strengths between biodiversity and ecosystem services (left) and human well-being (right), although we should note that it has proved to be impossible to evaluate these interaction strengths in practice.

For a long time, environmental policy in Europe was predominantly concerned with pollution remediation of soil, water, and air. In the United States, the Wilderness Act was passed in 1960s and all the major U.S. legislations for endangered species, air pollution, and toxicity were passed in the 1970s (even the Clean Water Act, enacted in 1948, was completely rewritten in 1972 and 1977). Since the 1970s, environmental legislation has broadened its remit and coverage of the major ecosystems, with a general progression from a focus on the immediate vicinity of human populations on land to more distant ecosystems, including the remote ocean depths. Worldwide, there are many historical examples of how freshwaters have been used and modified by humans for millennia (Palomo et al., 2016), although water pollution management came much later due to lack of appropriate monitoring tools (e.g. Friberg et al., 2011). When the first cases of soil pollution became apparent remediation was regarded as a minor operation that could be carried out by national governments, in contrast to transboundary air pollution, which demanded international cooperation, as in the classic case of identifying the causes and ecological consequences of nitrogen deposition and acid rain (De Vries et al., 2015; Friberg et al., 2011; Sala et al., 2000). International problems provided an impulse for international policy, at the same time that scientific cooperation and coordination of efforts were strengthened by disasters like Chernobyl. Within this globally changing environmental and legislative landscape, the MEA framework has become increasingly central to understanding how to couple ecological and social systems across many scales and how to evaluate the effects of resource degradation and mismanagement. Maintaining, enhancing, and, if necessary, restoring ESs have now become a high-level policy goal, leading to many large-scale projects, such as the drive to restore many river catchments across much of Europe (Feld et al., 2011), where the true societal and economic cost of centuries of pollution and habitat destruction are now recognized.

Unprecedented efforts have been made to document, analyze, and understand the effects of environmental change on ecosystems and human well-being, and to cast those effects as ESs within a cross-disciplinary conceptual framework that integrates environmental, social, and economic theory (Fig. 1). The first group of studies concentrates on local scales, identifying the relationships and connections between the diverse spectrum of ecological processes provided by ecosystems and social factors related to the core constituents of human well-being. The second group situates services and well-being within a direct and indirect context of drivers of environmental change (e.g. nitrogen deposition, elevated CO_2, biodiversity loss). These entities are primarily operational at a larger, even global scale, with deforestation and desertification being two classic examples of worldwide ES disruption (Ehrlich and Mooney, 1983). Daw et al. (2011) and Poppy et al. (2014) highlighted the need to understand the dimensional aggregation of these component groups, asking who benefits from different ESs and who takes decisions about different ESs. Such a (dis)aggregation requires effective visualization tools, like networks, and here we suggest possibilities to achieve this goal.

2. IMPACT OF THE MEA

Human health is (on average across the globe) better today than ever before, and, together with unprecedented population growth due to public sanitation improvements, health and wealth are arguably the main underlying factors behind the huge environmental impacts we see in almost all ecosystems (Whitmee et al., 2015). If we are to maintain and improve the well-being of the ever-increasing human population, we need to understand and manage the consequences of this growth for the natural ecosystems we interact with, both directly and indirectly. Stress ecology, social ecology, and sustainability science have received growing attention, especially in the light of projections that the global population could reach 10 billion by 2050, associated with sustained large-scale migrations from rural to urban areas.

To gain an overview of what, if anything, has changed noticeably within the relevant environmental sciences during the past two decades, following the MEA's publication, we conducted a literature search from 1995 to 2015 using Thomson-Reuters's ISI on the Web of Science core collection with a range of broad primary search terms (NUTRIENT CYCLING or SOIL FORMATION or PRIMARY PRODUCTION) as well as a suite of more specialized secondary terms ([FOOD or FRESHWATER or WOOD AND FIBER or FUEL or CLIMATE REGULATION or FLOOD REGULATION or DISEASE REGULATION or WATER PURIFICATION or AESTHETIC or SPIRITUAL or EDUCATIONAL or RECREATIONAL] and ['ECOS* SERVICE*' or 'ECOL* SERVICE*']). Together, these searches returned a total of 22,532 peer-reviewed articles, mostly from the subject areas: 'ENVIRONMENTAL SCIENCES ECOLOGY', 'MARINE FRESHWATER BIOLOGY', 'OCEANOGRAPHY', 'GEOLOGY', 'AGRICULTURE', 'FORESTRY', 'PLANT SCIENCES', 'BIODIVERSITY CONSERVATION', and 'METEOROLOGY ATMOSPHERIC SCIENCES'. An additional search conducted on (BIODIVERSITY and ['ECOS* SERVICE*' or 'ECOL* SERVICE*']) returned 4111 peer-reviewed papers from 1995 to 2015 (mostly from the subject areas: 'ENVIRONMENTAL SCIENCES ECOLOGY', 'BIODIVERSITY CONSERVATION', and 'AGRICULTURE') that were included in the final data set ($n=26{,}643$).

Assessing the difference in the number of publications on ESs before and after MEA revealed an almost exponential growth, manifested principally as interdisciplinary links that developed between environmental scientists, ecotoxicologists, and ecologists. This has resulted in a widespread adoption of ecological theory, much of which has been driven by the emergence of the ecosystem approach and a growing focus on provisioning of goods and sustainability (Figs. 2 and 3). The MEA, which in its various forms has itself been cited in the peer-reviewed literature over 10,000 ×, clearly contributed significantly to putting ES firmly on the agenda.

Building on early works by Costanza and Daly (1992), Perrings et al. (1992), and Daily (1997), the MEA recognized benefits that people receive from nature as goods and services. These include direct benefits (such as food), indirect benefits (such as regulating the climate), intangible benefits (such as a sense of well-being from knowing natural ecosystems exist),

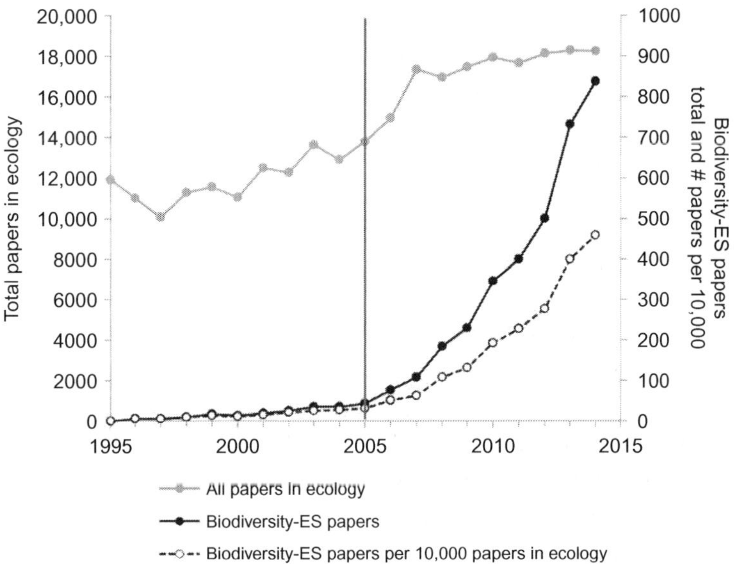

Figure 2 The number of papers published on the subject of biodiversity and ecosystem service(s) discovered in the Web of Science core collection (biodiversity–ES) has been rapidly increasing over the past decade (black trend, secondary axis), which is much greater than the overall increase in the number of ecological papers published (grey trend, primary axis). Therefore, the relative number of biodiversity–ES papers as a proportion of the total (number per 10,000 ecology papers; dotted line) still shows a marked increase. For reference, the publication of the MEA is shown with a vertical line.

and future benefits (belief that we continue to have the option to benefit from goods and services into the future) (Bateman et al., 2011). By catalyzing the ES approach at a global scale, the MEA boosted societal and political awareness that protecting ecosystem functioning and maintaining balance between supplies and demands of goods and services are essential prerequisites for human well-being. An intriguing example of how societal values are linked to regulating ESs is given by the case of water purification: clean water has become a *conditio sine qua non* of civilization since the ancient water and wastewater systems of Imperial Rome, but despite the huge knowledge accumulated in more than two millennia, it has been taken for granted in most societies. Its increasing shortage and the capacity of ecosystems to provide clean water have now turned it into a primary ES, in drylands and elsewhere (Fig. 3).

In the decade since its publication, the MEA has contributed to putting anthropogenic disturbance firmly on the political and scientific agendas. We

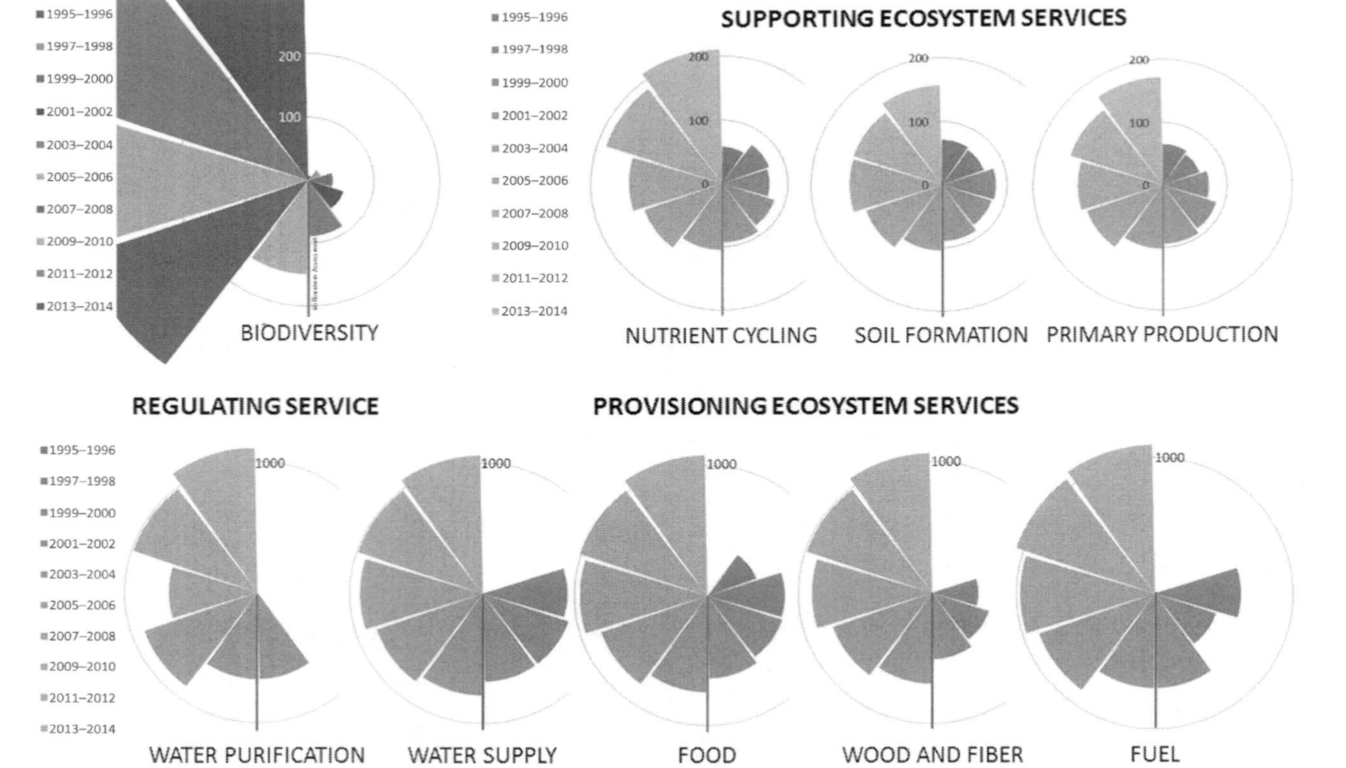

Figure 3 The rapid temporal increase in scientific peer-reviewed publications (the Web of Science was accessed August 11, 2015); relative reference (100%) is the average of the number of papers in 2004 (1 year before the Millennium Ecosystem Assessment) and 2005 (the boundary of the MEA is shown with a solid red (dark grey in the print version) line). Clockwise pies for each 2 years before the MEA (on the right) and after the MEA (on the left). Provisioning and regulating ESs (green (grey in the print version), lower panel) are plotted on a logarithmic scale, supporting ESs (orange (grey in the print version)) and biodiversity (grey) are plotted geometrically (upper panel). More details in the text.

conducted additional Web of Science surveys from 1995 to 2015 with the following search terms: ('NITROGEN DEPOSITION' or 'NITROGEN-DEPOSITION' or 'N DEPOSITION' or 'N-DEPOSITION'), ['LIGHT POLLUTION' and (BAT* or BIRD* or MOTH* or ECOL*)], (LANDSCAPE FRAGMENTATION), and [ECOL* and ('AGRICULTURE* INTENSIFICATION' or 'RURAL INTENSIFICATION')], mostly from the subject areas: 'ENVIRONMENTAL SCIENCES ECOLOGY', 'PLANT SCIENCES', 'AGRICULTURE', 'GEOLOGY', 'METEOROLOGY ATMOSPHERIC SCIENCES', 'FORESTRY', and 'BIODIVERSITY CONSERVATION.' Human-driven effects of landscape and habitat fragmentation ($n=1,137$, Fig. 4) and light pollution ($n=115$, Fig. 5, upper panel) exhibited a particularly rapid increase in publications, whereas global drivers like atmospheric deposition ($n=4679$, Fig. 5, lower panel) maintained the rate of increase (flatter trend). ESs as a whole have proven to be robust and (relatively) straightforward for dealing with otherwise overwhelmingly complex socio-ecological systems, and to do so in an integrative way that has grown in popularity among scientists and decision-makers (De Groot et al., 2010; De Vries et al., 2015; Paetzold et al., 2010; Stoll et al., 2015). This view is reflected in various environmental legislation of the European Union, such as the Habitats Directive, the Water Framework Directive (EU, 2000), and the European Marine Strategy Framework Directive (EU, 2008, 2010).

3. FUNCTIONAL ATTRIBUTES AND NETWORKS AS FRAMES FOR ECOSYSTEMS AND SOCIETIES

At the macroscale, ecosystems and human societies possess comparable attributes, insofar as they contain multiple interacting entities, such as individuals, species, or institutions, that respond both directly and indirectly to perturbations (Levin, 1998, 2000). Consider two instances: (1) any given ecosystem may incorporate continuous competition and facilitation among its species and functional groups, yet maintain ecological cohesion and (2) any given society may incorporate continuous competition and facilitation among its members and social groups, yet maintain cultural and economic cohesion. Both instances share horizontal diversity between subsets of similar entities and vertical diversity at different (energetic, cultural, economic) levels and layers. Although the usage of these terms is consistent with that employed in MEA (2005), our interpretation of (functional) entities and (horizontal and vertical) diversity is now much broader and also incorporates

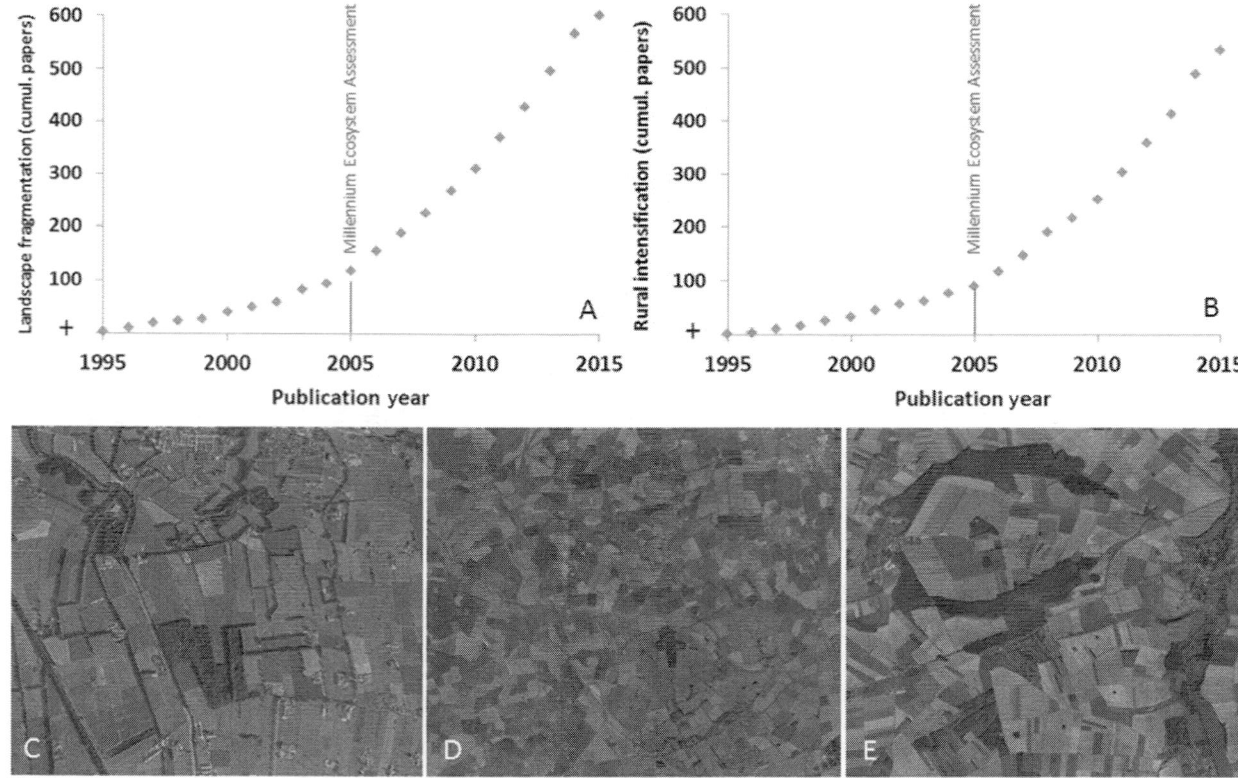

Figure 4 Temporal trends showing the cumulative growth of papers on landscape fragmentation (A) and rural intensification (B) for fragmented (C), independent (D), and mosaic landscapes (E).

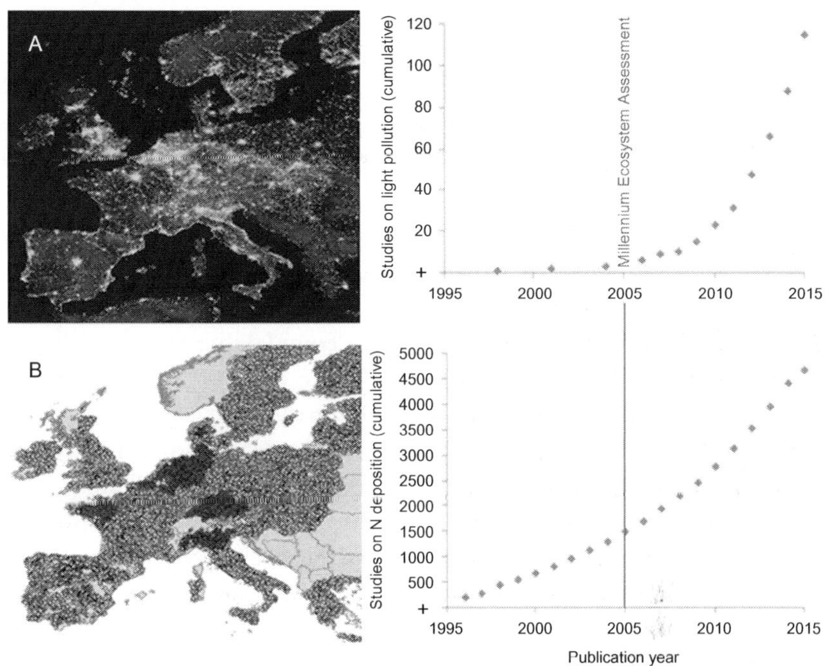

Figure 5 Temporal trends showing the cumulative growth of papers on light pollution (A: rapid increase since 2005) and nitrogen deposition (B: less rapid increase after 2005 but more constant growth). (B) Cluster analysis of ammonia deposition: the darker the colour the higher the NH_3 load (Mulder et al., 2015). *Photo credits (left part): (A) Radiance of the Earth by satellites, www.savethenight.eu, P. Cinzano and F. Falchi (University of Padova, Italy) and C.D. Elvidge (NOAA National Geophysical Data Center, Boulder, USA).*

other types of entities and diversities in general network theory. Understanding horizontal and vertical interrelationships among these entities is critical for management decisions, making appropriate tools necessary, especially as indirect responses to perturbations can be as strong as, or even stronger, than direct effects (Montoya et al., 2009; Moretti et al., 2013). Hence, tools to integrate such disparate repositories of knowledge and different forms of information are required; for instance, by identifying novel opportunities, assessing threats, or defining new issues (Sutherland et al., 2006, 2010, 2011). Importantly, natural ecosystems and human societies are not mutually exclusive, but are intimately connected—though they are still rarely studied with this perspective. As subsequently shown in this chapter, they are interdependent and dynamically connected, so to understand and predict the behaviour of one system requires an understanding of the other.

Similarities between ecological and social disciplines are often hard to identify, even though ecological and socio-economic disciplines are historically linked in their formation, if not always in their academic study. In their simplest form, cities, landscapes, and ecosystems are all open dissipative thermodynamic systems whose energy entrainment is (often assumed to be) maximized to confer stability against external disturbances (e.g. Bettencourt et al., 2007; Heal and Dighton, 1986; Kennedy et al., 2015). This leads to self-organizing structures requiring close integration of those units needing efficient servicing (Bettencourt et al., 2007; Kennedy et al., 2015), a continuous process whose apparent complexity reflects simple universal scaling laws (Bettencourt et al., 2007; Um et al., 2009). In the case of ecosystems, Carpenter (2003) suggests avoiding the term equilibrium, as this implies exclusion of the many other forms of steady-state dynamics seen in nature. Stability is not necessarily a static condition, whereas equilibrium is, but rather it is often a constrained or bounded dynamic process. Extremely low rates of change can resemble stability for many purposes, though not, technically, at equilibrium or even exhibiting stable dynamics (Holling, 1973).

Within this framework, even seemingly completely different data from ecological and socio-economic systems often appear to converge towards surprisingly similar phenomena. For instance, the frequency of sightings of bird species in the United States (e.g. a cultural or aesthetic ES) and the human population of cities (the ES recipients) in the United States share very plausible scaling laws (Clauset et al., 2009; but see also Stumpf and Porter, 2012, for caveats). Whether they are large cities, bird records or vegetation units, the huge amount of data available is useful for integration into ecological, social, and economic networks, although terminology can be rather confusing as too often the same term has rather different meanings in different fields. For these, and many other reasons, modelling of complex socio-ecological systems remains a major challenge in contemporary transdisciplinary research (Filatova et al., 2013). This task demands a comprehensive, interdisciplinary integration of ecological, social, and economic aspects with well-developed conceptual frameworks and theoretical as well as simulation models (An, 2012) and thus, demonstrates the pressing need for high-quality data, as well as a shared lexicon of terms (Wallace, 2007).

Many stakeholders aim to achieve stable system conditions to remain within their 'safe operating space': such a sustainable system is presumed to be persistent in the mathematical sense, if protecting against extinction (species loss or collapse of societies) and maintaining the same set of options by avoiding critical collapses. However, if we visualize complexity in just two information layers, fragile behaviours seem to reflect a disorganized complexity in simple models but an irreducible complexity in complex models (Alderson and Doyle, 2010; Weaver, 1948). Recent efforts towards standardization are providing new ways by which multiple information layers can be mapped onto one another for evaluation and management of stocks and flows, or ESs (Madin et al., 2007; Raffaelli and White, 2013; Raffaelli et al., 2014). Investigating responses at different scales can, therefore, allow a much better integration of research, an integration based upon the most universal and oldest language of scientists, mathematics (Cohen, 2004).

It is possible to elucidate social ties in space, such as characterizing how individuals (friends, relatives, and contacts) use their cities, as any urban space comprises a physical infrastructure and a social network (Wang et al., 2015). In this context, the perspective of a spatial network can be used to visualize the dynamic conditions of sustainability in different systems by optimized, space-filling, hierarchical branching networks (Bettencourt et al., 2007). Similarly, networks are also widely used in the medical world to identify 'disturbances' (e.g. Pichlmair et al., 2012). In the same way, it is possible to elucidate how entities in ecological networks are connected in space, for instance, how organisms (decomposers, producers, and consumers) separately breakdown, fix, or derive their own energy, as any food web is comprised of a chemical backbone and a constrained space (Hines et al., 2015; Mancinelli and Mulder, 2015). Any network can thus be seen as a simple data structure, a graph whose nodes identify the elements of a system and whose links identify their interactions where most of the structural information of social and ecological networks seems comparable to each other (Fig. 6). For instance, both the internet and the natural biosphere are promoted by an enormous variety of seemingly unrelated agents, and this could explain why both ecologists and social scientists have independently adopted network analysis as a common tool (Poulin, 2010): the challenge now is to use this common ground to help integrate these different disciplines more effectively.

4. NETWORK APPROACHES TO ESs AS A MEANS OF IMPLEMENTING THE MEA

The relationships between biodiversity, ecosystem functioning, and ecosystem services (B–EF–ES) have long been important gaps to address

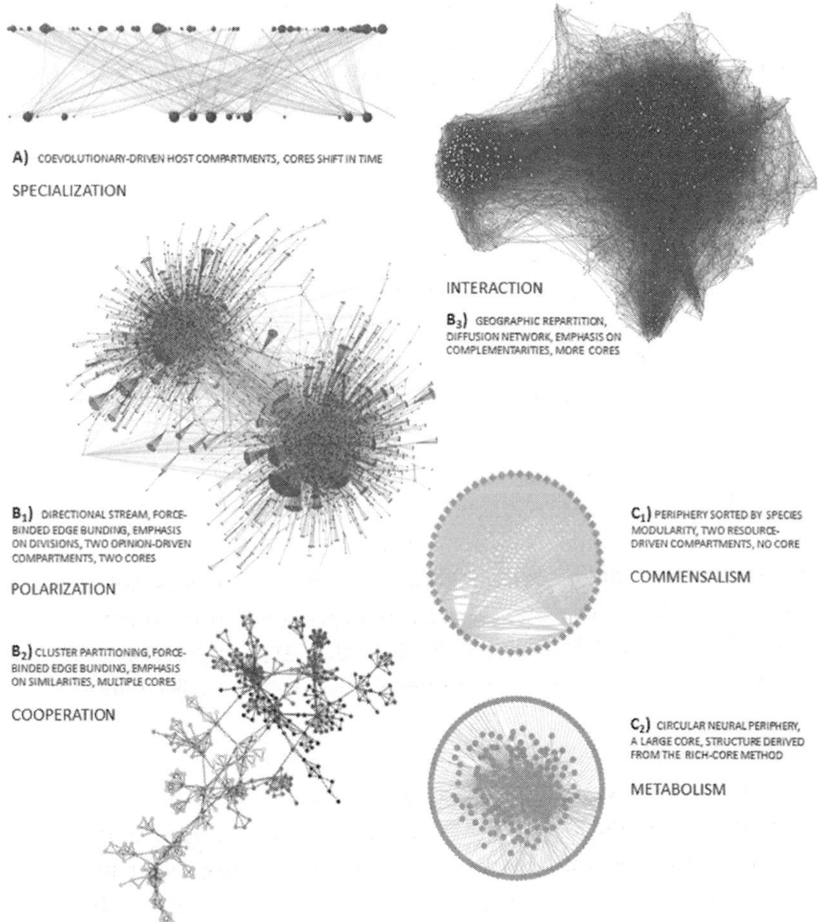

Figure 6 Examples of social, ecological, and evolutionary networks. (A) Bipartite interaction network from Fortuna et al. (2013), reproduced from PLOS under the terms of the Creative Commons Attribution License. (B) Communication networks—(B_1) Directional network generated by Twitter interactions: Each node is a single user, orange (grey in the print version) edges represent mentions and blue (dark grey in the print version) edges represent re-tweets and exemplify according to Vespignani (2012) the co-evolution of two communities (reproduced with permission of the author and of Nature Publishing Group); (B_2) cooperation network generated by scientific research: Each node is a single user, clusters exemplify common projects (giant component of scientists from Newman, 2006, defined as in Ma and Mondragón, 2012); (B_3) repartition network generated by phone calls in a large urban space: Each node is a single user, geographical complementarities exemplify local communities inhabiting different parts of the city (Wang et al., 2015). (C) Circular networks—(C_1) Detrital food web from a natural grassland: Functional traits determine the modularity of the periphery

(Continued)

in ES studies across scales, but remain poorly understood despite many efforts (e.g. Luck et al., 2009; Mace et al., 2012). Originally, the focus was on quantitative biodiversity-driven relationships, founded upon estimations of species richness at different scales: a specific ecosystem or habitat, a regional area, or even a whole continent (Balvanera et al., 2006; Butchart et al., 2010; Gotelli and Colwell, 2001; Hector and Bagchi, 2007; Hooper et al., 2005; Magurran, 2013). Accordingly, ranking procedures (scores) were often used: at ecosystem level, scores are commonly calculated as the deviation from reference conditions (i.e. an expected species list in undisturbed systems), a methodology that can be easily visualized by path analysis or structural equation modelling (SEM). SEM enables causal understanding to be inferred more strongly from observational data (Eisenhauer et al., 2015; Hines et al., 2015), but is also highly sensitive to both the intrinsic quality of the data set and the quantity of the records. In addition, SEM requires the standard assumptions of linear modelling: multivariate normality, additivity, and linear responses (Mitchell, 1992; Shipley, 2002). These assumptions (often contrasting the shapes of the B–EF–ES curves), as well as the strengths and weaknesses of SEM and path analysis, are discussed in detail by Pugesek et al. (2003), Martínez-López et al. (2013) and Westland (2015). Therefore, network approaches may be more appropriate for the large and heterogeneous B–EF–ES data sets.

Many components of networks theory have evolved separately: most theoretical biologists and computer engineers focused on mathematical metrics of networks (Jonsson, 2014; Wang et al., 2015), whereas ecologists tended to focus on structural changes along environmental gradients (e.g. Layer et al., 2010; Mulder and Elser, 2009; Woodward et al., 2010). Systems biology raises the intriguing prospect that some networks are inherently easier to control than others (Liu et al., 2011), which could have clear implications for sustainable management of ESs, especially if generic traits or indicators of the system can be identified that reveal this tendency. From this

Figure 6—cont'd (blue (grey in the print version) nodes) and the trophic links to the basal resources (green (grey in the print version) circles: fungi on the left, bacteria on the right) create two independent compartments ('Site F' from Mulder and Elser, 2009); (C_2) The 'small-world' neural network of *Caenorhabditis elegans*, together with *Escherichia coli* and *Drosophila melanogaster* one of the most widely investigated organisms (raw data from Watts and Strogatz, 1998; rich-core method in Ma and Mondragón, 2015). The network methodology can be used to visualize ongoing processes and hence to exemplify ESs, even benefitting from the rapid development of molecular ecology (see Vacher et al., 2016 for more network examples).

perspective, many powerful tools are already applicable to elucidate the importance of the network's topology and a certain degree of universality arises as soon as characters of a network are sufficient to quantify its features, such as scaling exponents that capture allometric and hydrological laws (Dodds and Rothman, 2000).

Networks can help provide the necessary understanding of relationships among entities as metrics to evaluate the improvement of ESs. The form of a network can help both academics and non-academics visualize many functions of a given organism or group of species in an ecosystem (Pocock et al., 2016). For instance, shifts in detrital organic material supply can cause dramatic changes in community structure and ecosystem functioning (e.g. Ibanez et al., 2013; Mancinelli and Mulder, 2015) thus affecting the supply of goods and services. Furthermore, the many groups of species that exploit in a similar manner the same class of environmental resources can be visualized, indicating levels of redundancy and ecosystem resilience capacity: e.g., if one node (or species) is lost, there are many alternative pathways in the interaction network through which the effects of its loss are essentially short-circuited. Also, developmental (successional) changes (Jonsson et al., 2005; Reiss et al., 2009), ecological stoichiometry (Mulder and Elser, 2009), overfishing (Jennings and Blanchard, 2004; Jennings et al., 1999), global warming (Yvon-Durocher et al., 2010), and fossil assemblages (Dunne et al., 2008) can be visualized by networks. Even at the level of individual variability in consumers' choice (Pettorelli et al., 2015; Tur et al., 2014), networks can be visualized and used to support conservation strategies based on resource requirements.

Ecological networks can be subdivided into three broad types: mutualistic plant–animal interactions, host–parasitoid, and prey–predator (trophic) webs (Bascompte and Jordano, 2007; Ings et al., 2009). Their increasing popularity has led to many open-source software packages, such as 'Pajek' (Batagelj, 1998), 'bipartite' (Dormann et al., 2008), 'Gephi' (Bastian et al., 2009), 'Cheddar' (Hudson et al., 2013), and 'Food Web Designer' (Sint and Traugott, 2015) to visualize the different aspects of networks. These software packages also allow the extraction of mathematical descriptors related to ecological properties and services (e.g. biodiversity of interactions, the trophic basis of production) that can be used for comparative analysis and could ultimately form a suite of indicators for monitoring responses to anthropogenic stressors (e.g. food chains should shorten and networks should simplify as stressors increase).

From an empirical standpoint, the number and quality of agricultural network studies is rapidly improving: the rate of growth in this field is even faster than in more traditional ecology and, if it continues apace, network-based approaches in managed ecosystems are surely bound shift from the sidelines into the mainstream (Bohan et al., 2013). The extension of metacommunity theory into metanetwork theory is now being pioneered in soil ecology and agroecology (Barberán et al., 2012; Pocock et al., 2012), largely due to the explicit recognition of the spatial and temporal patchiness of the landscape. This resonates with networks studies within the social sciences yet contrasts with much of traditional mainstream ecology, where spatiotemporal aspects are too often ignored as most studies are conducted in single, unreplicated systems, which are often (incorrectly) assumed to be isolated and closed systems.

The emerging field of eco-evolutionary dynamics is also being driven by studies of managed systems, in both fisheries science and agroecology, reflecting the extreme selective pressures being imposed by human activity and consequently the huge scope for ecological and evolutionary feedbacks to arise (Brennan et al., 2014). A good example of understanding feedback responses of human activities in managed systems is the use of pesticides: the rapid spread of pesticide resistance in commercial fisheries and the widespread alteration of freshwater community size-structures with attendant impacts on the food web are two pertinent examples that are attracting increasing attention. As pesticides cause regional biodiversity loss (Beketov et al., 2013) and erode different parts of the food web (Fig. 7), networks can visualize in a detailed yet intuitive manner the consequences of the environmental impacts of pesticide run-offs on non-target organisms and their ES delivery (Box 1).

Network theory can be applied to most kinds of complex self-organizing systems. These properties of being able to elucidate both the structure within complex systems and their metabolic scaling (Lentendu et al., 2014; Pawar et al., 2015) indicate that subnetworks, ecological networks, and network theory could be widely applied to practical problems, including management and decision-making processes. Examples include the design of nature reserves or the preservation of ESs in urban planning, as well as the management of commercial marine fish stocks for human consumption. While the study of networks is embedded in theoretical ecology, the application of such approaches to managed ecosystems has lagged behind. There are many reasons for this disconnection between pure and applied ecology, not least

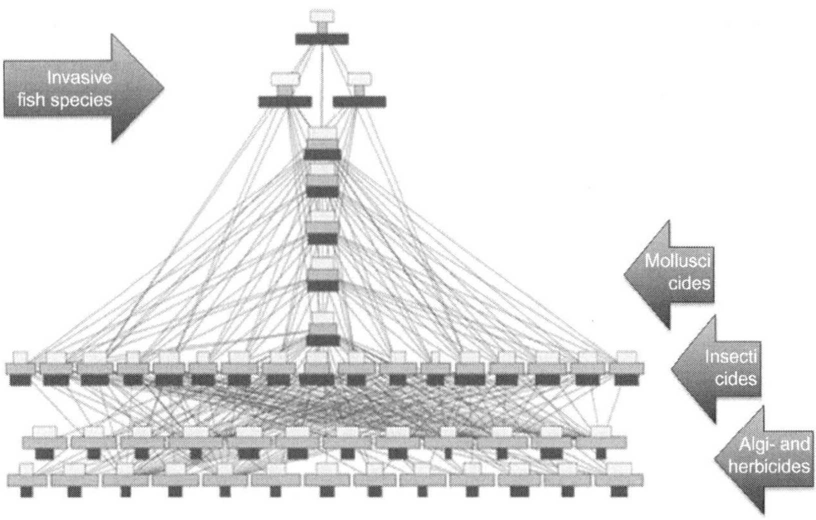

Figure 7 Network of an empirical aquatic food web in Tuesday Lake, MI, USA, arranged according to trophic height (Cohen et al., 2003; Jonsson et al., 2005). In 1985, the largemouth bass, a top predator formerly absent from the lake but native to the region, was deliberately introduced as a part of the first trophic cascade experiment (Carpenter and Kitchell, 1993; Carpenter et al., 1987). We mapped from top to bottom the adverse effects of comparable alien species (Cohen et al., 2009) and possible non-target effects of a family of pesticides (carbamates) on specific trophic guilds. Each node (species) is split in three log-scaled components, the population biomass (white bar), the numerical abundance (grey bar), and the average body mass (black bar). From the lower trophic level: phytoplankton (potentially affected by algicides or herbicides), zooplankton (potentially affected by insecticides or molluscicides), and fish (sensitive to top predators). According to the schematical application of ESs to aquatic food webs (Brennan et al., 2014), cultural and provisioning ESs may be provided by top predators (e.g. recreational angling), and regulating and supporting ESs (e.g. carbon sequestration) tend to be restricted to lower trophic levels.

being the long-held pervasive view that human-managed systems (e.g. agroecosystems and commercial fisheries) are not only different from supposedly pristine ecosystems but that they are also fundamentally artificial and thus not 'ecologically interesting' in a purely academic sense. This curious lack of investment in understanding the networks of managed systems is further highlighted by policy-driven environmental science tending to focus on disturbed or polluted ecosystems. Environmental policy thus exposes a general perception that natural systems, once perturbed, are somehow distinct from their natural counterparts. From this point of view, ESs provide a very suitable conceptual framework common to ecological science and policy, and network theory is a valuable tool common to multiple disciplines.

BOX 1 Science for Citizens and Citizen Science: Two Case Studies of Freshwater Ecosystems

Since the maintenance of an ecosystem is largely tied to the beneficiaries of ES provision, in particular situations, complex ecosystems are supposed to reduce human well-being (Lyytimäki and Sipilä, 2009). As an example, invasive zebra mussels in Lake Ontario and the St. Lawrence River in the United States were perceived to provide a positive contribution (benefit) by generating water clarity through filtration, as well as a negative contribution (disbenefit) by producing large amounts of nuisance algae (Limburg et al., 2015). The health risk associated with increasing water-associated pathogens (e.g. malaria) is another example where aquatic ecosystems are merely perceived to deliver a negative contribution to human well-being. These disadvantages (often defined as disservices, but see Section 6) are linked in human perception to disturbed aquatic systems, in which pathogen, pest, or parasite outbreaks are more likely to occur. To relate to the malaria example, in normally functioning wetlands populations of regulating predators significantly reduce mosquito populations, thereby also diminishing associated health risks as increasing mosquito populations are usually linked to artificial aquatic systems, such as reservoirs. It is recognized that agricultural influence from heavily fertilized agroecosystems causes substantial nitrogen leaching downward to the groundwater and laterally to the streams (Gordon et al., 2010; Verhoeven et al., 2006; Woodward et al., 2012). This increasing disturbance affects the functioning of wetlands and freshwater ecosystems, possibly leading to the disappearance of specialized species (e.g. Sterner and Elser, 2002). The hypothesis that healthy-functioning ecosystems overall deliver fewer disadvantages than disturbed ecosystems has yet to be tested. For instance, changes in land use and farming practices to feed rising populations have brought livestock animals increasingly close to rivers.

Citizen scientists, such as anglers, possess the skills to identify many macroinvertebrates and can monitor the status of ecosystems and report pollution incidents that threaten ES delivery. In the United Kingdom, biological tutors in conjunction with local agencies organize workshops to provide simple skills to citizen scientists whose data are valuable. Thompson et al. (2016) have recently shown how an insecticide spill in 2013 in the River Kennet altered the freshwater food-web structure and subsequently measured the resilience and recovery of the ecosystem across organizational levels from the structure of entire ecological network and ecosystem functioning (Fig. B1, right) to bacterial carbon substrate utilization and molecular ecology (Fig. B1, left). This provided a clear example of the close links between human society and natural ecosystems. The motivation for the citizen scientists to monitor the river's biota was a strong desire to ensure it was in a healthy condition.

Continued

BOX 1 Science for Citizens and Citizen Science: Two Case Studies of Freshwater Ecosystems—cont'd

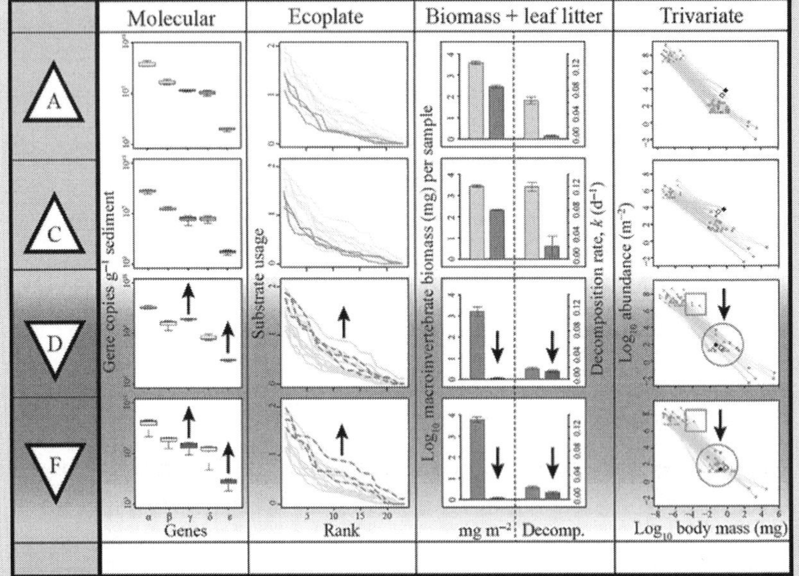

Figure B1 Impacted sites D and F shown with dark red (grey colour in the print version) background and control sites A and C with a light blue (grey in the print version) background. Six months after the incident, the shrimp *Gammarus pulex*, which is a key component of the diet of the commercial fishes on the river and the main driver of leaf-litter decomposition rates, was the slowest taxon to recover. The pesticide spill had a wide range of direct and indirect repercussions for both the 'brown' and 'green' pathways in the food web, resulting in altered ecosystem functioning and service provision (e.g. suppressed decomposition rates but also undesirable algal blooms and reduced prey availability as a supporting service for the commercial fisheries on the river).

However, despite their increasing popularity, socio-ecological networks are still ignored in ES studies and the generation of appropriate and compatible data remains a crucial step for the quantitative estimation of ESs (Feld et al., 2009; Wallace, 2007).

5. RESEARCH PRIORITIES ONE DECADE AFTER THE MEA

A priorities-listing exercise was designed to give a broad overview of research priorities for scientists and stakeholders, based on the expert knowledge of the co-authors, who represent a range of expertise across different disciplines and countries. We do not claim that this set of views is representative for the global situation, as any such survey is inevitably biased by the topic of expertise, geographical location, ecosystem type, or other variables that cannot be controlled. Nevertheless, we aimed to collate and sift the views of this set of experts in the field to explore some of the major trends in the field since the publication of the MEA. The master list of headings we circulated is not exhaustive, but simply a broadly representative coverage of the main topics covered in the original MEA, divided into subheadings. The list was sent to a common pool of researchers by email and then discussed subsequently following receipt of the responses.

Responses were collated for groups of broad topical questions: a first block of five fundamental questions was derived from general trends in the existing literature (Fig. 2 and 3), while a second block of five questions was more applied and narrower in focus. These two blocks reflected Holling's basic functions (1987), given that exploitation, conservation, and release have been mostly addressed within the supporting, regulating, and provisioning ESs, while cultural services and constituents of human well-being were somehow the recurrent background beyond the second block (reorganization). On average, 182 words were added to the master list by each participant to define the 'top-five' priorities. All the answers were aggregated within into a single file (repetitions due to the use of the same template text and associated references were removed).

A striking pattern is that all the replies are reasonably evenly distributed across the original 10 categories, although not sufficiently evenly to obtain an equal number of issues in each category. Screening the word cloud of all the text supplied by the participants showed a common focus on particular aspects (Fig. 8). Due to the different ways used by different contributors to formulate the same issue, the overlap was often high and responses could be merged. During this phase, a new category 'What is the role of global connections in ESs delivery, and how should this impact our management and understanding/prediction of future provision?' was added. Overall, we identified 36 key scientific issues that, if answered, we felt would drive future advances in the field ESs. The resulting issues (as bullets) are discussed in the following part, starting with the more generic issues listed from Section 5.1 onwards.

5.1 Underpinning Knowledge: From Functioning to Services

Biodiversity (species richness and functional diversity) and ecosystem functioning are not independent, and the debate regarding the identification of crucial aspects of the relationship between them is still open (Cardinale et al., 2012; Huston et al., 2000; Jax, 2010; Tilman et al., 1997). Effects of

Figure 8 A word cloud generated from the research priorities collected during our internal survey. The cloud provides greater prominence (the numbers of entries determine the relative sizes) to the words that appear more frequently in the collated text. For instance, the word 'service' has been used 5 × as often as 'biodiversity' and the word 'ecosystem' has been used 2 × as often as 'species'.

dominant species (whether a certain community assemblage is necessary to form and support a given ecosystem) matter at several levels (Ospina et al., 2012; Perrings et al., 1992) and can question the amount of biodiversity needed to maintain a function or provide a service (Kleijn et al., 2015). The discussion as to whether we need phylogenetic versus taxonomic versus functional levels of biodiversity (Gerhold et al., 2015; Mouchet et al., 2010; Mouquet et al., 2012) is ongoing. Many recent papers have attempted to gauge whether phylogenetic diversity is a useful proxy for community assembly or functioning but it does not seem to be the case all the time. Network approaches to the question may give more emphasis on the impact of ecosystem complexity, with the structure of ecological interaction networks appearing as key to our understanding of the dynamic and the functioning of ecosystems (e.g. Fontaine, 2013; Thébault and Loreau, 2003). Despite the widespread non-linearity in relationships between biodiversity and ESs (e.g. Bennett et al., 2009; Carpenter et al., 2009; Grêt-Regamey et al., 2014; Koch et al., 2009; Lester et al., 2013) that fuels debates on methodologies to quantify biodiversity and its spatial and temporal variations, ES is still clearly a useful conceptual framework to bridge the typically isolated disciplines of the social and natural sciences.

Figure 9 addresses the extent to what various ESs depend on biodiversity: How many species are 'needed' for the service delivery? But which process rate is desirable, and at which scale? The table in Cardinale et al. (2012) has a set of comparisons of correlates of ESs with various measures of diversity,

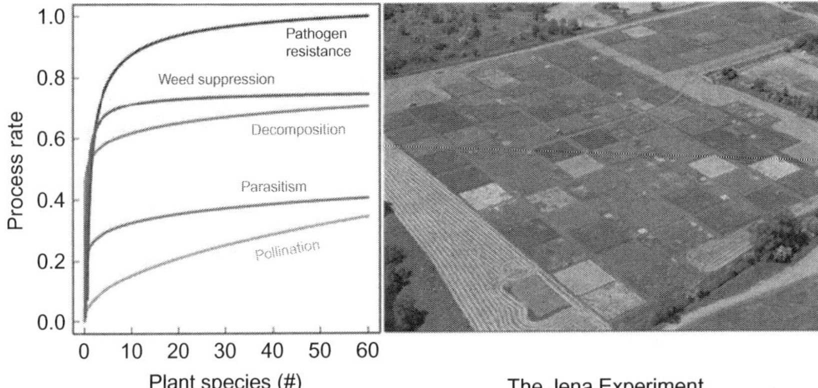

Figure 9 In a 10-ha experimental field near Jena (Germany), the number of vascular plant species were controlled and ecological processes were measured (Scherber et al., 2010). In that experiment, processes and ESs reacted rapidly to the initial increase in plant species diversity, after which some services like 'weed suppression' (here as inverse of invasion) tend to saturate, while supporting services like 'pollination' and—to a lesser extent—'decomposition' remain enhanced by biodiversity. Scherber et al. (2010) also exposed experimental nesting sites for wild bees and measured parasitism rates as proxies for occurring top–down control: parasitism rates increased with the number of plant species, resulting in potential biological control in species-rich ecosystems. *Figure recomputed with the original data by C. Scherber; photo credit W. Voigt.*

and the picture is much more mixed (cf. Mace et al., 2012). Given the large amount of available data, methods of statistical reduction have been rapidly replaced by interaction metrics and graph theory (Poisot et al., 2013; Thébault and Fontaine, 2010), allowing to visualize the relationships between ecosystems and ESs in a more intuitive way. Networks enable the integration of metabarcoding and interactions in graphs (e.g. Ji et al., 2013; Pocock et al., 2016; Vacher et al., 2016, and references therein) and allow visualization when a single function or a group of functions are needed (or not) for a specific ES, a group of ESs, or a category of ESs. Currently, estimation of global ES values remains crude (Naidoo et al., 2010) and collecting ES metrics varies enormously in cost and complexity (Naeem et al., 2015). ESs are at present still largely studied independently and networks will enable a complementary service-lead approach to map ESs onto functions rather than *vice versa*. In the following sections, we have compiled the major sets of questions under umbrella terms (in italics) that group them together within a recognizable recurrent theme:

- *Scales (#1)*: Which spatial scale is important for which ES? How to choose the resolution and how to set the grid size of the underlying abiotic environment? What scales are relevant for ES versus other scales like animal movement or conservation management? Does the relationship between biodiversity and ecosystem functioning (B–EF hereafter) and services change across scales? What are the most appropriate spatio-temporal scales and resolutions to study the links between B and ES? How does regional habitat loss influence ES of local patches and how can we scale up the knowledge from B–EF experiments to management or global scale?

- *Trade-offs and synergies (#1)*: What are the shapes of the B–EF–ES curves and how much redundancy does biodiversity provide? How few dimensions of ES can we measure: are multiple services correlated or orthogonal? When ESs interact, how do we cope with the non-linearity that ensues? Why do the most productive ecosystems (Douglas fir, redwoods, beech woods, tidal and freshwater marshes, bamboo forests) typically have low plant diversity? Which ESs are strongly correlated with biodiversity and which are not? Can we use molecular techniques to quantify microbial diversity (DNA) and microbial functioning (RNA)?
- *Metrics*: Which biodiversity metrics are the best descriptors of service delivery? As power law functions can take almost any shape (saturating, linear, concave, convex) and exponents are easy to compare both across systems and within ecosystems, are allometric exponents suitable for comparison?
- *Dimensions*: What are the dimensions of biodiversity that most matter in the delivery of ESs? Are there different mechanisms driving relationship between biodiversity and categories of ESs? Must we weigh the impact of non-native species, for instance, by proportionating invasive species on ESs (as in urban ecosystems) and/or is the regenerative capacity and redundancy of ES depending on seasonal changes in the ecosystem structure?

5.2 Regulating Services

Mutualistic symbioses, commensalism, parasitism, and amensalism (e.g. whereby parasites might change the animal behaviour and either contribute or impede the delivery of specific services) are of general importance to regulating services. A large number of studies have been dedicated to understanding how a single pathogen agent interacts with its host, without taking into account the role of the overall biotic environment. This reductionist approach of pathogenesis has, however, evolved considerably in the past decade, triggered by the development of network ecology (Hudson et al., 2006; Lafferty et al., 2008; Vacher et al., 2008) followed by that of meta-omics (Berendsen et al., 2012; Hacquard and Schadt, 2015; Vayssier-Taussat et al., 2014). The transition to a more holistic understanding of diseases has led to the recent emergence of the 'pathobiome' concept, which represents the agent integrated within its wider biotic environment (Vayssier-Taussat et al., 2014). Our understanding of the relationship between network properties and disease regulation is still in its infancy (Vacher et al., 2016). The idea revolves around the fact that symbiotic interactions might be key for functioning (e.g. the black queen hypothesis in microbial communities) and, hence, for services (Carroll, 1988; Jackson et al., 2012; Kiers et al., 2007; Polin et al., 2014; Rapparini and Peñuelas, 2014). Regulating services delivered through different co-production processes mostly benefit human well-being locally, although many regulating ESs are influenced by local-to-regional management and global changes (climate warming, pollution, landscape, fragmentation) spanning multiple scales (Gill et al., 2016; Hein et al., 2006; Stephens et al., 2015). Overall, service diversity coupled to biodiversity seems to be a good and reliable predictor for the delivery of regulating ESs (Raudsepp-Hearne et al., 2010a). We grouped questions under emerging themes, as before:

- *Scales (#2)*: How do we accommodate the more global scale that regulating ESs operate over with the more local scale of provisioning and cultural services? Under which circumstances can these be provided globally, without regard to location, and when are they location specific? What are the implications for management of these differences? What is the minimal/maximal/optimalsize of ecosystems in respect to different regulating ESs? What are the differences between vegetation types, such as different tree species, on the efficiency of service delivery? At which scale should we measure microbial community structure in order to predict ecosystem health?
- *Technological control*: To what extent can bioengineering or (non-natural) capital be used? Can bioengineering help to maintain the output/endpoint of regulating services in the depauperate biota of disturbed systems? Can we improve the use of microbial organisms for climate regulation and waste processing? What are the consequences of planting crops/trees for biofuel on adjacent and connected ecosystems and what are the consequences on ES dynamics?
- *Biological control*: What are the main biotic (community) drivers of disease control? How will rhizosphere microbial clusters (indirectly) interact

with each other during plant competition, and will this create synergies (e.g. disease suppression) in relation to ESs? There seems to be—at least in aphids—a trade-off between assimilation of symbionts giving resistance to parasitoids versus resistance to predators: can pest control be enhanced, and can we manipulate that to assist these 'pest controllers'?

- *Disease control*: Is there a relationship between the structure of the residential microbiota within a host and its susceptibility to disease? Is disease susceptibility accounted for by the presence of a few species or by the structure of the whole microbial community? How to use networks to highlight the specific microorganisms and/or the properties of the whole microbial community that regulate disease?

5.3 Provisioning Services

Biodiversity is the outcome of countless ecological and evolutionary events that occur over many scales in time and space: when humans genetically modify organisms, changes driven by (often local) economic interests are added onto >10,000 years of both artificial and natural selection. Regardless of the way we may define 'nature', natural capital stocks are in some ways analogous to financial capital in bank accounts. For instance, the financial systems have a high modularity (Haldane and May, 2011), like trait-mediated networks (Fig. 6C). While it is possible for us to extract high yields of resources from (natural and managed) ecosystems, if more than the interests yielded on that capital become extracted, then any system crashes (Raffaelli, 2016). The well-regulated forest management in many European countries is a good stock-and-flow example of an ecosystem-yield approach that aimed for a sustainable timber provision. An example of one that is far less effective is that of the traditional species-centred (as opposed to ecosystem-based) approach to managing global commercial fisheries, which has been implicated in the crashes of many stocks around the world. Flows between domestic banks and across technological or biological networks are comparable, as shown by small-world similarities between financial, technological, and biological models (Newman, 2003; Raffaelli, 2016; Watts and Strogatz, 1998). If these similarities are as general as suggested in the literature, constrained regularities can be identified and extended to ESs. Three main groupings of questions emerged under the heading of provisioning services:

- *Monitoring*: What are the best ways to monitor provisioning ESs, seen the low priority given to long-term change by governments and agencies?

Can citizen science help us filling in the gap by bird-watching, butterfly counts, or vegetation surveys? Do alien species that are common in urbanized systems enhance provisioning services?

- *Modelling*: How can we model how harvesting of animals cascade through ecological networks affects other taxa and how important is the diversity of available resources?
- *Emergy*: Can stocks of natural capital and flows of environmental resources capture the full value of provisioning ESs within the concept of embodied energy (emergy)?

5.4 Supporting Services

Urban systems are growing faster than any other land cover type (Meyer and Turner, 1992). Maintaining agricultural yield at a sustainable level requires that the regenerative capacity of driving subsystems is sufficiently strong despite fragmentation. Hence, economic trade-offs arise between management practices (e.g. conventional and no-tillage agriculture) and between rural and urbanized systems, and dealing with these has become a growing challenge. Large-scale anthropogenic disturbances, like the effects of nitrogen deposition on the diversity of mycorrhizal fungi (Chung et al., 2009; Cotton et al., 2015) and of atmospheric pollution in general on the genetic pool of pollinating insects (Gill et al., 2016), have also been investigated. The latter authors conclude that studies carried out during a single year may be difficult to generalize, as data across large environmental gradients and over long time spans are needed for a strong analysis, and these limitations apply to much of the field, where long-term large-scale empirical data are scarce (cf. Tylianakis and Coux, 2014). This makes the mechanistic interpretation of co-occurring ESs more difficult and a major challenge for future research, especially given the tendency for research funding to focus on short-term novelty, rather than monitoring the same set of model systems for many years (Box 2).

- *Trade-offs and synergies (#2)*: What are the relative contributions of community biomass, species richness, or trait diversity to biogeochemical ESs (e.g. hydrologic infiltration, soil stabilization, carbon sequestration, microclimate amelioration)? Why are some relationships between species diversity and productivity in natural ecosystems opposite of the relationships typically found in B–EF experiments?
- *Balance*: To what extent do ES-providing species or groups of species depend on non-service-related species? Non-crop plants provide habitat

BOX 2 Biodiversity and Ecosystem Functioning: Productivity in Terrestrial Ecosystems

Services, especially in agriculture, can be strongly trait mediated (Wood et al., 2015). If we consider the several types of plant–insect interactions (DeAngelis and Mooij, 2005; Fægri and Van der Pijl, 1979), a selective chemical pressure due to co-occurring direct effects of pollutants on plants and secondary effects of polluted hosts on their pollinators is likely to occur. Fægri and Van der Pijl (1979) introduced the so-called pollination syndrome to classify the pollination strategy according to the agents (wind or pollinators) by which pollen is transferred, and they showed that many insect-pollinated plants in agroecosystems have multiple pollination strategies, making them less dependent on a specific invertebrate, in contrast to plants in tropical forests and plantations. The trait-mediated disservices will be then different according to the geographical location of the site, as tropical ecosystems suffer the most by massive deforestation and landscape fragmentation and temperate ecosystems are often endangered by pollution. In a case study conducted in the Netherlands, the nectar plants for butterflies were the only showing stress from heavy metals, whereas the nectar plants for moths were the most tolerant to heavy metal pollution (Mulder et al., 2005). Hence, only the pollination service provided by adult butterflies was indirectly affected by pollution (but see Gill et al. (2016) for more case studies).

ESs are strongly influenced by shifts in land-use practices. Transformation of productive (species-poor) into less-productive (species-rich) grasslands remains a current practice in conservation and restoration ecology (Bakker, 1989). Such a transformation is a typical example of how some services are unpredictable in the soil: Wardle et al. (2004) coupled a relative fungal dominance in soils to nitrogen poor litter, although it is not always the case as shown by empirical evidence for effective competition between microbes and plants for nitrogen uptake (e.g. Laakso et al., 2000; Setälä et al., 1998). This phenomenon determines the structure of entire ecological networks. In particular, the distribution and length of trophic links are essential in the categorization of the food-web structure, and we may expect that by evaluating their trophic links ecological networks might provide a tool to better forecast supporting and provisioning ES. For instance, changes in weeds and invertebrates between the herbicide management of spring-sown maize, beet and oilseed rape, and winter-sown oilseed rape, and the herbicide management of genetically modified herbicide-tolerant varieties was evaluated across Britain (Bohan et al., 2011; Firbank et al., 2003), and the trophic links between each prey species and consumer species were given a probability score and weighted by logic-based machine learning (Bohan et al., 2011; Pocock et al., 2016). Such metawebs demonstrate that the long trophic links deviated more from the community response than the short (often intraguild) links, as most functional groups were found not to overlap each other (Sechi et al., 2015). In general, processes, functions, and services result from a complex interplay of (a)biotic interactions (e.g. Hines et al., 2015).

For example, the plant biodiversity (centre of the Fig. B2) translates at the same time into changes in the aboveground and the belowground networks, each of

BOX 2 Biodiversity and Ecosystem Functioning: Productivity in Terrestrial Ecosystems—cont'd

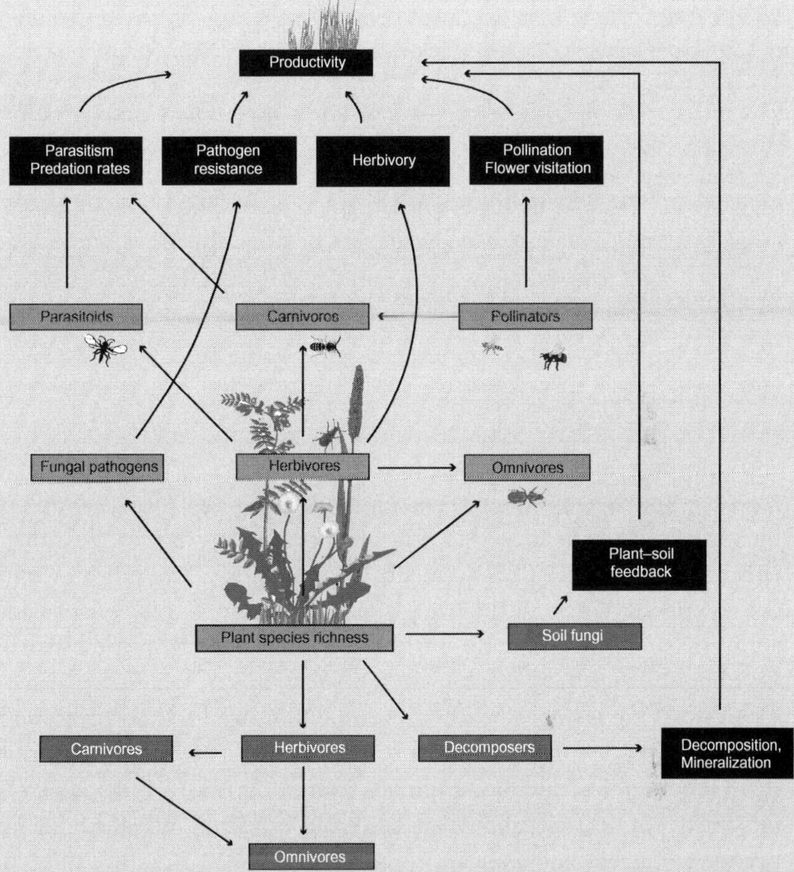

Figure B2 How above- and belowground multitrophic interactions may translate into ecosystem services. Most functional guilds (herbivores, omnivores, carnivores) of the detrital soil food web (brown pathway) are mirroring those of the aboveground food web (green pathway) in this conceptual graph. Their close synergy is shown by supporting ESs like nutrient cycling (decomposition, mineralization) and provisioning ESs like pollination, jointly determining the primary productivity of the entire ecosystem, here as system's output (top black box). Like in the aforementioned freshwater system, pesticides have, also in terrestrial systems, a wide range of repercussions for both the 'brown' (dark grey in the print version) and 'green' (grey in the print version) pathways in any food web. For instance, direct effects of fungicides on pathogens living on the phyllosphere, hence belonging to the 'green pathway', are often linked to indirect effects on the rhizosphere fungi and mycorrhizae of the 'brown pathway'.

Continued

BOX 2 Biodiversity and Ecosystem Functioning: Productivity in Terrestrial Ecosystems—cont'd

which results in different processes (examples in the black boxes) that affect the system's output (here as productivity). Even wild microherbivores contribute to such biogeochemical cycles (Belovsky and Slade, 2000). Network-based approaches are required if we want to understand multiple ESs, allowing predictions to be made for more organizational levels. Furthermore, responses of interacting components to interacting drivers can be predicted using such frameworks. For instance, homeostasis of C:N:P ratios (the so-called Redfield Ratio) strongly changes the food quality by CO_2 enrichment (Loladze, 2002). All such biodiversity-induced changes will cause, both above- and belowgrounds, structural shifts in ecological networks (e.g. Eisenhauer et al., 2013; Mulder et al., 2012, 2013; Reuman et al., 2008; Scherber et al., 2010).

and resources for pollinators, but can be a source of competition for resources (nutrients and light) and harbour crop pests. How should pollinator habitats be managed to enhance populations but not be a major competitor for crops? Can we identify tipping points and can we exploit ecosystems to increase/manage their ESs? How can we balance agroecosystem ESs and trade-offs? How should a habitat be best spatially distributed for greatest gain?

- *Corridors*: What is the value of vegetated buffer strips to in-stream fungal leaf-decomposers, and how is the ecological quality influenced by non-managed buffer strips along surface water corridors (side effects of increased connectivity between land and stream)? What is the link between nutrient and toxicant removal by flooded riparian zones and the terrestrial vegetation that supports pollinators and other insects?

5.5 Cultural and Aesthetic Services

Network theory may improve knowledge of relationships between biodiversity and its functions on one hand, and driving subsystems on the other. This is also true for cultural ESs in general and 'charismatic fauna' in particular, especially as the latter are mostly towards the top of the food web. Cultural ESs tend to be fund-service (non-consumptive) in nature and are by definition subjective. If the endpoint is 'to maintain beauty or scarcity' of a particular ecosystem, then monetization itself is not an issue *per se* that can be made operational in environmental or economic contexts. If the endpoint is to maintain or increase the price value of housing that overlooks a neighbouring ecosystem of aesthetic value, then monetization may be a useful economic proxy. To a certain extent, the values of some of these ESs seem to reflect the level of the local social organization (ES granted to a nation, provided to a community, and even to an individual): for instance, zoological and botanical gardens are known to positively influence the attitudes of the visitors (Williams et al., 2015). Probably therefore ornamental plants—almost always introduced—are more influential in this service provision than native plants. Interestingly, 'charismatic species' or totemic animals (e.g. the North American bald eagle) are often large, rare species high in the food chain, though there are plenty of exceptions; in general, they could simply be defined as organisms whose presence causes emotional changes in humans.

- *Emotional value*: How can we do better than 'willingness to pay' to assess the relative value of these services, e.g., using information offered via social media? What are biases in 'offered' information via social media? What is the meaning of spiritual value(s)?
- *Historical preservation*: Are there similarities between the protection of cultural heritage (e.g. a church or a painting) and natural heritage (e.g. a national park or seashore)? What does 'nature' mean and which parts are mostly appreciated? How is nature 'used' by mankind? Can we (and/or should we) monetize these values?
- *Graduality*: What is the best way to assess and monitor changes in cultural and aesthetic services? Can we define trade-offs where humans perceive a change in cultural/aesthetic ESs or is the threshold gradually reached? How quickly do humans get used/adapt to a decrease in cultural/aesthetic services of their home environments?

5.6 Synergies Among Services and Multiple Drivers: How Can We Quantify Main Effects and Interactions Among ESs and Their Drivers in the Real World?

There is a need to increase the capacity to measure and model the factors that currently lack in ES assessments (the dispossessed, the incommensurable, the unquantifiable—*sensu* Daw et al., 2015). Lavorel et al. (2011) mapped ES delivery using plant traits and soil abiotics, showing that trait distribution across landscapes is helpful to understand the mechanisms underlying ES delivery. However, although some taxa played a more major role, would the same be the case in human-dominated systems? Moreover, the main ES categories 'behave' differently. Provisioning ESs are typically based on stock-flow resources, unlike regulating and cultural ESs, which are typically fund-service based. How do we ensure we can capture these in the same way if interactions apparently change through time (Bennett et al., 2009)?

- *Stress*: How many dimensions of stressors are involved—e.g., is it always the large, rare species high in the food web that are the most strongly affected, as seems to be the case for climate warming, habitat fragmentation, acidification, and drought? Do certain combinations of stressors amplify or modulate the effects of others? Which services are likely to be diminished and which enhanced by climate change or the spread of invasive species? Can interactions persist even though one or more stressors (e.g. summer drought events, severe fires, or pesticide run-offs) are temporary?
- *Traits*: Beyond mapping: we need to advance the science for analyzing and projecting change in multiple ESs in location-based studies. Can we produce functional models of ES delivery that accommodate ecosystem condition, ecosystem change as a response to multiple factors, thresholds, and uncertainty, and that can inform management decisions? Can trait and species distribution databases be merged and can the derived trait distribution maps be used to predict multiple ES provisioning at large spatial scales?
- *Trade-offs and synergies (#3)*: How can we improve tools for measuring and modelling joint behaviour of multiple ESs (trade-offs, synergies, etc.)? What causes relationships between services to be either trade-offs or synergies? Are they caused only by response to the same driver, or are there cases where ESs are truly interacting through ecological processes? What steps can we take to either reduce or enhance these effects, for instance, by manipulation of network structure?

5.7 How Are Services Linked in Different Realms?

When valuing ES changes, we must account for the complexity and connectedness of ecosystems in order to enhance the accuracy of values across different layers (Wegner and Pascual, 2011). Supporting ESs are the foundation of provisioning, regulating, and cultural ESs (e.g. De Groot et al., 2002; Naeem et al., 2009; Wallace, 2007). This makes the linkage between categories of ESs and separate ecosystems (surely if they belong to different realms) difficult within the existing conceptual framework. Moreover, within the freshwater–marine–terrestrial realms connections between ecological processes can be fundamentally different, affecting coupling of ecological processes or linking of services (Krumins et al., 2013; Mancinelli and Mulder, 2015) and the importance of the aquatic–terrestrial ecotones has been well addressed (Polis et al., 1997, 2004).

But can we make sure that we will be able to identify and locate all the beneficiaries across such large domains, sometimes even across political borders (López-Hoffman et al., 2010)? Protected ecosystems are classical examples for large (transboundary) domains: such areas exhibit a large number of important and valuable ESs (biodiversity, fisheries, recreational), yet they protect also invasive species and often act as reservoirs (Burfeind et al., 2013; Hiley et al., 2014; Pejchar and Mooney, 2009). In some aquatic ecosystems, like peatlands, the protective role with respect to flooding is rapidly vanishing, and as a result, the biodiversity and associated recreational and educational ESs are also decreasing (Lamers et al., 2015).

- *Holism*: How do trade patterns affect ecosystem management, and how do changes in ecosystems that accompany these trade patterns likely affect future ES delivery? In ES delivery, should this impact our management and prediction of future provision?
- *Landscape planning*: To what extent can we use ESs for defence against flooding due to rising sea-level? Does land-sharing versus land-sparing better optimize ES delivery? As urban systems are 'loose' in their energetics and flows, would that evidently cascade down to adjacent systems with unwanted consequences?
- *Tipping points and whole-system shifts*: Can we identify pinchpoints (such as the extent to which freshwater fisheries of migratory species, like salmon, are dependent on coastal fisheries)? What is the importance of ontogenetic niche shift—turning one ES provider into another, or turning a neutral process into an ES provider?

5.8 How Do We Prioritize the 'Value' of Services?

Scaling can be a problem for provisioning, regulating, and sustaining ESs, but not necessarily from a biophysical and mathematical perspective. Identifying the conditions that enable important changes, the drivers, what is reversible and what is not in ESs (cf. Davies et al., 2014), is now an urgent concern. An effective currency for measuring ESs in standardized and comparable ways will be a key issue (Bennett et al., 2015; Howe et al., 2014). It is a further concern that the value of some important ESs (nature) is difficult to quantify, but ESs should not be assumed to have zero value simply because they are harder to measure.

Valuing natural capital appears central to bringing conservation into the main stream of modern societies (Daily et al., 2009), but Palomo et al. (2016) show how quantity and quality of delivered ESs depend on different kinds of capital, which will also create different trade-offs that affect ES sustainability (Bateman et al., 2011; Reyers et al., 2013). Well-being may increase as certain ESs degrade (Raudsepp-Hearne et al., 2010b), but paradoxically it is also true that the environmental degradation reflects increased human well-being (at least in the short-term intragenerational scale). There are always winners and losers and we need to know more about who will win and lose where and when. This makes the prioritization of important ESs difficult, especially if we have to consider the social equity in rapidly growing economies (Pascual et al., 2014), but a focus on biophysical ES modelling approaches, such as in ARIES (ARtificial Intelligence for ESs) leaves the translation of ES to economic values to the end user (Villa et al., 2014).

- *Values*: Can hypothetical (stated preference-based) and experimental valuation approaches (a form of non-monetary valuation based on choice experiments) versus people's revealed preference approaches, be a more effective means of valuing ESs with no direct monetary value, especially given intangible values such as cultural service values? Since ESs are always provided in bundles, does it make sense to value one ES, or should we only measure bundles of ESs? How can we realistically quantify ESs in terms of money including all hidden costs?
- *Priorities*: Can we prioritize ESs according to decreasing human needs and how does such a ranking change in different cultural/educational domains? Should values be based on the direct economic benefits for human society or adjusted according to the scarcity or vulnerability of the service impacted by human society?
- *Fairness*: How, when, and where are ESs co-produced by social-ecological systems? How to achieve fairness in the governance and policy instruments, such as payments for ecosystem services, to support the delivery of ES? How do ES values match with the notion of environmental fairness/justice which in turn is based on the institutional (both formal—such as policies—and informal—such as collective action norms and principles) settings? What are the culturally legitimate means of linking beneficiaries and providers to ensure ES delivery?
- *Trade-offs and synergies (#4)*: How do we balance the values of benefits and burdens of delivering different services to different members of a

community? What are the social trade-offs in ES, and what are the 'injustices' and 'inequalities' associated with the distribution of benefits and burdens of ES delivery? Why do lowland tropical regions with high biodiversity typically have more problems with disease, malnutrition, and human health than higher-elevation tropics and the temperate zone? How can equality of ESs be achieved in the face of gross global inequalities?

5.9 Coupling Models to Data: How Do We Develop a Better Predictive Understanding?

Most data sets we currently have are heterogeneous and there are often strong limitations to their access (e.g. in the case of GMOs). But we need many more freely available databases and the community urgently needs to continue building a universal open-source database for traits, records, services, and trades. At the moment, we have one huge annotated collection of all publicly available DNA sequences (Benson et al., 2013) and some smaller databases in part available upon request, like that for vascular plant traits (Kattge et al., 2011). The stimulating suggestions of early investigations (e.g. Montoya et al., 2003) indicating relationships between food-web structure and the ESs provided by terrestrial ecosystems have been repeatedly confirmed, suggesting that food-web properties can explain some ESs not only across land-use systems (e.g. De Vries et al., 2013) but possibly even at much larger spatial scales (cf. Hudson et al., 2014; Kissling et al., 2012; Thomas et al., 2015).

Hence, land-use history matters in the ES delivery, making the urgency and the value of such a database greater. Even so, are there legacies of past provision that will matter in future provision? For example, the way that we harvest timber (how much, how often, how wide) can influence not only the immediate provision of other services (wild berries, carbon storage, greenery harvest for floral use), but the way services recover over time, which influences future ES delivery and, importantly, even future timber provision. While we know part of this pattern for some services in some locations, we are still far from having a general understanding of the role of legacies of past use on future ES delivery.

- *Data mining*: Can data on human well-being be incorporated to models to better predict the value of ESs? How do we incorporate abiotic factors as nodes into network theory? Do we have to distinguish between old and new stressors? How are ESs impacted by the increasing occurrence of extreme events?

- *Parameterization*: To what extent can we exploit existing data to parameterize models? How complex do models have to be to get sufficient power, and how can we link terrestrial and freshwater models? Can we obtain better predictive understanding by using multifaceted approaches?
- *Scales (#3)*: What about the concept of multifunctionality? Do we accommodate scale dependencies when combining local-level data with global-level models? How much biodiversity can we afford to lose in future scenarios (2020, 2050, 2100) before services become unsustainable?

5.10 How Can We Manage Systems for Sustainable Delivery of ESs?

We have to accept a continuous management of agroecosystems to obtain sustainable and deliverable ESs. A good example is that methods to produce food (such as ploughing and fertilizer use) can affect water quality now and, through accumulation of nutrients in the soil, also dramatically affect it far into the future, even after farming has ceased (Bennett et al., 2009; Carpenter, 2005). It will also be essential to improve communication and decision-support tools for public understanding of alternative options for managing multiple ESs (Mace et al., 2015).

- *Network of networks*: Does the concept 'multiple ESs' make the issue too complex to be manageable? Given that biodiversity is distributed across spatial scales, should we identify and conserve 'umbrella services' that will effectively promote services regulated by species at smaller spatial scales? And if so, how does the functioning of neighbouring ecosystems affect the delivery of a given ES?
- *Conflicts of interest*: How do we deal with trade-offs and conflicts of interest (e.g. shallow lakes rich in macrophytes have clear water and high biodiversity—but may also be difficult to use for rowing and fishing)? What is the appropriate management unit to maximize the delivery of ESs across a landscape? How to optimize transboundary policies to protect ESs? How to deal with possible contrasting ESs in restoration projects? How can the health industry be persuaded that ecosystem restoration is cost-effective?
- *Stocks and flows*: How can the demand for ESs, and the way that it varies over space and time, be linked to the supply side analyzes that ecologists most often undertake? ESs are 'flows' that often depend upon a source, or a 'stock'. How can we ensure that analyzes incorporate stock

depletion and its potentially non-linear impacts on service delivery? Can we quantify trade-offs by maximizing connectivity between (sub)systems (intragroup homogeneity) where possible and intergroup heterogeneity otherwise? Can we understand how trade-offs shift over time, with feedbacks, impact, etc.?

- *Scales (#4)*: Can ES sustainability be managed at a local scale, or are large-scale approaches such as the catchment approach necessary? What are the policy instruments (e.g. payment for services) that would benefit both biodiversity conservation and ES delivery? How do we get workers from different disciplines to work together? How do we manage (or not) urban habitats as they return to a healthy state?

5.11 What is the Role of Global Connections in ESs Delivery, and How Should This Impact Our Management and Understanding/Prediction of Future Provision?

In an era of global connections (Liu et al., 2013, 2015), high-income countries meet demand for some ESs through international trade (e.g. Perfecto and Armbrecht, 2003; Perfecto and Vandermeer, 2008), allowing them to protect biodiversity and ES delivery that are more easily produced locally and less easily traded (e.g. recreation). The exponential growth of global population and exponential economic growth (directly correlated with the growth of physical production of consumable goods) of countries such as China and India are driving an increase of the global human trophic level (Bonhommeau et al., 2013), also causing an intensification of the exploitation of marine food webs (Roopnarine, 2014). Such a worldwide increase is paralleled by a simultaneous 'fishing down effect' (Pauly et al., 1998) of finfish species. Networks allow computing social, economic, and ecological aspects, making a focus on ES in different realms (marine, freshwater, terrestrial) possible, and examples of this application already being used in marine systems can be found in the EcoPath software widely used as a basis for gauging anthropogenic and environmental change on the production of commercial fish species within food webs.

- *Trade patterns*: How likely do trade patterns affect ecosystem management, and how will the environmental changes that often accompany these trade patterns affect future ES delivery?
- *Willingness to pay*: Do premium prices for food like coffee need to come exclusively from market forces? In other words, are consumers willing to pay higher prices to alleviate poverty, mitigate biodiversity loss, and hence paying for ESs elsewhere?

6. PRELIMINARY CONCLUSIONS

Our 36 research priorities, as defined in the bulleted subheadings, are broad and diverse, yet there are some similarities among the questions and even across the topical categories: for instance, both the 'scales' and the 'trade-offs and synergies' subheadings are addressed 4 ×. Regardless of the type of ES, most pleas and open questions address our concerns with *dimensions*. Spatial and temporal scales, stocks and flows, and costs and benefits are in fact nothing other than dimensional values in a particular unit. In addition, the plea for fine-resolution data reflects our concerns with defining appropriate dimensions. On the one hand, the coarser resolution of environmental grids derived from satellite imagery is appropriate to predict distributional shifts of species and ecosystems in response to climate change, invasive species, and land overexploitation (e.g. Pettorelli et al., 2014; Verbruggen et al., 2009). On the other hand, studies aiming to understand microhabitats and local variables that vary over small geographic distances should use interpolated grids that are either fine enough to reflect properties *in situ* (e.g. Martínez et al., 2012) or remotely via the use of drones. As such, the choice of a specific unit is strongly linked to the available data resolution and the delineation of any service-providing unit is depending on the considered ES (Luck et al., 2003). As soon as we accept that different units will be appropriate for different groups of ESs, it will become possible to reach a broader and more workable consensus. From that perspective, the scientific community needs to provide the evidence for the appropriate units for any ES quantification.

Another issue, indirectly related to dimensions but directly reflecting ES quantification, is that of so-called ecosystem *disservices* (the negative or unintended consequences according to Pataki et al., 2011, or more simply the costs—being services the benefits—as in Escobedo et al., 2011). Losses of biodiversity or wildlife habitat, sedimentation of waterways, emissions of greenhouse gases, and pesticide run-off seem to be, for Power (2010) and Rasmussen et al. (2012), typical disadvantages of agroecosystems. Disadvantages seem to be widespread also in urbanized areas as the term 'disservices' is increasing in urban planning (Von Döhren and Haase, 2015). Interestingly, in ISI Web of Science, the term 'disservice' seems to be used in Life Sciences twice as frequently as in Social Sciences and 4 × as much as in Health Sciences. There has been a tendency to study human–nature systems as separate entities and with unidirectional connections between human and natural systems (An, 2012), although the conceptualization of social-ecological systems is growing (Liu et al., 2007; Ostrom, 2007, 2009). However, it is beyond the scope of this overview to delve further into the philosophical issues surrounding ESs.

ESs are on the rise in their use in environmental management. The traditional functional ecology point of view quickly evolves towards a societal-needs perspective rooted in the classical social sciences. This route of thoughts pointed out some open issues, and those in agriculture seem to be particularly challenging. Agriculture can be seen as the longest running field experiment ever conducted, and understanding how artificial crop selection and land-use practices have moulded much of the Earth's surface can help us gain a better picture of how to manage these complex systems to maximize the return of the goods and services they provide. It is becoming increasingly apparent that these systems are not the barren monocultures they have long been assumed to be. Even oil palm plantations in the tropics and intensively farmed arable fields in temperate regions, although they may not be as diverse as the surrounding habitats, possess complex interaction networks. Understanding these ecological networks could help us to assess unintended consequences of the loss or relocation of species and to improve sustainable management of our future ecosystems and also, ultimately, of the wider biosphere.

ACKNOWLEDGEMENTS

We thank the 111 people who participated in workshops and conversations that resulted in the initial master list of 25 issues and are indebted to Ton De Nijs, Michael A. Huston, Theodora Petanidou, and Thomas Tscheulin who contributed additional information for our survey. Georgina M. Mace and an anonymous referee also provided valuable comments on our previous drafts.

REFERENCES

Alderson, D.L., Doyle, J.C., 2010. Contrasting views of complexity and their implications for network-centric infrastructures. IEEE Trans. Syst. Man Cybern. Part A: Syst. Hum. 40, 839–852.

An, L., 2012. Modeling human decisions in coupled human and natural systems: review of agent-based models. Ecol. Model. 229, 25–36.

Annan, K.A., 2000. We The Peoples – The Role of the United Nations in the 21st Century. United Nations, New York, NY.

Bakker, J.P., 1989. Nature Management by Cutting and Grazing. Kluwer Academic Publishers, Dordrecht, The Netherlands.

Balvanera, P., Pfisterer, A.B., Buchmann, N., He, J.S., Nakashizuka, T., Raffaelli, D., Schmid, B., 2006. Quantifying the evidence for biodiversity effects on ecosystem functioning and services. Ecol. Lett. 9, 1146–1156.

Barberán, A., Bates, S.T., Casamayor, E.O., Fierer, N., 2012. Using network analysis to explore co-occurrence patterns in soil microbial communities. ISME J. 6, 343–351.

Bascompte, J., Jordano, P., 2007. Plant–animal mutualistic networks: the architecture of biodiversity. Annu. Rev. Ecol. Syst. 38, 567–593.

Bastian, M., Heymann, S., Jacomy, M., 2009. Gephi: an open source software for exploring and manipulating networks. In: Proceedings of the International Conference on Weblogs and Social Media (ICWSM) 17–20 May 2009, San Jose, CA.

Batagelj, M., 1998. Pajek—program for large network analysis. Connections 21, 47–57.

Bateman, I.J., Mace, G.M., Fezzi, C., Atkinson, G., Turner, K., 2011. Economic analysis for ecosystem service assessments. Environ. Resour. Econ. 48, 177–218.

Beketov, M., Kefford, B.J., Schäfer, R.B., Liess, M., 2013. Pesticides reduce regional biodiversity of stream invertebrates. Proc. Natl. Acad. Sci. U. S. A. 110, 11039–11043.

Belovsky, G.E., Slade, J.B., 2000. Insect herbivory accelerates nutrient cycling and increases plant production. Proc. Natl. Acad. Sci. U. S. A. 97, 14412–14417.

Bennett, E.M., Peterson, G.D., Gordon, L., 2009. Understanding relationships among multiple ecosystem services. Ecol. Lett. 12, 1–11.

Bennett, E.M., Cramer, W., Begossi, A., Cundill, G., Díaz, S., Egoh, E., Geijzendorffer, I.R., Krug, C.B., Lavorel, S., Lazos, E., Lebel, L., Martín-Lopez, B., et al., 2015. Linking ecosystem services to human well-being: three challenges for designing research for sustainability. Curr. Opin. Environ. Sustain. 14, 76–85.

Benson, D.A., Cavanaugh, M., Clark, K., Karsch-Mizrachi, I., Lipman, D.J., Ostell, J., Sayers, E.W., 2013. GenBank. Nucleic Acids Res. 41, D36–D42.

Berendsen, R.L., Pieterse, C.M.J., Bakker, P.A.H.M., 2012. The rhizosphere microbiome and plant health. Trends Plant Sci. 17, 478–486.

Bettencourt, L.M.A., Lobo, J., Helbing, D., Kuhnert, C., West, G.B., 2007. Growth, innovation, scaling and the pace of life in cities. Proc. Natl. Acad. Sci. U. S. A. 104, 7301–7306.

Bohan, D.A., Caron-Lormier, G., Muggleton, S., Raybould, A., Tamaddoni-Nezhad, A., 2011. Automated discovery of food webs from ecological data using logic-based machine learning. PLoS ONE 6, e29028.

Bohan, D.A., Raybould, A., Mulder, C., Woodward, G., Tamaddoni-Nezhad, A., Bluthgen, N., Pocock, M., Muggleton, S., Evans, D.M., Astegiano, J., Massol, F., Loeuille, N., et al., 2013. Networking agroecology: integrating the diversity of agroecosystem interactions. Adv. Ecol. Res. 49, 1–67.

Bonhommeau, S., Dubroca, L., Le Pape, O., Barde, J., Kaplan, D.M., Chassot, E., Nieblas, A.E., 2013. Eating up the world's food web and the human trophic level. Proc. Natl. Acad. Sci. U. S. A. 110, 20617–20620.

Brennan, A., Woodward, G., Seehausen, O., Muñoz-Fuentes, V., Moritz, C., Guelmami, A., Abbott, R.J., Edelaar, P., 2014. Hybridization due to changing species distributions: adding problems or solutions to conservation of biodiversity during global change? Evol. Ecol. Res. 16, 475–491.

Burfeind, D.D., Pitt, K.A., Connolly, R.M., Byers, J.E., 2013. Performance of non-native species within marine reserves. Biol. Invasions 15, 17–28.

Butchart, S.H.M., Walpole, M., Collen, B., Van Strien, A., Scharlemann, J.P.W., Almond, R.E.A., Baillie, J.E.M., Bomhard, B., Brown, C., Bruno, J., 2010. Global biodiversity: indicators of recent declines. Science 328, 1164–1168.

Cardinale, B.J., Duffy, J.E., Gonzalez, A., Hooper, D.U., Perrings, C., Venail, P., Narwani, A., Mace, G.M., Tilman, D., Wardle, D.A., et al., 2012. Biodiversity loss and its impact on humanity. Nature 486, 59–67.

Carpenter, S.R., 2003. Regime shifts in lake ecosystems: pattern and variation. Excel. Ecol. Ser. 15, 1–199.

Carpenter, S.R., 2005. Eutrophication of aquatic ecosystems: bistability and soil phosphorus. Proc. Natl. Acad. Sci. U. S. A. 102, 10002–10005.

Carpenter, S.R., Kitchell, J.F., 1993. Experimental lakes, manipulations and measurements. In: Carpenter, S.R., Kitchell, J.F. (Eds.), The Trophic Cascade in Lakes. Cambridge University Press, Cambridge, UK, pp. 15–25.

Carpenter, S.R., Kitchell, J.F., Hodgson, J.R., Cochran, P.A., Elser, J.J., Elser, M.M., Lodge, D.M., Kretchmer, D., He, X., Von Ende, C.N., 1987. Regulation of lake primary productivity by food web structure. Ecology 68, 1863–1876.

Carpenter, S.R., Mooney, H.A., Agard, J., Capistrano, D., DeFries, R.S., Díaz, S., Dietz, T., Duraiappah, A.K., Oteng-Yeboah, A., Pereira, H.M., Perrings, C., Reid, W.V., et al., 2009. Science for managing ecosystem services: beyond the Millennium Ecosystem Assessment. Proc. Natl. Acad. Sci. U. S. A. 106, 1305–1312.

Carroll, G., 1988. Fungal endophytes in stems and leaves: from latent pathogen to mutualistic symbiont. Ecology 69, 2–9.

Chung, H., Zak, D.R., Reich, P.B., Ellsworth, D.S., 2009. Plant species richness, elevated CO_2, and atmospheric nitrogen deposition alter soil microbial community composition and function. Glob. Cha. Biol. 13, 980–989.

Clauset, A., Shalizi, C.R., Newman, M.E.J., 2009. Power-law distributions in empirical data. SIAM Rev. 51, 661–703.

Cohen, J.E., 2004. Mathematics is biology's next microscope, only better; biology is mathematics' next physics, only better. PLoS Biol. 2, 2017–2023.

Cohen, J.E., Jonsson, T., Carpenter, S.R., 2003. Ecological community description using the foodweb, species abundance, and body size. Proc. Natl. Acad. Sci. U. S. A. 100, 1781–1786.

Cohen, J.E., Schittler, D.N., Raffaelli, D.G., Reuman, D.C., 2009. Food webs are more than the sum of their tritrophic parts. Proc. Natl. Acad. Sci. U. S. A. 106, 22335–22340.

Costanza, R., Daly, H.E., 1992. Natural capital and sustainable development. Conserv. Biol. 6, 37–46.

Costanza, R., Cumberland, J., Daly, H., Goodland, R., Norgaard, R., 1997. An Introduction to Ecological Economics. CRC Press, Boca Raton, FL.

Cotton, T.E.A., Fitter, A.H., Miller, R.M., Dumbrell, A.J., Helgason, T., 2015. Fungi in the future: interannual variation and effects of atmospheric change on arbuscular mycorrhizal fungal communities. New Phytol. 205, 1598–1607.

Daily, G.C. (Ed.), 1997. Nature's Services. Island Press, Washington, DC.

Daily, G.C., Polasky, S., Goldstein, J., Kareiva, P.M., Mooney, H.A., Pejchar, L., Ricketts, T.H., Salzman, J., Shallenberger, R., 2009. Ecosystem services in decision making: time to deliver. Front. Ecol. Environ. 7, 21–28.

Davies, T.E., Fazey, I.R.A., Cresswell, W., Pettorelli, N., 2014. Missing the trees for the wood: why we are failing to see success in pro-poor conservation. Anim. Conserv. 17, 303–312.

Daw, T.M., Brown, K., Rosendo, S., Pomeroy, R., 2011. Applying the ecosystem services concept to poverty alleviation: the need to disaggregate human well-being. Environ. Conserv. 38, 370–379.

Daw, T.M., Coulthard, S., Cheung, W.W.L., Brown, K., Abunge, C., Galafassi, D., Peterson, G.D., McClanahan, T.R., Omukoto, J.O., Munyi, L., 2015. Evaluating taboo trade-offs in ecosystems services and human well-being. Proc. Natl. Acad. Sci. U. S. A. 112, 6949–6954.

DeAngelis, D.L., Mooij, W.M., 2005. Individual-based modeling of ecological and evolutionary processes. Annu. Rev. Ecol. Evol. Syst. 36, 147–168.

De Groot, R.S., Wilson, M.A., Boumans, R.M.J., 2002. A typology for the classification, description and valuation of ecosystem functions, goods and services. Ecol. Econ. 41, 393–408.

De Groot, R.S., Alkemade, R., Braat, L., Hein, L., Willemen, L., 2010. Challenges in integrating the concept of ecosystem services and values in landscape planning, management and decision making. Ecol. Complex. 7, 260–272.

De Vries, F.T., Thébault, E., Liiri, M., Birkhofer, K., Tsiafouli, M.A., Bjørnlund, L., Jørgensen, H.B., Brady, M.V., Christensen, S., De Ruiter, P.C., 2013. Soil food web properties explain ecosystem services across European land use systems. Proc. Natl. Acad. Sci. U. S. A. 110, 14296–14301.

De Vries, W., Posch, M., Reinds, G.J., Bonten, L.T.C., Mol-Dijkstra, J.P., Wamelink, G.W.W., Hettelingh, J.-P., 2015. Integrated assessment of impacts of atmospheric deposition and climate change on forest ecosystem services in Europe. Environ. Pollut. Ser. 25, 589–612.

Díaz, S., Fargione, J., Chapin III, F.S., Tilman, D., 2006. Biodiversity loss threatens human well-being. PLoS Biol. 4, e277.

Dodds, P.S., Rothman, D.H., 2000. Scaling, universality and geomorphology. Annu. Rev. Earth Planet. Sci. 28, 1–41.

Dormann, C.F., Gruber, B., Fründ, J., 2008. Introducing the bipartite package: analysing ecological networks. R News 8, 8–11.

Dunne, J.A., Williams, R.J., Martinez, N.D., Wood, R.A., Erwin, D.H., 2008. Compilation and network analyses of Cambrian food webs. PLoS Biol. 6, 693–708.

Ehrlich, P.R., Mooney, H.A., 1983. Extinction, substitution, and ecosystem services. Bioscience 33, 248–254.

Eisenhauer, N., Dobies, T., Cesarz, S., Hobbie, S.E., Meyer, R.J., Worm, K., Reich, P.B., 2013. Plant diversity effects on soil food webs are stronger than those of elevated CO_2 and N deposition in a long-term grassland experiment. Proc. Natl. Acad. Sci. U. S. A. 110, 6889–6894.

Eisenhauer, N., Bowker, M.A., Grace, J.B., Powell, J.R., 2015. From patterns to causal understanding: structural equation modeling (SEM) in soil ecology. Pedobiologia 58, 65–72.

Escobedo, F.J., Kroeger, T., Wagner, J.E., 2011. Urban forests and pollution mitigation: analyzing ecosystem services and disservices. Environ. Pollut. 159, 2078–2087.

EU, 2000. Directive 2000/60/EC of the European Parliament and of the Council of 23 October 2000, establishing a framework for Community action in the field of water policy (Water Framework Directive). Off. J. Eur. Union L327, 1–72.

EU, 2008. Directive 2008/56/EC of the European Parliament and of the Council of 17 June 2008, establishing a framework for community action in the field of marine environmental policy (Marine Strategy Framework Directive). Off. J. Eur. Union L164, 19–40.

EU, 2010. Commission decision of 1 September 2010 on criteria and methodological standards on good environmental status of marine waters. Off. J. Eur. Union L232, 14–24.

Fægri, K., Van der Pijl, L., 1979. The Principles of Pollination Ecology, third rev. ed. Pergamon Press, New York, NY.

Feld, C.K., Martins da Silva, P., Sousa, P.J., De Bello, F., Bugter, R., Grandin, U., Hering, D., Lavorel, S., Mountford, O., Pardo, I., 2009. Indicators of biodiversity and ecosystem services: a synthesis across ecosystems and spatial scales. Oikos 118, 1862–1871.

Feld, C.K., Birk, S., Bradley, D.C., Hering, D., Kail, J., Marzin, A., Melcher, A., Nemitz, D., Pedersen, M.L., Pletterbauer, F., Pont, D., Verdonschot, P.F.M., Friberg, N., 2011. From natural to degraded rivers and back again: a test of restoration ecology theory and practice. Adv. Ecol. Res. 44, 119–209.

Filatova, T., Verburg, P.H., Parker, D.C., Stannard, C.A., 2013. Spatial agent-based models for socio-ecological systems: challenges and prospects. Environ. Model. Softw. 45, 1–7.

Firbank, L.G., Heard, M.S., Woiwod, I.P., Hawes, C., Haughton, A.J., Champion, G.T., Scott, R.J., Hill, M.O., Dewar, A.M., Squire, G.R., May, M.J., Brooks, D.R., et al.,

2003. An introduction to the farm-scale evaluations of genetically modified herbicide-tolerant crops. J. Appl. Ecol. 40, 2–16.

Fisher, B., Turner, R.K., 2008. Ecosystem services: classification for valuation. Biol. Conserv. 141, 1167–1169.

Fisher, B., Turner, K., Zylstra, M., Brouwer, R., De Groot, R., Farber, S., Ferraro, P., Green, R., Hadley, D., Harlow, J., Jefferiss, P., Kirkby, C., et al., 2008. Ecosystem services and economic theory: integration for policy-relevant research. Ecol. Appl. 18, 2050–2067.

Fontaine, C., 2013. Ecology: abundant equals nested. Nature 500, 411–412.

Fortuna, M.A., Zaman, L., Wagner, A.P., Ofria, C., 2013. Evolving digital ecological networks. PLoS Comput. Biol. 9, e1002928.

Friberg, N., Bonada, N., Bradley, D.C., Dunbar, M.J., Edwards, F.K., Grey, J., Hayes, R.B., Hildrew, A.G., Lamouroux, N., Trimmer, M., Woodward, G., 2011. Biomonitoring of human impacts in freshwater ecosystems: the good, the bad and the ugly. Adv. Ecol. Res. 44, 1–68.

Gerhold, P., Cahill, J.F., Winter, M., Bartish, I.V., Prinzing, A., 2015. Phylogenetic patterns are not proxies of community assembly mechanisms (they are far better). Funct. Ecol. 29, 600–614.

Gill, R.J., Baldock, K.C.R., Brown, M.J.F., Cresswell, J.E., Dicks, L.V., Fountain, M.T., Garratt, M.P.D., Gough, L.A., Heard, M.S., Holland, J.M., Ollerton, J., Stone, G.N., et al., 2016. Protecting an ecosystem service: approaches to understanding and mitigating threats to wild insect pollinators. Adv. Ecol. Res. 54, in press.

Gordon, L.J., Finlayson, C.M., Falkenmark, M., 2010. Managing water in agriculture for food production and other ecosystem services. Agric. Water Manag. 97, 512–519.

Gotelli, N.J., Colwell, R.K., 2001. Quantifying biodiversity: procedures and pitfalls in the measurement and comparison of species richness. Ecol. Lett. 4, 379–391.

Grêt-Regamey, A., Rabe, S.E., Crespo, R., Lautenbach, S., Ryffel, A., Schlup, B., 2014. On the importance of non-linear relationships between landscape patterns and the sustainable provision of ecosystem services. Landsc. Ecol. 29, 201–212.

Gunderson, L.H., Holling, C.S. (Eds.), 2002. Pararchy: Understanding Transformations in Human and Natural Systems. Island Press, Washington, DC.

Hacquard, S., Schadt, C.W., 2015. Towards a holistic understanding of the beneficial interactions across the Populus microbiome. New Phytol. 205, 1424–1430.

Haldane, A.G., May, R.M., 2011. Systemic risk in banking ecosystems. Nature 469, 351–355.

Heal, O.W., Dighton, J., 1986. Nutrient cycling and decomposition of natural terrestrial ecosystems. Dev. Biogeochem. 3, 14–73.

Hector, A., Bagchi, R., 2007. Biodiversity and ecosystem multifunctionality. Nature 448, 188–190.

Hein, L., Van Koppen, K., De Groot, R.S., Van Ierland, E.C., 2006. Spatial scales, stakeholders and the valuation of ecosystem services. Ecol. Econ. 57, 209–228.

Hiley, J.R., Bradbury, R.B., Thomas, C.D., 2014. Introduced and natural colonists show contrasting patterns of protected area association in UK wetlands. Divers. Distrib. 20, 943–951.

Hines, J., van der Putten, W.H., De Deyn, G.B., Wagg, C., Voigt, W., Mulder, C., Weisser, W.W., Engel, J., Melian, C., Scheu, S., Birkhofer, K., Ebeling, A., et al., 2015. Towards an integration of biodiversity–ecosystem functioning and food web theory to evaluate relationships between multiple ecosystem services. Adv. Ecol. Res. 53, 161–199.

Holling, C.S., 1973. Resilience and stability of ecological systems. Annu. Rev. Ecol. Syst. 4, 1–23.

Holling, C.S., 1987. Simplifying the complex: the paradigms of ecological function and structure. Eur. J. Oper. Res. 30, 139–146.

Hooper, D.U., Chapin III, F.S., Ewel, J.J., Hector, A., Inchausti, P., Lavorel, S., Lawton, J.H., Lodge, D.M., Loreau, M., Naeem, S., 2005. Effects of biodiversity on ecosystem functioning: a consensus of current knowledge. Ecol. Monogr. 75, 3–35.

Howe, C., Suich, H., Vira, B., Mace, G.M., 2014. Creating win-wins from trade-offs? Ecosystem services for human well-being: a meta-analysis of ecosystem service trade-offs and synergies in the real world. Glob. Environ. Change 28, 263–275.

Hudson, P.J., Dobson, A.P., Lafferty, K.D., 2006. Is a healthy ecosystem one that is rich in parasites? Trends Ecol. Evol. 21, 381–385.

Hudson, L.N., Emerson, R., Jenkins, G.B., Layer, K., Ledger, M.E., Pichler, D.E., Thompson, M.S.A., O'Gorman, E.J., Woodward, G., Reuman, D.C., 2013. Cheddar: analysis and visualisation of ecological communities in R. Methods Ecol. Evol. 4, 99–104.

Hudson, L.N., Newbold, T., Contu, S., Hill, S.L.L., Lysenko, I., De Palma, A., Phillips, H.R.P., Senior, R.A., Bennett, D.J., Booth, H., Choimes, A., Correia, D.L.P., et al., 2014. The PREDICTS database: a global database of how local terrestrial biodiversity responds to human impacts. Ecol. Evol. 4, 4701–4735.

Huston, M.A., Aarssen, L.W., Austin, M.P., Cade, B.S., Fridley, J.D., Garnier, E., Grime, J.P., Hodgson, J., Lauenroth, W.K., Thompson, K., Vandermeer, J.H., Wardle, D.A., 2000. No consistent effect of plant diversity on productivity. Science 289, 1255a.

Ibanez, S., Bernard, L., Coq, S., Moretti, M., Lavorel, S., Gallet, C., 2013. Herbivory differentially alters litter dynamics of two functionally contrasted grasses. Funct. Ecol. 27, 1064–1074.

Ings, T.C., Montoya, J.M., Bascompte, J., Blüthgen, N., Brown, L., Dormann, C.F., Edwards, F., Figueroa, D., Jacob, U., Jones, J.I., Lauridsen, R.B., Ledger, M.E., et al., 2009. Ecological networks—beyond food webs. J. Anim. Ecol. 78, 253–269.

Jackson, L.E., Bowles, T.M., Hodson, A.K., Lazcano, C., 2012. Soil microbial-root and microbial-rhizosphere processes to increase nitrogen availability and retention in agroecosystems. Curr. Opin. Environ. Sustain. 4, 517–522.

Jax, K., 2010. Ecosystem Functioning. Cambridge University Press, New York, NY.

Jennings, S., Blanchard, J.L., 2004. Fish abundance with no fishing: predictions based on macroecological theory. J. Anim. Ecol. 73, 632–642.

Jennings, S., Greenstreet, S.P.R., Reynolds, J.D., 1999. Structural change in an exploited fish community: a consequence of differential fishing effects on species with contrasting life histories. J. Anim. Ecol. 68, 617–627.

Ji, Y., Ashton, L., Pedley, S.M., Edwards, D.P., Tang, Y., Nakamura, A., Kitching, R., Dolman, P.M., Woodcock, P., Edwards, F.A., Larsen, T.H., Hsu, W.W., et al., 2013. Reliable, verifiable and efficient monitoring of biodiversity via metabarcoding. Ecol. Lett. 16, 1245–1257.

Jonsson, T., 2014. Trophic links and the relationship between predator and prey body sizes in food webs. Community Ecol. 15, 54–64.

Jonsson, T., Cohen, J.E., Carpenter, S.R., 2005. Food webs, body size, and species abundance in ecological community description. Adv. Ecol. Res. 36, 1–84.

Kattge, J., Díaz, S., Lavorel, S., Prentice, I., Leadley, P., Bonisch, G., Garnier, E., Westoby, M., Reich, P., Wright, I., Cornelissen, J., Violle, C., et al., 2011. TRY—a global database of plant traits. Glob. Cha. Biol. 17, 2905–2935.

Kennedy, C.A., Stewart, I., Facchini, A., Cersosimo, I., Mele, R., Chen, B., Uda, M., Kansal, A., Chiu, A., Kim, K., Dubeux, C., La Rovere, E.L., et al., 2015. Energy and material flows of megacities. Proc. Natl. Acad. Sci. U. S. A. 112, 5985–5990.

Kiers, E.T., Hutton, M.G., Denison, R.F., 2007. Human selection and the relaxation of legume defences against ineffective rhizobia. Proc. R. Soc. Lond. B 274, 3119–3126.

Kissling, W.D., Dormann, C.F., Groeneveld, J., Hickler, T., Kühn, I., McInerny, G.J., Montoya, J.M., Römermann, C., Schiffers, K., Schurr, F.M., Singer, A.,

Svenning, J.-C., et al., 2012. Towards novel approaches to modelling biotic interactions in multispecies assemblages at large spatial extents. J. Biogeogr. 39, 2163–2178.

Kleijn, D., Winfree, R., Bartomeus, I., Carvalheiro, L.G., Henry, M., Isaacs, R., Klein, A.-M., Kremen, C., M'Gonigle, L.K., Rader, R., Ricketts, T.H., Williams, N.M., et al., 2015. Delivery of crop pollination services is an insufficient argument for wild pollinator conservation. Nat. Commun. 6, 7414.

Koch, E.W., Barbier, E.B., Silliman, B.R., Reed, D.J., Perillo, G.M.E., Hacker, S.D., Granek, E.F., Primavera, J.H., Muthiga, N., Polasky, S., 2009. Non-linearity in ecosystem services: temporal and spatial variability in coastal protection. Front. Ecol. Environ. 7, 29–37.

Krumins, J.A., Van Oevelen, D., Bezemer, T.M., De Deyn, G.B., Hol, W.H.G., Van Donk, E., De Boer, W., De Ruiter, P.C., Middelburg, J.J., Monroy, F., Soetaert, K., Thébault, E., et al., 2013. Soil and freshwater and marine sediment food webs: their structure and function. Bioscience 63, 35–42.

Laakso, J., Setälä, H., Palojärvi, A., 2000. Influence of decomposer food web structure and nitrogen availability on plant growth. Plant Soil 225, 153–165.

Lafferty, K.D., Allesina, S., Arim, M., Briggs, C.J., De Leo, G., Dobson, A.P., Dunne, J.A., Johnson, P.T.J., Kuris, A.M., Marcogliese, D.J., Martinez, N.D., Memmott, J., et al., 2008. Parasites in food webs: the ultimate missing links. Ecol. Lett. 11, 533–546.

Lamers, L.P.M., Vile, M.A., Grootjans, A.P., Acreman, M.C., Van Diggelen, R., Evans, M.G., Richardson, C.J., Rochefort, L., Kooijman, A.M., Roelofs, J.G.M., Smolders, A.J.P., 2015. Ecological restoration of rich fens in Europe and North America: from trial and error to an evidence-based approach. Biol. Rev. 90, 182–203.

Lavorel, S., Grigulis, K., Lamarque, P., Colace, M.P., Garden, D., Girel, J., 2011. Using plant functional traits to understand the landscape distribution of multiple ecosystem services. J. Ecol. 99, 135–147.

Layer, K., Riede, J.O., Hildrew, A.G., Woodward, G., 2010. Food web structure and stability in 20 streams across a wide pH gradient. Adv. Ecol. Res. 42, 265–299.

Lentendu, G., Wubet, T., Chatzinotas, A., Wilhelm, C., Buscot, F., Schlegel, M., 2014. Effects of long-term differential fertilization on eukaryotic microbial communities in an arable soil: a multiple barcoding approach. Mol. Ecol. 23, 3341–3355.

Lester, S.E., Costello, C., Halpern, B.S., Gaines, S.D., White, C., Barth, J.A., 2013. Evaluating tradeoffs among ecosystem services to inform marine spatial planning. Mar. Policy 38, 80–89.

Levin, S.A., 1998. Ecosystems and the biosphere as complex adaptive systems. Ecosystems 1, 431–436.

Levin, S.A., 2000. Fragile Dominion: Complexity and the Commons. Princeton University Press, Princeton, NJ.

Limburg, K.E., Luzadis, V.A., Ramsey, M., Schulz, K.L., Mayer, C.M., 2010. The good, the bad, and the algae: perceiving ecosystem services and disservices generated by zebra and quagga mussels. J. Great Lakes Res. 36, 86–94.

Liss, K.N., Mitchell, M.G.E., MacDonald, G.K., Mahajan, S., Méthot, J., Jacob, A.L., Maguire, D., Metson, G., Ziter, C., Dancose, K., Martins, K., Terrado, M., et al., 2013. Variability in ecosystem service measurement: a case study of pollination service studies. Front. Ecol. Environ. 11, 414–422.

Liu, J., Dietz, T., Carpenter, S.R., Alberti, M., Folke, C., Moran, E., Pell, A.N., Deadman, P., Kratz, T., Lubchenco, J., Ostrom, E., Ouyang, Z., et al., 2007. Complexity of coupled human and natural systems. Science 317, 1513–1516.

Liu, Y.-Y., Slotine, J.-J., Barabási, A.-L., 2011. Controllability of complex networks. Nature 473, 167–173.

Liu, J., Hull, V., Batistella, M., DeFries, R., Dietz, T., Fu, F., Hertel, T.W., Izaurralde, R.C., Lambin, E.F., Li, S., Martinelli, L.A., McConnell, W.J., et al., 2013. Framing sustainability in a telecoupled world. Ecol. Soc. 18, 26.

Liu, J., Mooney, H., Hull, V., Davis, S.J., Gaskell, J., Hertel, T., Lubchenco, J., Seto, K.C., Gleick, P., Kremen, C., Li, S., 2015. Systems integration for global sustainability. Science 347, 1258832.

Loladze, I., 2002. Rising atmospheric CO_2 and human nutrition: toward globally imbalanced plant stoichiometry? Trends Ecol. Evol. 17, 457–461.

López-Hoffman, L., Varady, R.G., Flessa, K.W., Balvanera, P., 2010. Ecosystem services across borders: a framework for transboundary conservation policy. Front. Ecol. Environ. 8, 84–91.

Luck, G.W., Daily, G.C., Ehrlich, P.R., 2003. Population diversity and ecosystem services. Trends Ecol. Evol. 18, 331–336.

Luck, G.W., Harrington, R., Harrison, P.A., Kremen, C., Berry, P.M., Bugter, R., Dawson, T.R., De Bello, F., Díaz, S., Feld, C.K., Haslett, J.R., Hering, D., et al., 2009. Quantifying the contribution of organisms to the provision of ecosystem services. Bioscience 59, 223–235.

Lyytimäki, J., Sipilä, M., 2009. Hoping on one leg—the challenge of ecosystem disservices for urban green management. Urban For. Urban Green. 8, 309–315.

Ma, A., Mondragón, R., 2012. Evaluation of network robustness using a node tearing algorithm. Phys. A 391, 6674–6681.

Ma, A., Mondragón, R., 2015. Rich-cores in networks. PLoS ONE 10, e0119678.

Mace, G.M., Norris, K., Fitter, A.H., 2012. Biodiversity and ecosystem services: a multilayered relationship. Trends Ecol. Evol. 27, 19–26.

Mace, G.M., Hails, R.S., Cryle, P., Harlow, J., Clarke, S.J., 2015. Towards a risk register for natural capital. J. Appl. Ecol. 52, 641–653.

Madin, J., Bowers, S., Schildhauer, M., Krivov, S., Pennington, D., Villa, F., 2007. An ontology for describing and synthesizing ecological observation data. Ecol. Inform. 2, 279–296.

Magurran, A.E., 2013. Measuring Biological Diversity. John Wiley & Sons, Malden, MA.

Mancinelli, G., Mulder, C., 2015. Detrital dynamics and cascading effects on supporting ecosystem services. Adv. Ecol. Res. 53, 97–160.

Martínez, B., Viejo, R.M., Carreno, F., Aranda, S.C., 2012. Habitat distribution models for intertidal seaweeds: responses to climatic and non-climatic drivers. J. Biogeogr. 39, 1877–1890.

Martínez-López, F.J., Gázquez-Abad, J.C., Sousa, C.M.P., 2013. Structural equation modelling in marketing and business research: critical issues and practical recommendations. Eur. J. Mark. 47, 115–152.

MEA, 2005. Ecosystems and Human Well-being: Biodiversity Synthesis. Island Press, Washington, DC.

Mesarovic, M.D., 1984. Complexity and the Global Future. Proc. Club Rome, Helsinki, p. 192.

Meyer, W.B., Turner II, B.L., 1992. Human population growth and global land-use/cover change. Annu. Rev. Ecol. Syst. 23, 39–61.

Mitchell, R.J., 1992. Testing evolutionary and ecological hypotheses using path analysis and structural equation modelling. Funct. Ecol. 6, 123–129.

Montoya, J.M., Rodríguez, M.A., Hawkins, B.A., 2003. Food web complexity and higher-level ecosystem services. Ecol. Lett. 6, 587–593.

Montoya, J.M., Woodward, G., Emmerson, M.C., Solé, R.V., 2009. Press perturbations and indirect effects in real food webs. Ecology 90, 2426–2433.

Moretti, M., De Bello, F., Ibanez, S., Fontana, S., Pezzatti, G.B., Dziock, F., Rixen, C., Lavorel, S., 2013. Linking traits between plants and invertebrate herbivores to track functional effects of land-use changes. J. Veg. Sci. 24, 949–962.

Mouchet, M.A., Villeger, S., Mason, N.W.H., Mouillot, D., 2010. Functional diversity measures: an overview of their redundancy and their ability to discriminate community assembly rules. Funct. Ecol. 24, 867–876.

Mouquet, N., Devictor, V., Meynard, C.N., Munoz, F., Bersier, L.-F., Chave, J., Couteron, P., Dalecky, A., Fontaine, C., Gravel, D., Hardy, O.J., Jabot, F., et al., 2012. Ecophylogenetics: advances and perspectives. Biol. Rev. 87, 769–785.

Mulder, C., Elser, J.J., 2009. Soil acidity, ecological stoichiometry and allometric scaling in grassland food webs. Glob. Cha. Biol. 15, 2730–2738.

Mulder, C., Aldenberg, T., De Zwart, D., Van Wijnen, H.J., Breure, A.M., 2005. Evaluating the impact of pollution on plant–Lepidoptera relationships. Environmetrics 16, 357–373.

Mulder, C., Boit, A., Mori, S., Vonk, J.A., Dyer, S.D., Faggiano, L., Geisen, S., González, A.-L., Kaspari, M., Lavorel, S., Marquet, P.A., Rossberg, A.G., et al., 2012. Distributional (in) congruence of biodiversity-ecosystem functioning. Adv. Ecol. Res. 46, 1–88.

Mulder, C., Ahrestani, F.S., Bahn, M., Bohan, D.A., Bonkowski, M., Griffiths, B.S., Guicharnaud, R.A., Kattge, J., Krogh, P.H., Lavorel, S., Lewis, O.T., Mancinelli, G., et al., 2013. Connecting the green and brown worlds: allometric and stoichiometric predictability of above- and belowground networks. Adv. Ecol. Res. 49, 69–175.

Mulder, C., Hettelingh, J.-P., Montanarella, L., Pasimeni, M.R., Posch, M., Voigt, W., Zurlini, G., 2015. Chemical footprints of anthropogenic nitrogen deposition on recent soil C:N ratios in Europe. Biogeosciences 12, 4113–4119.

Naeem, S., Bunker, D.E., Hector, A., Loreau, M., Perrings, C., 2009. Can we predict the effects of global change on biodiversity loss and ecosystem functioning? In: Naeem, S., Bunker, D.E., Hector, A., Loreau, M., Perrings, C. (Eds.), Biodiversity, Ecosystem Functioning, and Human Wellbeing. Oxford University Press, Oxford, pp. 290–298.

Naeem, S., Duffy, J.E., Zavaleta, E., 2012. The functions of biological diversity in an age of extinction. Science 336, 1401–1406.

Naeem, S., Ingram, J.C., Varga, A., Agardy, T., Barten, P., Bennett, G., Bloomgarden, E., Bremer, L.L., Burkill, P., Cattau, M., Ching, C., Colby, C., et al., 2015. Get the science right when paying for nature's services. Science 347, 1206–1207.

Naidoo, R., Balmford, A., Costanza, R., Fisher, B., Green, R.E., Lehner, B., Malcolm, T.R., Ricketts, T.H., 2010. Global mapping of ecosystem services and conservation priorities. Proc. Natl. Acad. Sci. U. S. A. 105, 9495–9500.

Newman, M., 2003. The structure and function of complex networks. SIAM Rev. 45, 167–256.

Newman, M., 2006. Finding community structure in networks using the eigenvectors of matrices. Phys. Rev. E 74, 036104.

Ospina, S., Rusch, G.M., Pezo, D.A., Casanoves, F., Sinclair, F.L., 2012. More stable productivity of semi natural grasslands than sown pastures in a seasonally dry climate. PLoS ONE 7, e35555.

Ostrom, E., 2007. A diagnostic approach for going beyond panaceas. Proc. Natl. Acad. Sci. U. S. A. 104, 15181–15187.

Ostrom, E., 2009. A general framework for analyzing sustainability of social–ecological systems. Science 35, 419–422.

Paetzold, A., Warren, P.H., Maltby, L.L., 2010. A framework for assessing ecological quality based on ecosystem services. Ecol. Complex. 7, 273–281.

Palomo, I., Felipe-Lucia, M.R., Bennett, E.M., Martín-López, B., Pascual, U., 2016. Disentangling the pathways and effects of ecosystem service co-production. Adv. Ecol. Res. 54, in press.

Pascual, U., Phelps, J., Garmendia, E., Brown, K., Corbera, E., Martin, A., Gomez-Baggethun, E., Muradian, R., 2014. Social equity matters in payments for ecosystem services. Bioscience 64, 1027–1036.

Pataki, D.E., Carreiro, M.M., Cherrier, J., Grulke, N.E., Jennings, V., Pincetl, S., Pouyat, R.V., Whitlow, T.H., Zipperer, W.C., 2011. Coupling biogeochemical cycles in urban environments: ecosystem services, green solutions, and misconceptions. Front. Ecol. Environ. 9, 27–36.

Pauly, D., Christensen, V., Dalsgaard, J., Froese, R., Torres, F., 1998. Fishing down marine food webs. Science 279, 860–863.

Pawar, S., Dell, A.I., Savage, V.M., 2015. From metabolic constraints on individuals to the dynamics of ecosystems. In: Belgrano, A., Woodward, G., Jacob, U. (Eds.), Aquatic Functional Biodiversity: An Ecological and Evolutionary Perspective. Academic Press, London, pp. 3–36.

Pejchar, L., Mooney, H.A., 2009. Invasive species, ecosystem services and human well-being. Trends Ecol. Evol. 24, 497–504.

Perfecto, I., Armbrecht, I., 2003. The coffee agroecosystem in the Neotropics: combining ecological and economic goals. In: Vandermeer, J. (Ed.), Tropical Agroecosystems. CRC, Boca Raton, FL, pp. 159–194.

Perfecto, I., Vandermeer, J., 2008. Biodiversity conservation in tropical agroecosystems: a new conservation paradigm. Ann. N.Y. Acad. Sci. 1134, 173–200.

Perrings, C., Folke, C., Mäler, K.G., 1992. The ecology and economics of biodiversity loss: the research agenda. Ambio 21, 201–211.

Perrings, C., Naeem, S., Ahrestani, F.S., Bunker, D.E., Burkill, P., Canziani, G., Elmqvist, T., Fuhrman, J.A., Jaksic, F.M., Kawabata, A., Kinzig, A., Mace, G.M., et al., 2011. Ecosystem services, targets, and indicators for the conservation and sustainable use of biodiversity. Front. Ecol. Environ. 9, 512–520.

Pettorelli, N., Laurance, W.F., O'Brien, T.G., Wegmann, M., Nagendra, H., Turner, W., 2014. Satellite remote sensing for applied ecologists: opportunities and challenges. J. Appl. Ecol. 51, 839–848.

Pettorelli, N., Hilborn, A., Duncan, C., Durant, S.M., 2015. Individual variability: the missing component to our understanding of predator–prey interactions. Adv. Ecol. Res. 52, 19–44.

Pichlmair, A., Kandasamy, K., Alvisi, G., Mulhern, O., Sacco, R., Habjan, M., Binder, M., Stefanovic, A., Eberle, C.-A., Goncalves, A., Bürckstümmer, T., Müller, A.C., et al., 2012. Viral immune modulators perturb the human molecular network by common and unique strategies. Nature 487, 486–490.

Pocock, M.J.O., Evans, D.M., Memmott, J., 2012. The robustness and restoration of a network of ecological networks. Science 335, 973–977.

Pocock, M.J.O., Evans, D.M., Fontaine, C., Harvey, M., Julliard, R., McLaughlin, Ó., Silvertown, J., Tamaddoni-Nezhad, A., White, P.C.L., Bohan, D.A., 2016. The visualisation of ecological networks, and their use as a tool for engagement, advocacy and management. Adv. Ecol. Res. 54, in press.

Poisot, T., Mouquet, N., Gravel, D., 2013. Trophic complementarity drives the biodiversity–ecosystem functioning relationship in food webs. Ecol. Lett. 16, 853–861.

Polin, S., Simon, J.-C., Outreman, Y., 2014. An ecological cost associated with protective symbionts of aphids. Ecol. Evol. 4, 836–840.

Polis, G.A., Anderson, W.B., Holt, R.D., 1997. Toward an integration of landscape and food web ecology: the dynamics of spatially subsidized food webs. Annu. Rev. Ecol. Syst. 28, 289–316.

Polis, G.A., Power, M.E., Huxel, G.R. (Eds.), 2004. Food Webs at the Landscape Level. University of Chicago Press, Chicago, MI.

Poppy, G.M., Chiotha, S., Eigenbrod, F., Harvey, C.A., Honzák, M., Hudson, M.D., Jarvis, A., Madise, N.J., Schreckenberg, K., Shackleton, C.M., Villa, F., Dawson, T.P., 2014. Food security in a perfect storm: using the ecosystem services framework to increase understanding. Philos. Trans. R. Soc. Lond. B Biol. Sci. 369, 20120288.

Poulin, R., 2010. Network analysis shining light on parasite ecology and diversity. Trends Parasitol. 26, 492–498.

Power, A.G., 2010. Ecosystem services and agriculture: tradeoffs and synergies. Philos. Trans. R. Soc. Lond. B Biol. Sci. 365, 2959–2971.

Pugesek, B.H., Tomer, A., Von Eye, A., 2003. Structural Equation Modeling: Applications in Ecological and Evolutionary Biology. Cambridge University Press, Cambridge, UK.

Raffaelli, D., 2016. Ecosystem structures and processes: characterising natural capital stocks and flows. In: Potschin, M., Haines-Young, R., Fish, R., Turner, R.K. (Eds.), Routledge Handbook of Ecosystem Services. Routledge: Taylor & Francis Group, New York, NY, in press.

Raffaelli, D., White, P.C.L., 2013. Ecosystems and their services in a changing world: an ecological perspective. Adv. Ecol. Res. 48, 1–70.

Raffaelli, D., Bullock, J.M., Cinderby, S., Durance, I., Emmett, B., Harris, J., Hicks, K., Oliver, T.H., Patersonk, D., White, P.C.L., 2014. Big data and ecosystem research programmes. Adv. Ecol. Res. 51, 41–77.

Rapparini, F., Peñuelas, J., 2014. Mycorrhizal fungi to alleviate drought stress on plant growth. In: Miransari, M. (Ed.), In: Use of Microbes for the Alleviation of Soil Stresses vol. 1. Springer, New York, NY, pp. 21–42.

Rasmussen, J.J., Wiberg-Larsen, P., Baattrup-Pedersen, A., Friberg, N., Kronvang, B., 2012. Stream habitat structure influences macroinvertebrate response to pesticides. Environ. Pollut. 164, 142–149.

Raudsepp-Hearne, C., Peterson, G.D., Bennett, E.M., 2010a. Ecosystem service bundles for analyzing tradeoffs in diverse landscapes. Proc. Natl. Acad. Sci. U. S. A. 107, 5242–5247.

Raudsepp-Hearne, C.M.T., Peterson, G.D., Bennett, E.M., Holland, T., Benessaiah, K., MacDonald, G.K., Pfeifer, L., 2010b. Untangling the environmentalist's paradox: why is human well-being increasing as ecosystem services degrade? Bioscience 60, 576–589.

Reiss, J., Bridle, J.R., Montoya, G.M., Woodward, G., 2009. Emerging horizons in biodiversity and ecosystem functioning research. Trends Ecol. Evol. 24, 505–514.

Reuman, D.C., Mulder, C., Raffaelli, D., Cohen, J.E., 2008. Three allometric relations of population density to body mass: theoretical integration and empirical tests in 149 food webs. Ecol. Lett. 11, 1216–1228.

Reyers, B., Biggs, R., Cumming, G.S., Elmqvist, T., Hejnowicz, A.P., Polasky, S., 2013. Getting the measure of ecosystem services: a social–ecological approach. Front. Ecol. Environ. 11, 268–273.

Roopnarine, P.D., 2014. Humans are apex predators. Proc. Natl. Acad. Sci. U. S. A. 111, E796.

Sala, O.E., Chapin III, F.S., Armesto, J.J., Berlow, E., Bloomfield, J., Dirzo, R., Huber-Sanwald, E., Huenneke, L.F., Jackson, R.B., Kinzig, A., Leemans, R., Lodge, D.M., et al., 2000. Global biodiversity scenarios for the year 2100. Science 287, 1770–1774.

Scherber, C., Eisenhauer, N., Weisser, W.W., Schmid, B., Voigt, W., Fischer, M., Schulze, E.-D., Roscher, C., Weigelt, A., Allan, E., Beßler, H., Bonkowski, M., et al., 2010. Bottom-up effects of plant diversity on multitrophic interactions in a biodiversity experiment. Nature 468, 553–556.

Sechi, V., Brussaard, L., De Goede, R.G.M., Rutgers, M., Mulder, C., 2015. Choice of resolution by functional trait or taxonomy affects allometric scaling in soil food webs. Am. Nat. 185, 142–149.

Setälä, H., Laakso, J., Mikola, J., Huhta, V., 1998. Functional diversity of decomposer organisms in relation to primary production. Appl. Soil Ecol. 9, 25–31.

Shipley, B., 2002. Cause and Correlation in Biology: A User's Guide to Path Analysis, Structural Equations and Causal Inference. Cambridge University Press, Cambridge, UK.

Sint, D., Traugott, M., 2015. Food Web Designer: a flexible tool to visualize interaction networks. J. Pest. Sci. http://dx.doi.org/10.1007/s10340-015-0686-7.

Stephens, P.A., Pettorelli, N., Barlow, J., Whittingham, M.J., Cadotte, M.W., 2015. Management by proxy? The use of indices in applied ecology. J. Appl. Ecol. 52, 1–6.

Sterner, R.W., Elser, J.J., 2002. Ecological Stoichiometry: The Biology of Elements from Molecules to the Biosphere. Princeton University Press, Princeton, NJ.

Stoll, S., Frenzel, M., Burkhard, B., Adamescu, M., Augustaitis, A., Baeßler, C., Bonet, F.J., Carranza, M.L., Cazacu, C., Cosor, G.L., 2015. Assessment of ecosystem integrity and service gradients across Europe using the LTER Europe network. Ecol. Model. 295, 75–87.

Stumpf, M.P.H., Porter, M.A., 2012. Critical truths about power laws. Science 335, 665–666.

Sutherland, W.J., Armstrong-Brown, S., Armsworth, P.R., Brereton, T., Brickland, J., Campbell, C.D., Chamberlain, D.E., Cooke, A.I., Dulvy, N.K., Dusic, N.R., Fitton, M., Freckleton, R.P., et al., 2006. The identification of 100 ecological questions of high policy relevance in the UK. J. Appl. Ecol. 43, 617–627.

Sutherland, W.J., Albon, S.D., Allison, H., Armstrong-Brown, S., Bailey, M.J., Brereton, T., Boyd, I.L., Carey, P., Edwards, J., Gill, M., Hill, D., Hodge, I., et al., 2010. The identification of priority policy options for UK nature conservation. J. Appl. Ecol. 47, 955–965.

Sutherland, W.J., Fleishman, E., Mascia, M.B., Pretty, J., Rudd, M.A., 2011. Methods for collaboratively identifying research priorities and emerging issues in science and policy. Methods Ecol. Evol. 2, 238–247.

Sutherland, W.J., Bardsley, S., Clout, M., Depledge, M.H., Dicks, L.V., Fellman, L., Fleishman, E., Gibbons, D.W., Keim, B., Lickorish, F., Margerison, C., Monk, K.A., et al., 2013. A horizon scan of global conservation issues for 2013. Trends Ecol. Evol. 28, 16–22.

Thébault, E., Fontaine, C., 2010. Stability of ecological communities and the architecture of mutualistic and trophic networks. Science 329, 853–856.

Thébault, E., Loreau, M., 2003. Food-web constraints on biodiversity–ecosystem functioning relationships. Proc. Natl. Acad. Sci. U. S. A. 100, 14949–14954.

Thomas, R.Q., Brookshire, E.N.J., Gerber, S., 2015. Nitrogen limitation on land: how can it occur in Earth system models? Glob. Cha. Biol. 21, 1777–1793.

Thompson, M.S.A., Bankier, C., Bell, T., Dumbrell, A.J., Gray, C., Ledger, M.E., Lehmann, K., McKew, B.A., Sayer, C.D., Shelley, F., Trimmer, M., Warren, S.L., et al., 2016. Gene-to-ecosystem impacts of a catastrophic pesticide spill: testing a multilevel bioassessment approach in a river ecosystem. Freshw. Biol., http://dx.doi.org/10.1111/fwb.126760.

Tilman, D., Knops, J., Wedin, D., Reich, P., Ritchie, M., Siemann, E., 1997. The influence of functional diversity and composition on ecosystem processes. Science 277, 1300–1302.

Tur, C., Vigalondo, B., Trøjelsgaard, K., Olesen, J.M., Traveset, A., 2014. Downscaling pollen–transport networks to the level of individuals. J. Anim. Ecol. 83, 306–317.

Tylianakis, J.M., Coux, C., 2014. Tipping points in ecological networks. Trends Plant Sci. 19, 281–283.

Um, J., Son, S.-W., Lee, S.-I., Jeong, H., Kim, B.J., 2009. Scaling laws between population and facility densities. Proc. Natl. Acad. Sci. U. S. A. 106, 14236–14240.

Vacher, C., Piou, D., Desprez-Loustau, M.-L., 2008. Architecture of an antagonistic tree/fungus network: the asymmetric influence of past evolutionary history. PLoS ONE 3, e1740.

Vacher, C., Tamaddoni-Nezhad, A., Kamenova, S., Peyrard, N., Moalic, Y., Sabbadin, R., Schwaller, L., Chiquet, J., Smith, M.A., Vallance, J., Fievet, V., Jakuschkin, B., Bohan, D.A., 2016. Learning ecological networks from next-generation sequencing data. Adv. Ecol. Res. 54, in press.

Vayssier-Taussat, M., Albina, E., Citti, C., Cosson, J.-F., Jacques, M.-A., Lebrun, M.-H., Le Loir, Y., Ogliastro, M., Petit, M.-A., Roumagnac, P., Candresse, T., 2014. Shifting the paradigm from pathogens to pathobiome: new concepts in the light of meta-omics. Front. Cell. Infect. Microbiol. 4, 29.

Verbruggen, H., Tyberghein, L., Pauly, K., Vlaeminck, C., Van Nieuwenhuyze, K., Kooistra, W.H.C.F., Leliaert, F., De Clerck, O., 2009. Macroecology meets macroevolution: evolutionary niche dynamics in the seaweed Halimeda. Glob. Ecol. Biogeogr. 18, 393–405.

Verhoeven, J.T.A., Arheimer, B., Yin, C., Hefting, M.M., 2006. Regional and global concerns over wetlands and water quality. Trends Ecol. Evol. 21, 96–103.

Vespignani, A., 2012. Modelling dynamical processes in complex socio-technical systems. Nat. Phys. 8, 32–39.

Villa, F., Bagstad, K.J., Voigt, B., Johnson, G.W., Portela, R., Honzák, M., Batker, D., 2014. A methodology for adaptable and robust ecosystem services assessment. PLoS ONE 9, e91001.

Von Döhren, P., Haase, D., 2015. Ecosystem disservices research: a review of the state of the art with a focus on cities. Ecol. Indic. 52, 490–497.

Walker, B., Salt, D., 2006. Resilience Thinking. Sustaining Ecosystems and People in a Changing World. Island Press, Washington, DC.

Wallace, K.J., 2007. Classification of ecosystem services: problems and solutions. Biol. Conserv. 139, 235–246.

Wang, Y., Kang, C., Bettencourt, L.M.A., Liu, Y., Andris, C., 2015. Linked activity spaces: embedding social networks in urban space. Geotechnol. Environ. 13, 313–336.

Wardle, D.A., Bardgett, R.D., Klironomos, J.N., Setälä, H., Van der Putten, W.H., Wall, D.H., 2004. Ecological linkages between aboveground and belowground biota. Science 304, 1629–1633.

Watts, D.J., Strogatz, S.H., 1998. Collective dynamics of 'small-world' networks. Nature 393, 440–442.

Weaver, W., 1948. Science and complexity. Am. Sci. 36, 536–544.

Wegner, G., Pascual, U., 2011. Cost-benefit analysis in the context of ecosystem services for human well-being: a multidisciplinary critique. Glob. Environ. Change 21, 492–504.

Westland, J.C., 2015. Structural Equation Models. Springer, Heidelberg.

Whitmee, S., Haines, A., Beyrer, C., Boltz, F., Capon, A.G., Ferreira de Souza Dias, B., Ezeh, A., Frumkin, H., Gong, P., Head, P., Horton, R., Mace, G.M., et al., 2015. Safeguarding human health in the Anthropocene epoch: report of The Rockefeller Foundation–Lancet Commission on planetary health. Lancet. http://dx.doi.org/10.1016/S0140-6736(15)60901-1.

Williams, S.J., Jones, J.P.G., Gibbons, J.M., Clubbe, C., 2015. Botanic gardens can positively influence visitors' environmental attitudes. Biodivers. Conserv. 24, 1609–1620.

Wood, S.A., Karp, D.S., DeClerck, F., Kremen, C., Naeem, S., Palm, C.A., 2015. Functional traits in agriculture: agrobiodiversity and ecosystem services. Trends Ecol. Evol. 30, 531–539.

Woodward, G., Blanchard, J., Lauridsen, R.B., Edwards, F.K., Jones, I.W., Figueroa, D., Warren, P.H., Petchey, O.L., 2010. Individual-based food webs: species identity, body size and sampling effects. Adv. Ecol. Res. 43, 211–266.

Woodward, G., Gessner, M.O., Giller, P.S., Gulis, V., Hladyz, S., Lecerf, A., Malmqvist, B., McKie, B.G., Tiegs, S.D., Cariss, H., Dobson, M., Elosegi, A., et al., 2012. Continental-scale effects of nutrient pollution on stream ecosystem functioning. Science 336, 1438–1440.

Yvon-Durocher, G., Jones, J.I., Trimmer, M., Woodward, G., Montoya, J.M., 2010. Warming alters the metabolic balance of ecosystems. Philos. Trans. R. Soc. Lond. B Biol. Sci. 365, 2117–2126.

CHAPTER TWO

Linking Biodiversity, Ecosystem Functioning and Services, and Ecological Resilience: Towards an Integrative Framework for Improved Management

Amélie Truchy*, David G. Angeler*, Ryan A. Sponseller†, Richard K. Johnson*, Brendan G. McKie*,[1]

*Department of Aquatic Sciences and Assessment, Swedish University of Agricultural Sciences, Uppsala, Sweden
†Department of Ecology and Environmental Sciences, Umeå University, Umeå, Sweden
[1]Corresponding author: e-mail address: brendan.mckie@slu.se

Contents

1. Introduction 56
Glossary 58
2. Drivers of Ecosystem Functioning 60
   2.1 Abiotic Factors 60
   2.2 Biodiversity as a Driver of Ecosystem Functioning at Multiple Scales of Organisation 62
3. Adding Spatiotemporal Dimensions 70
   3.1 Metacommunities and Meta-Ecosystems 70
   3.2 Resilience and the Stability of Functioning Over Multiple Spatiotemporal Scales 72
   3.3 Incorporating Resilience Aspects into Portfolio Theory for Management of Ecosystem Services 76
4. Extending and Parameterising a Trait-Based Framework for Predicting Functional Redundancy and Outcomes for Ecosystem Functioning and Services 77
5. Concluding Remarks 82
Acknowledgements 85
References 86

Abstract

Final ecosystem services (i.e. services that directly benefit humanity) depend fundamentally upon the various processes, regulated by organisms, which underpin ecosystem functioning and maintain ecosystem structures. Such processes include *inter alia* primary productivity, detritus decomposition, pollination, soil formation, and nutrient

Advances in Ecological Research, Volume 53
ISSN 0065-2504
http://dx.doi.org/10.1016/bs.aecr.2015.09.004

uptake and fixation. Insights into the abiotic, biotic, and spatial factors regulating these "supporting ecosystem processes" have arisen from within multiple fields of ecology which have not always been well integrated, including research on biodiversity–ecosystem functioning (B-EF) and biodiversity–ecosystem service (B-ES) relationships, meta-ecosystem ecology, and ecological resilience. Here, we draw together insights from these fields towards a framework suitable for addressing impacts of human disturbances on ecosystem processes and the services they support. We further discuss application of portfolio theory and a trait-based framework as unifying approaches in the assessment and management of ecosystem functioning and services, and identify a set of "resilience attributes" useful for assessing the resilience of ecosystem structure, functioning, and service delivery. Finally, we discuss future research challenges and opportunities, including uncertainties involved in linking species traits and interactions with ecosystem functioning and services. We conclude that the necessary theory and tools are already in place to begin the unification of B-EF, B-ES, meta-ecosystem, and resilience frameworks and to test their application in the assessment and management of ecosystem services.

1. INTRODUCTION

A key component of the ecosystem services framework is the notion that the final goods and benefits humanity derives from nature depend fundamentally upon the various processes, largely regulated by organisms, which underpin ecosystem functioning and maintain ecosystem structures. This precept is clear in Daily's (1997) definition of ecosystem services as "the conditions and processes through which natural ecosystems, and the species that make them up, sustain and fulfil human life" and was ultimately refined in The Millennium Ecosystem Assessment's (MEA) division of ecosystem services into four different categories (Fig. 1) (UNEP, 2005). The first three of these categories are provisioning, regulating, and cultural services and are regarded as "final" ecosystem services that directly benefit human populations (Fig. 1). The fourth category in the MEA is "supporting ecosystem services", largely comprising various ecosystem-level processes which underpin the delivery of final services (UNEP, 2005). These include processes involved in the cycling of nutrients and energy (e.g. primary productivity, organic matter decomposition, nutrient fixation, nutrient uptake), generation and maintenance of ecosystem structures (e.g. soil formation, reef construction), or the maintenance of populations (Smith et al., 2013). These ecosystem processes are regulated by both abiotic and biotic drivers operating over multiple spatiotemporal scales and are sensitive to human

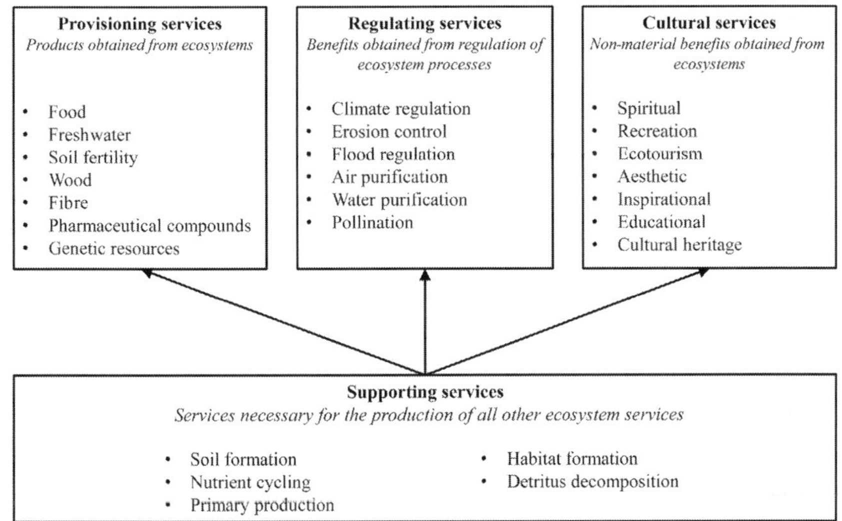

Figure 1 Categories (UNEP, 2005) and examples of ecosystem services (Daily, 1997; Daily et al., 2000; Malmqvist and Rundle, 2002; UNEP, 2005), according to the Millennium Ecosystem Assessment framework.

disturbances (Frainer and McKie, 2015; Wardle and Jonsson, 2014). Insights into these drivers and the impacts of human activities have arisen from multiple fields of ecological enquiry—although these have not always been well integrated—including biodiversity–ecosystem functioning (B-EF) and biodiversity–ecosystem service (B-ES) research, meta-ecosystem research, and resilience theory. There is now a need to draw together insights from these fields towards a comprehensive framework suitable for addressing impacts of human disturbances on ecosystem processes and services over multiple scales, and for investigating uncertainties related to global environmental change. Such a synthesis will ultimately assist decision making in environmental policy and management aiming for resilient ecosystems and the sustainable use of ecosystem services (Durance et al., 2016; Gill et al., 2016; Mancinelli and Mulder, 2015).

Recently, the need to explicitly assess all putative ecosystem services, including supporting ecosystem processes, relative to socio-economic values as well as ecological values has been reemphasised (see Glossary), if they are to be fully incorporated into the ecosystem services framework (Boyd and Banzhaf, 2007; Haines-Young and Potschin, 2010; Palomo et al., 2016). This is not always straightforward, not only because appropriate economic or social valuations of specific ecosystem processes are often not available

GLOSSARY: GUIDING DEFINITIONS OF KEY TERMS

Diversity effect The "contribution of a community to a process rate that cannot be explained by summing the weighted individual contributions of the constituent species" (Gessner et al., 2010). Diversity effects are thus non-additive and can arise from the interplay of multiple species (e.g. from facilitative interactions or complementary resource use), or from the "selection effect", reflecting the greater probability that diverse communities will include particular species with strong influences on ecosystem processes.

Ecological resilience The amount of disruption that is required to transform a system from being maintained by one set of reinforcing processes and structures to a different set of processes and structures (Holling, 1973). This represents a multi-equilibrium concept of resilience, in contrast with the single-equilibrium concept underlying *Engineering resilience*.

Ecosystem functioning "The joint effects of all processes that sustain an ecosystem" (Reiss et al., 2009).

Ecosystem process A process emerging at the ecosystem level and involving interactions between species within their food web, and with their environment, often involving transformations of nutrients and energy (e.g. primary production), generation of habitat structures (e.g. reef building), or maintenance of populations (e.g. pollination) (Gessner and Chauvet, 2002; Smith et al., 2013).

Ecosystem services "The conditions and processes through which natural ecosystems, and the species that make them up, sustain and fulfil human life" (Daily, 1997). Ecosystem services should always be defined relative to human needs and thus should be assessed according to their socio-economic as well as ecological values.

Engineering resilience A commonly used resilience definition that emphasises the speed with which a perturbed ecosystem returns to previous food web configurations and levels of ecosystem functioning following release of stressors (Gunderson, 2000; Pimm, 1991). This represents a single-equilibrium concept of resilience, in contrast with the multi-equilibrium concept of *Ecological resilience*.

Effect traits Characteristics of an organism's phenotype, such as resource acquisition and biomass production rates, which affect both its fitness and its effects on ecosystem processes (Verberk et al., 2013; Violle et al., 2007).

Meta-ecosystem "A set of ecosystems connected by spatial flows of energy, materials, and organisms across ecosystem boundaries" (Loreau et al., 2003).

Response traits Characteristics of an organism's phenotype that regulate its environmental responses, reflecting especially its environmental tolerances and ecological flexibility (Violle et al., 2007).

(Durance et al., 2016; Marbuah et al., 2014) but also because once such valuations are applied the distinction between these processes as supporting or regulating/provisioning services often becomes unclear (Bastian et al., 2015; Mace et al., 2012). For example, the primary productivity of a field is often regarded as a supporting service underpinning provisioning and regulating services such as food production or carbon sequestration. However, when harvested and given a monetary value—for example, as hay or a grain crop—the primary productivity of a field is normally more usefully regarded as a "final" provisioning ecosystem service (Bastian et al., 2015; Mace et al., 2012). In this contribution, we follow the MEA in regarding individual ecosystem processes as key ecosystem-level attributes that support provisioning, regulating, and cultural services, and thus constituting an essential component of any research or management programme focussed on ecosystem services (Bastian et al., 2008; Haines-Young and Potschin, 2010). Where they can appropriately be related to societal needs or economic values, we further discuss these processes as services in their own right (e.g. see Astegiano et al., 2015; Mancinelli and Mulder, 2015), or as components of a suite of ecosystem attributes comprising the ecosystem service "portfolio" of a given habitat (Griffiths et al., 2014).

A common underpinning of research on not only ecosystem processes but also meta-ecosystems, ecological resilience, and ecosystem services is the idea that ecosystems should be characterised not only by their taxonomic composition but also according to how they function. Ecosystem functioning is defined as "the joint effects of all processes (fluxes of energy and matter) that sustain an ecosystem" over time and space through biological activities (Naeem and Wright, 2003; Reiss et al., 2009). Crucially, in understanding variability in ecosystem functioning, including that induced by humans, it is not enough to study taxonomic changes alone, because species composition can change without concomitant functional changes, and vice versa (Bunn and Davies, 2000; Dirzo et al., 2014). For example, after drought events, the loss of species in a plant community might not affect overall plant productivity (an ecosystem-level process) if other tolerant species are able to increase their growth rates and compensate for those losses (Tilman and Downing, 1994). Likewise, functioning can change even when species are unaffected (for example, reflecting changed interactions or behaviours by the resident species; McKie and Malmqvist, 2009).

The importance of non-additive relationships between species, species traits, and ecosystem processes is most clearly recognised in B-EF research, but remains to be fully incorporated into related fields, such as meta-ecosystem and resilience research, or in the development of a more extensive B-ES framework. The potential importance of this is seen when B-EF research is linked with ecosystem services (Daily et al., 2000; Reiss et al., 2009; UNEP, 2005). For instance, declines in the diversity of invertebrates that process detritus have been shown to non-additively alter decomposition rates and nutrient cycling (Dirzo et al., 2014; Gessner et al., 2010), while losses of algal diversity may interact to affect an ecosystem's capacity to sequester nutrients (Cardinale, 2011), all processes potentially considered as either supporting or regulating services. Most of these non-additive effects of species loss have been observed at local scales, but an improved understanding of how they are dampened or amplified across larger spatiotemporal scales would enhance prospects for managing ecological networks for sustainable and resilient delivery of ecosystem services (Durance et al., 2016).

In this review, we link ecosystem services with key concepts related to ecosystem functioning and then explore different drivers of functioning, including abiotic (e.g. oxygen, light), biotic (e.g. species and their traits, biodiversity), and spatial drivers (e.g. metacommunity processes), as well as human impacts on these linkages, to move towards a more rigorous B-ES framework. We develop an expanded spatiotemporal perspective on these linkages by incorporating aspects of food web and resilience theory, and consider the potential application of portfolio theory in the management of ecosystem functioning and services. While we draw on examples from many different ecosystem types, we have a particular focus on freshwater habitats. Lakes and streams are notable not only as key providers of multiple ecosystem services, including provisioning of drinking water, mitigation of pollutants, and multiple aesthetic and recreational values, but also as focal ecosystems for much cutting edge research on B-EF, ecological connectivity, and resilience (Durance et al., 2016). We finish by extending an earlier framework linking individuals, species, and traits with ecosystem processes to also incorporate supporting ecosystem processes and non-additive effects arising from species interactions. Finally, we consider challenges facing future research, particularly in quantifying uncertainties in the monitoring and assessment of ecosystem functioning, resilience, and ecosystem services.

2. DRIVERS OF ECOSYSTEM FUNCTIONING

2.1 Abiotic Factors

Understanding variability in the ecosystem processes which underpin ecosystem services generally (e.g. soil formation, detritus decomposition), and which in some cases may constitute ecosystem services in their own right (e.g. pollination, primary production in agriculture, nutrient uptake in enriched catchments), is a sound starting point for understanding the variability underpinning service delivery. By definition, an "ecosystem-level" process involves interactions between species within their food web and with their environment (Gessner and Chauvet, 2002). Accordingly, abiotic factors that affect organisms are also potentially important drivers of functioning and ecosystem service delivery. This includes temperature as a basic driver of metabolic processes (Brown et al., 2004), with photosynthesis, respiration, and organismal growth all key processes that follow Van't Hoff's rule, whereby the rate of underlying chemical reactions increases with temperature (Myers, 2003). Light and nutrient availability are two further abiotic factors of particular importance for primary producers (Bott, 2006; Hauer and Hill, 2006), with nutrients also important for decomposers (e.g. Burrows et al., 2015; Rosemond et al., 2015). In streams, a substantial proportion of variability in functioning can be explained by variability in these three drivers alone. This is seen in the ecosystem-level changes associated with removal of stream riparian vegetation, which typically increases light, nutrients, and temperature (Caissie, 2006; McKie and Cranston, 2001; Peierls et al., 1991), influencing biodiversity (Hlúbiková et al., 2014; McKie and Cranston, 2001), and affecting multiple supporting ecosystem processes and ecosystem services (Hladyz et al., 2011), including provisioning of clean water (Burrell et al., 2014), gross primary production and ecosystem respiration (Burrell et al., 2014; Sabater et al., 1998; Young and Huryn, 1999), and decomposition of leaf litter (Gessner and Chauvet, 2002; Hladyz et al., 2011; Lagrue et al., 2011; McKie and Malmqvist, 2009). Other abiotic drivers can also have profound influences on functioning, including relative humidity and soil moisture content in terrestrial and semi-aquatic habitats (Gessner et al., 2010), substrate composition, and—in streams—sediment loading (which can decrease light availability and limit primary production, regardless of nutrient concentrations; Bunn and Davies, 2000; Young and Huryn, 1999). Moreover, in many aquatic systems, hydrological regimes are fundamental organisers of temporal patterns in biotic structure (Townsend and Hildrew, 1994) and ecosystem process rates (Stanley et al., 2010).

Often, interactions among these various abiotic drivers may have synergistic (e.g. combined positive effects of temperature and nutrients on algal productivity) or antagonistic (e.g. counteracting effects of nutrients and sediment deposition on algal productivity) effects on a given process rate (Ferreira and Chauvet, 2011; Piggott et al., 2015), resulting in non-linear changes in ecosystem functioning along broad environmental gradients (Woodward et al., 2012). Crucially, such effects rarely arise solely from direct interactions between the drivers themselves (as when two chemical stressors counteract one another in an antagonistic abiotic interaction), but most often involve *complex interactions and feedbacks* between the abiotic drivers, biotic interactions, and changes in biodiversity within ecological communities (McKie et al., 2009; Townsend et al., 2008). Accordingly, an understanding of ecosystem functioning based only on abiotic factors would rarely be complete—multiple biotic drivers also need to be incorporated, including the particular roles of individual species and their traits, multitrophic interactions, and biodiversity itself.

2.2 Biodiversity as a Driver of Ecosystem Functioning at Multiple Scales of Organisation

Biodiversity represents "all heritability-based variation at all levels of organisation, from the genes within a single local population, to the species composing all or part of a local community, and finally to the communities themselves that compose the living parts of the multifarious ecosystems of the world" (Wilson, 1992). As with many topics in biology, Darwin identified the potential importance of biodiversity for ecosystem functioning: "It has been experimentally proved that if a plot of ground be sown with one species of grass, and a similar plot be sown with several distinct genera of grasses, a greater number of plants and a greater weight of dry herbage can thus be raised" (Darwin, 1859). This idea has resurfaced in recent decades, with much research attention paid to biodiversity of not only species but also key species "traits" as key biotic drivers of ecosystem processes, and therefore of the ecosystem services supported by these processes (Astegiano et al., 2015; Reiss et al., 2009).

2.2.1 Species Traits: A Crucial Link Between Diversity and Ecosystem Functioning

The "traits" of an organism are the components of its phenotype that regulate its responses to environmental factors such as temperature, soil or water conditions, precipitation and resource availability, and its influences on ecosystem processes (Naeem and Wright, 2003; Petchey and Gaston, 2006; Violle et al., 2007). Accordingly, traits are typically divided into two non-exclusive categories: (1) *response traits* that regulate the responses of species to environmental conditions, reflecting especially their environmental tolerance and ecological flexibility, and (2) functional *effect traits*, such as resource acquisition and biomass production rates, which influence both individual fitness (Verberk et al., 2013; Violle et al., 2007) and the effects of organisms on ecosystem processes (Fig. 2; Hooper et al., 2002; Lavorel and Garnier, 2002; Naeem and Wright, 2003). In practice, information on true effect traits (e.g. species-specific resource assimilation rates) is rarely available for all, or even some, species in a given functional guild. Consequently, traits easily quantified at the individual or species level are often used as

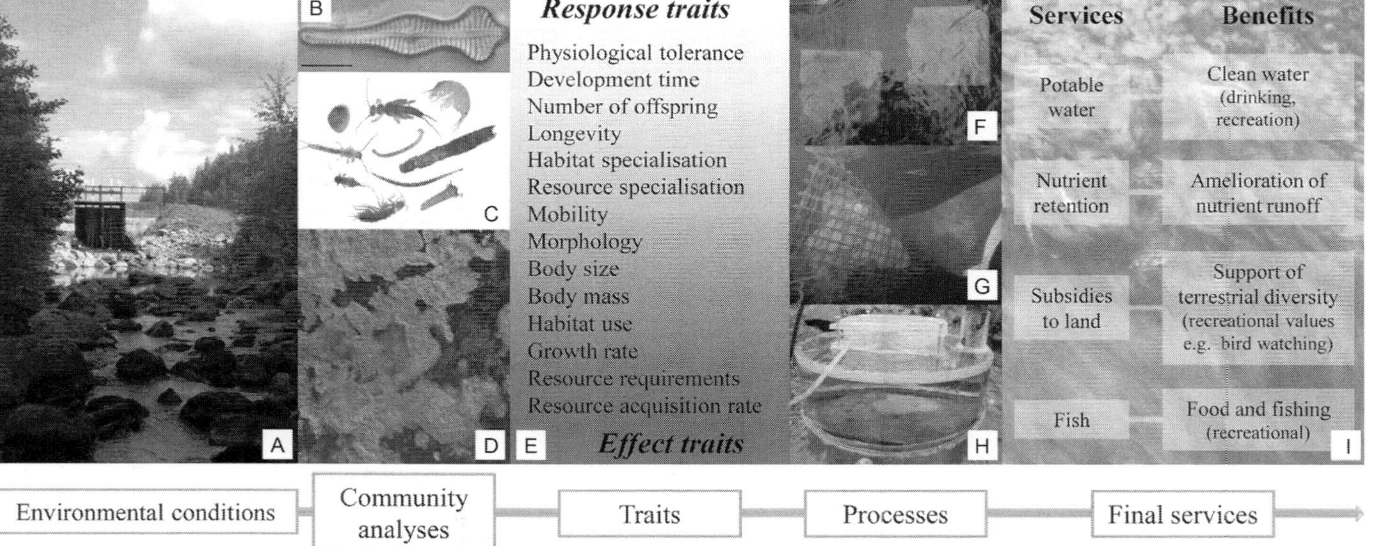

Figure 2 Linking anthropogenic stressors, species, and traits with supporting and final ecosystem services in stream habitats. Environmental conditions and stressors in a degraded stream (A) can have impacts on the composition of key organism groups, including algae (B), invertebrates (C), fish, and aquatic plants (D; seen here in an aerial view of a lake). This causes shifts in the composition of *response traits*, with stress often favouring species that are tolerant, generalist, develop rapidly, and short-lived. Associated functional *effect traits* also shift, as reflected in measures of ecosystem processes such as algal productivity, detritus decomposition, and the respiration of biofilms (F–H); and ultimately altering final ecosystem services supported by these processes (I). Methods for quantifying algal productivity (F—the tile method), leaf decomposition (G—the litter-bag method), and biofilm performance (H—respiration chambers) are illustrated. Quantitative information on true *functional effect* traits (e.g. growth and consumption rates) is rarely available, hence appropriate response trait proxies are often used, such as body size—both a key response (stress favours smaller size) and effect (due to mass-specific metabolic demands) trait. *Photographs courtesy of (B) Dr. Steffi Gottschalk, (D) Dr. Frauke Ecke, and (H) Dr. Jon Benstead.*

proxies for both response and effect traits, depending on the research questions asked (Fig. 2; Frainer and McKie, 2015; Frainer et al., 2014). For example, body size and growth rates are clearly traits that can both influence the capacity of an organism to *respond* to the environment (with stress often favouring small-sized, fast growing taxa), and also their *effects* on processing of resources and community-level productivity (Fig. 2). Identifying the specific biological traits that have the strongest influences on particular ecosystem processes and contribute most to functional diversity (Tilman, 2001) provides a crucial link in the development of a broader framework for understanding how species composition and diversity affect ecosystem functioning and, by extension, the delivery of ecosystem services (Astegiano et al., 2015). For instance, identifying traits that characterise "winning" species in the current global crises would assist in forecasting future changes in ecosystem functioning and service delivery (Dirzo et al., 2014).

Coarse schemes for classifying organisms into groups according to their traits have existed for some time, such as the division of aquatic insects into functional feeding groups (Cummins, 1974), including predators, collectors, and detritus shredders. Coarse classification schemes are also often applied to plants, distinguishing, e.g., nitrogen fixers, C-3 grasses, C-4 grasses, C-3 forbs, and C-4 forbs (Díaz and Cabido, 2001; Tilman, 2001). Unfortunately, these classification schemes often have limited value in representing variability in the true functional diversity of assemblages, and variability in functioning, since species rarely fall neatly into single, broadly defined functional groupings (Hooper et al., 2002; Reiss et al., 2009). Moreover, some species may also express some functional traits only in a specific environmental context or life stage, and other traits under different environmental conditions or life stages (Frainer et al., 2014; Naeem and Wright, 2003). Finally, some species may be difficult to allocate to any, broadly defined, functional group, because they possess a high number of unique traits (Mouillot et al., 2013; Naeem and Wright, 2003).

These issues have led to the development of species-level trait classifications for many different organism groups, including aquatic macroinvertebrates (Poff et al., 2006; Schmidt-Kloiber and Hering, 2015) and terrestrial plants (Bonan et al., 2012; Kattge et al., 2011), as well as sophisticated methodologies for objectively identifying trait clusters in communities. These include dendrogram-based approaches that use coding to account for functional redundancy (Mouchet et al., 2008; Petchey and Gaston, 2007; see Petchey and Gaston, 2006; Villeger et al., 2008 for additional approaches). Many of these approaches account not only for trait composition and richness but also fluctuations in the absolute density and relative abundance of traits within an assemblage (Hillebrand and Matthiessen, 2009; Petchey and Gaston, 2006). Process rates in an assemblage strongly dominated by one species are likely to reflect the traits of that single species, in line with Grime's *mass ratio* hypothesis (Dangles and Malmqvist, 2004; Frainer and McKie, 2015; Grime, 1998). In contrast, species richness *per se* might be more important in an assemblage where distinct traits are represented by similar numbers of individuals (Frainer et al., 2014; Mouillot et al., 2005; Petchey and Gaston, 2006). It is thus important to evaluate trait distributions, as a dominant trait might reflect the average of the assemblage or, alternatively, might be an outlier trait compared to the combined traits of the other species (Hillebrand and Matthiessen, 2009).

Recently, key concepts from trait-based ecology, the mass-ratio hypothesis, and the metabolic theory of ecology have been integrated into the framework of *trait-driver theory*, which links traits, community assembly, and processes within a predictive framework that can be applied across broad environmental gradients (Enquist et al., 2015). Underlying this is the principle that the robust quantification of trait identity and diversity can be used to both predict and explain variation in ecosystem functioning, and even ecosystem services (Enquist et al., 2015). Potential application of this is seen, for example, in studies addressing how the diversity and composition of freshwater invertebrate traits (Frainer and McKie, 2015; Frainer et al., 2014) and of plant-litter traits (Handa et al., 2014; Heemsbergen et al., 2004) influence the breakdown of detritus, as well as in studies linking above- and below-ground plant traits to rates of primary production by terrestrial plants (Comas et al., 2013; Roscher et al., 2012). Nevertheless, there are some risks in basing predictions on variability in functioning purely on the composition and diversity of traits. These arise from the potential for non-additive interactions among species, and between species and the environment, to alter patterns of trait expression and outcomes for ecosystem processes. A framework for investigating such non-additive effects has been extensively developed within the field of B-EF research.

2.2.2 Biodiversity and the Effects of Species Interactions on Ecosystem Functioning

In response to the global biodiversity crisis, an intensive research effort developed from the early 1990s to investigate links between biodiversity and ecosystem functioning (Reiss et al., 2009; Srivastava, 2002). The most common biodiversity index investigated in B-EF research is the number of species within a habitat, i.e., α species richness, at a very local scale (Díaz and Cabido, 2001; Tilman, 2001). Early B-EF studies mainly focussed on whether ecosystem processes such as plant productivity or litter decomposition are consistently enhanced by increasing species richness *per se*, or whether variation in process rates is primarily regulated by changes in the identity of species comprising the ecosystem (Huston, 1997; Johnson et al., 1996; Naeem et al., 2002). Positive effects of diversity *per se*—arising from the interplay of multiple species—may arise from complementary niche partitioning (Cardinale et al., 2002; Díaz and Cabido, 2001; Loreau and Hector, 2001; Mulder et al., 2001), which occurs when several species coexist at a given site and complement each other spatially and temporally in their patterns of resource use (Cardinale et al., 2004; Díaz and Cabido, 2001; Loreau and Hector, 2001; Vaughn, 2010). For example, shallow-rooted grasses and deep-rooted shrubs in cold steppes partition the soil profile and thereby use different resources at the same site, while cool-season and warm-season grasses in prairies use the same resources but at different times of year (Díaz and Cabido, 2001). Facilitation is another mechanism involving the interplay among multiple species, and occurs when activities of some species enhance or facilitate activities of others and, in turn, ecosystem process rates (Gessner and Chauvet, 2002; Jonsson and Malmqvist, 2003; Tiunov and Scheu, 2005). For instance, within the suite of processes underpinning water purification in freshwaters, facilitation is seen when diverse assemblages of filter-feeder caddisflies capture more suspended material than they could do in monoculture, due to "current shading" (Cardinale et al., 2002).

Implicit in both these mechanisms is the idea that it is not merely the traits of species that are important but also the way those species interact to influence trait expression. Thus, effects of complementarity are more likely when sufficient spatiotemporal complexity exists for species to finely partition resources without extensive interference competition, and facilitative interactions lift community performance above expectations based on individual species traits. In other situations, interference competition or antagonistic interactions may undermine these effects, and result in the dominance of functioning by single species which may or may not be able to maintain processing rates at the same level as a more even community, depending on their specific functional traits (Cardinale, 2011; McKie et al., 2008). Given that trait expression and the outcomes of species interactions can both vary according to environmental context (e.g. McKie and Pearson, 2006; O'Connor and Donohue, 2013; Törnroos et al., 2015), there is strong potential for variation in the form of B-EF relationships, with positive, negative, and neutral relationships all observed within the same functional guild (Cardinale et al., 2012; Gessner et al., 2010). In one illustrative example, both positive and negative relationships between detritivore richness and leaf decomposition have been observed for the same detritivore assemblages in different environmental settings (Jonsson, 2006; McKie et al., 2009), highlighting the potential for environmental factors to alter the outcome of species interactions (from complementary to antagonistic), and hence B-EF, and even B-ES (Cardinale, 2011) relationships.

It has long been recognised that single species can also have strong effects on ecosystem-level attributes, including ecosystem services, as captured in the "keystone species", "ecosystem engineer", and "foundation species" concepts (Ellison et al., 2005; Jones et al., 1997; Power et al., 1996). Within B-EF research, a mechanism that has been proposed to account for strong contributions of particular species to diversity–functioning relationships is the sampling or selection effect (Huston, 1997; Loreau, 2000). This model hypothesises that species are different in terms of competitive traits (e.g. stress tolerance, nitrogen-fixation ability, high seed germination rate) and that better competitors are also more productive. Accordingly, richer communities should be on average more productive because they have a greater chance of comprising and being dominated by the most productive species (Cardinale et al., 2006; Thompson and Starzomski, 2007; Tilman, 2001), though species identity effects can also drive negative relationships between species richness and ecosystem function (Creed et al., 2009; McKie et al., 2009). In ecosystem service management, it may be particularly useful to distinguish B-ES relationships that depend strongly on single species from those driven by relationships among multiple species, since services that are largely dependent on keystone or foundational species (e.g. fisheries dependent on only one or a few top predator species) may often be relatively tractable to management narrowly targeting that service (Durance et al., 2016). In contrast, management of services dependent on the cumulative activities of multiple, and often cryptic, species (as for water purification services strongly dependent microbial organisms) may be more challenging to optimise (Durance et al., 2016).

Presently, the weight of evidence from two decades of predominantly experimental research suggests that increasing species richness is often, but not universally, associated with enhanced rates of supporting ecosystem processes and services (Balvanera et al., 2006; Cardinale, 2011; Cardinale et al., 2006, but see Baulch et al., 2011), and that many of these effects are driven by changes in diversity *per se*, and not just by the occurrence of particular species (Srivastava and Vellend, 2005; Thompson and

Starzomski, 2007). Notably, evidence for the importance of biodiversity for ecosystem functioning tends to strengthen when assessed over larger spatiotemporal scales (Cardinale et al., 2007; Jonsson, 2006; Matthiessen and Hillebrand, 2006; Stachowicz et al., 2008; Tilman et al., 2001), and as more ecosystem processes are considered simultaneously (Gamfeldt et al., 2008; Perkins et al., 2015). These findings highlight the likely importance of biodiversity in underpinning not only ecosystem multifunctionality but also for supporting ecosystem services (e.g. Maestre et al., 2012; Gamfeldt et al., 2013), given that provisioning, cultural, and regulating services all typically rest on the activities of multiple organism groups, and multiple ecosystem processes, distributed across multiple habitat compartments, and even linking across ecosystem boundaries (Durance et al., 2016).

To what extent can variability in relationships between biodiversity and these ecosystem services be understood using the concepts and findings from B-EF research, which although theoretically and empirically rich is often narrowly focussed on single processes over short time scales? When an ecosystem process is clearly valued as a final ecosystem service (such as productivity of agricultural fields or pollination), we expect that predictions from B-EF research should be directly applicable for investigating B-ES relationships. However, for services that emerge as amalgams of multiple ecosystem processes, often driven by the actions of microbial communities, the links between biodiversity of single functional guilds at small spatiotemporal scales and ecosystem services are less clear. The observation that individual B-EF relationships can switch from positive to negative depending on outcomes of species interactions and environmental context (e.g. McKie et al., 2009) points to the potential for substantial uncertainties when predicting outcomes for ES of variability in multiple, underlying processes. It is also possible that thresholds in B-ES relationships exist, whereby contingencies in multiple B-EF relationships occurring at low levels of richness become less important as more species contribute to more processes (a form of "statistical averaging", *sensu* Doak et al., 1998). Indeed, the evidence that biodiversity generally increases in importance as more processes and services are considered (Gamfeldt et al., 2008, 2013; Perkins et al., 2015), or that multiple ecosystem processes and services are strongly compromised when food webs are overly simplified (Bohan et al., 2013; Thompson et al., in press), while still limited does indicate that biodiversity *per se* may be even more important, and possibly consistent, as a driver of ecosystem services than of individual underlying processes.

2.2.3 Integrating Food Web Theory

The capacity of an ecosystem to deliver a range of both supporting and final ecosystem services depends on more than just biodiversity within single trophic levels (the focus of most B-EF research) but also on interactions among trophic levels, and the structure of food webs (Carpenter et al., 1985, 1987; Montoya et al., 2003). It is well known that biodiversity changes within one trophic level can have impacts on species at other levels, whether directly through consumer–resource interactions, or indirectly via behavioural changes (Carpenter et al., 1985; Hines et al., 2015; McKie and Pearson, 2006; Paine, 1966; Thébault and Loreau, 2003). Both direct and indirect pathways can also affect ecosystem processes, due to direct impacts of species on the processes, or as a result of indirect changes in species interactions, potentially arising both top-down or bottom-up within the trophic web (Gessner et al., 2010; Raffaelli et al., 2002). Losses of top predators often have particularly strong effects, as seen in the classic trophic cascade, where intermediate consumers, freed from predation pressure, suppress primary production rates (Carpenter et al., 1985; Paine, 1966). Such cascades might similarly have implications for services associated with detritus-based systems (Mancinelli and Mulder, 2015). However, in other cases, the presence of predators can enhance processes at the base of the food web, when predation frees more efficient species from competition with less-efficient species that would otherwise dominate the process (Gessner et al., 2010; Jabiol et al., 2013).

Anthropogenic stressors often result in changed food web structure (Massol and Petit, 2013; Tamaddoni-Nezhad et al., 2013), with decreased connectance (Gilbert, 2009), losses of biological interactions (Valiente-Banuet et al., 2015), shorter food chain lengths associated with the extinction of top predators (Petchey et al., 1999), and other alterations in community composition and size structure (Jonsson et al., 2015a). Insights into the consequences of such food web simplification for ecosystem functioning and services can be drawn from the few experimental studies in which both the horizontal and vertical components of food webs were considered (Bastian et al., 2008; Downing and Leibold, 2002; Duffy et al., 2007; Jabiol et al., 2013; O'Connor and Donohue, 2013), or from catastrophes impacting food web structure *in situ* (Thompson et al., in press). For example, Jabiol et al. (2013) in a study of a model detrital food web found that rates of a supporting ecosystem process—leaf-litter decomposition—were maximal in the most complex food web considered (when all trophic levels were present with maximal species richness), with the cumulative effects of species loss within and across trophic levels reducing process rates. However, in another study, the response of primary productivity to losses of trophic complexity was more idiosyncratic (Downing and Leibold, 2002). Overall, considering both multitrophic interactions and effects of environmental factors (e.g. nutrient enrichment; O'Connor and Donohue, 2013; also see Bastian et al., 2008) suggests that changes in ecosystem functioning (Hines et al., 2015; Jabiol et al., 2013; Thébault and Loreau, 2003, 2006) and ecosystem services (Durance et al., 2016; Mancinelli and Mulder, 2015) resulting from widespread biodiversity loss could be more complicated, and possibly more severe, than previously suggested from single trophic level studies.

3. ADDING SPATIOTEMPORAL DIMENSIONS

3.1 Metacommunities and Meta-Ecosystems

Most methods for quantifying the ecosystem processes that underpin ecosystem services are most easily applied at local scales—a patch of leaves or a field plot or, at best, a stream reach (e.g. for methods quantifying reach-scale respiration and nutrient uptake in running waters). However, single habitat patches are typically connected with other patches by multiple flows of organisms and materials, which may strongly influence patterns of functioning at local scales. For example, the movement of organisms (i.e. immigration and emigration) among habitat patches has strong potential to influence the composition and diversity of species traits, and species interactions at local scales, and hence ecosystem functioning and the delivery of ecosystem services at local and broader spatial scales (Cardinale et al., 2004; Gill et al., 2016; Hagen et al., 2012; Loreau et al., 2003; Massol and Petit, 2013).

In recognition of this, the earlier concept of the "metacommunity", developed to understand how regional connectivity can affect local biodiversity, has been extended to the concept of the "meta-ecosystem" by Loreau et al. (2003), which also accounts for spatial flows of energy and materials such as inorganic nutrients or detritus (Gravel et al., 2010; Loreau et al., 2003). Thus, just as a metacommunity represents "a set of local communities that are linked by dispersal of multiple potentially interacting species" (Leibold et al., 2004), so can a meta-ecosystem be defined as "a set of ecosystems connected by spatial flows of energy, materials, and organisms across ecosystem boundaries" (Loreau et al., 2003). The ecosystem compartments may comprise trophic levels, species, or functional groups. Four paradigms underlie metacommunity theory—patch dynamics, neutral theory, species sorting, and mass effects (Leibold et al., 2004; Mouquet and

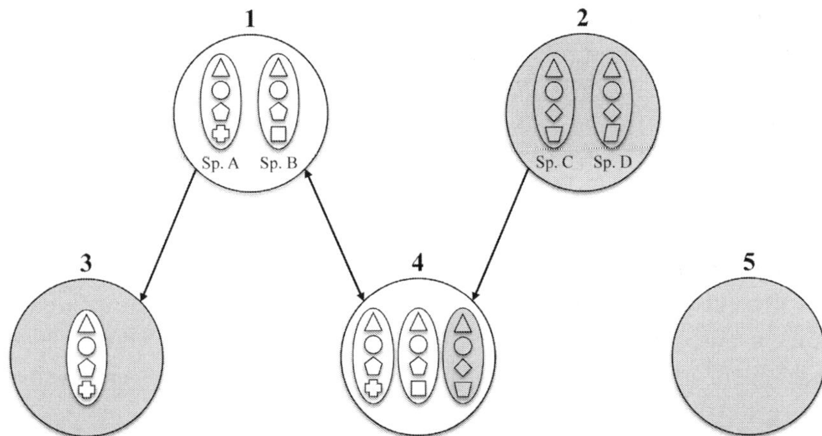

Figure 3 Linking the species sorting and mass-effect paradigms in metacommunity theory to outcomes for species traits and ecosystem functioning. In this figure, orange (white in the print version) patches represent acidic ponds and are the preferred habitat of two species of leaf-eating consumer, represented by the red (white in the print version) ovals and denoted A and B, while blue (dark grey in the print version) patches are alkaline ponds and are preferred by species C and D (the purple (grey in the print version) ovals). The traits characteristic of each species are represented by symbols overlaid onto the ovals (after Reiss et al., 2009). The distribution of species between ponds 1 and 2 is in line with their environmental preferences, reflecting the species sorting paradigm of metacommunity theory. Diversity in the other ponds is further regulated by dispersal, with the arrows between ponds representing open dispersal pathways. In this example, there is an open dispersal pathway to the alkaline pond 3 from the acidic pond 1, but not from either of the other alkaline ponds. Species A has high-dispersal ability, and a steady flow of colonisers from pond 1 maintains its presence in pond 3 despite its low fitness in the alkaline environment. Species C similarly persists in the acid pond 4, despite competition from the more acid tolerant species A and B, because of constant recolonisation from pond 2. Thus, ecosystem functioning is supported in pond 3 and functional diversity enhanced in pond 4 by a source–sink dynamic, reflecting the mass-effect paradigm. A strong dispersal barrier prevents colonisation of acidic pond 5 by any of the species.

Loreau, 2003)—and these also provide the basis for understanding how organism and energy flows influence ecosystem functioning at multiple scales (Fig. 3). The four models differ in their assumptions about species and the quality of habitat patches, making very different predictions about the relative importance of dispersal versus local environmental features for diversity and ecosystem functioning at habitat and regional scales. Indeed, these models explicitly address the fact that diversity can be both patterned and structured differently at different spatial scales, ranging from α-diversity, which is the species diversity within a patch, to β-diversity, which represents the heterogeneity among patches, and γ-diversity, which describes the diversity of the entire metacommunity (France and Duffy, 2006;

Whittaker, 1960, 1972; see also Hagen et al., 2012). Of these, α-diversity has been the focus of most previous B-EF research, although some studies have assessed the role of dispersal on trophic cascades and the stability of ecosystem functioning in metacommunities (Howeth and Leibold, 2010).

Over the last decade, the body of studies investigating the importance of a spatial dimension for understanding how biotic distributions influence ecosystem functioning has grown (Hagen et al., 2012; Maestre et al., 2005; Matthiessen et al., 2007; Venail et al., 2010). The responses of different species to habitat fragmentation are often highly complex, complicating predictions of outcomes for ecosystem functioning and service delivery (Hagen et al., 2012). Nevertheless, there is empirical support for positive relationships between β- and γ-diversities and ecosystem functioning, in well-connected ecological networks (Matthiessen and Hillebrand, 2006). Local ecosystems may serve as sources and/or sinks for species traits and thereby contribute to the maintenance of different functions at a broader scale (β-diversity) (Loreau et al., 2003; Mouquet and Loreau, 2003; Öckinger and Smith, 2007). For example, in agricultural mosaics, field margins or semi-natural grasslands harbour a higher diversity of pollinators (and thus their associated traits) than crops, and thereby allow the maintenance of pollination services (Massol and Petit, 2013; Öckinger and Smith, 2007). Moreover, species traits related to dispersal and colonisation abilities (Bommarco et al., 2010; Ewers and Didham, 2006; Montoya et al., 2008; Öckinger et al., 2010) can strongly influence ecosystem services (e.g. pollination or biocontrol services in agroecosystems) at the regional and landscape scales. Losses of species at local scales may have stronger effects on functioning in isolated habitat patches dominated by species with low dispersal abilities compared to well-connected habitat networks (Hagen et al., 2012; Loreau et al., 2003; Matthiessen and Hillebrand, 2006), since the maintenance of key functional traits at local scales is more likely in well-connected, high-dispersal networks, providing "spatial insurance" for ecosystem functioning against variations in environmental conditions (Fig. 3; Loreau et al., 2003). This further points to the potential importance of dispersal in maintaining not only short-term stability of ecosystem processes but also the resilience and stability of ecosystem services in the face of environmental changes.

3.2 Resilience and the Stability of Functioning Over Multiple Spatiotemporal Scales

The meta-ecosystem concept has great potential to advance our basic understanding of how biodiversity affects ecosystem functioning across scales of space and time, and thus can provide a strong bridge for scaling up to B-ES relationships in human-modified landscapes. However, research addressing the roles of habitat connectivity in regulating ecosystem functioning, food web complexity and stability, and ecosystem service delivery over realistic spatiotemporal scales still remains limited (Calcagno et al., 2011; Gravel et al., 2011a,b; Pillai et al., 2011). While experimental quantification of the effects of species richness at habitat patch scales and over short time periods has been tractable for many ecosystem processes, these are generally not capable of capturing the true complexity of biodiversity and ecosystem functioning relationships in natural systems (Duffy, 2009), and especially over larger spatiotemporal scales (Cardinale et al., 2007; Stachowicz et al., 2008; Tilman et al., 2001). The challenges in experimentally studying multiple patches arranged with realistic spatial configurations and flows of organisms and materials among them, and over meaningful time scales, are substantial (Cardinale et al., 2007; Stachowicz et al., 2008; Tilman et al., 2001). In response, some researchers are moving away from manipulative experiments towards investigation of variation in ecosystem functioning along natural biodiversity gradients or "natural experiments" *in situ* (e.g. Dangles and Malmqvist, 2004; Dangles et al., 2011; Frainer and McKie, 2015; Frainer et al., 2014; Sundqvist et al., 2013; Thompson et al., in press; Tolkkinen et al., 2013). For example, along a gradient of increasing habitat modification, Tylianakis et al. (2007) linked changes in food web structure and elevated parasitism rates in bees, with potential implications for the pollination and biocontrol services in which these insects are involved in. However, although investigation of existing B-EF and B-ES gradients *in situ* increases realism, temporal scope often remains limited, except where space-for-time substitutive designs are employed, as in studies of successional chronosequences (Jonsson et al., 2015b; Wardle et al., 2011). An alternative approach is offered by the field of ecological resilience research, which has developed a framework based on utilisation of species traits as proxies for ecosystem "functions", within which larger-scale and longer-term consequences of biodiversity and ecological connectivity for food web stability, ecosystem functioning, and ecosystem services can be investigated. While there are important limitations in the use of traits as proxies for functioning, application of this approach in combination with key insights from the meta-ecosystem, B-EF and B-ES perspectives can assist in the assessment, forecasting, and management of structural and functional vulnerabilities in ecological networks.

Ecological resilience, following Holling's (1973) definition, is a measure of the amount of disruption that is required to transform a system from being maintained by one set of reinforcing processes and structures to a different set of processes and structures. This represents a multi-equilibrium concept of resilience, which is most appropriate when a system can reorganise into different alternative states or regimes (i.e. shift from one stability domain to another; Durance et al., 2016; Holling, 1973; Mancinelli and Mulder, 2015). An important feature of such transitions is that the emergence of new reinforcing processes creates a stable equilibrium and prevents easy transitions between (or among) alternative states (i.e. the transitions become hysteretic; Scheffer et al., 2001). In contrast, the commonly used single-equilibrium definition of resilience emphasises the speed with which a perturbed ecosystem is able to return to previous food web configurations and/or levels of ecosystem functioning following release of some or all stressors (Gunderson, 2000; Pimm, 1991). This is analogous to the idea of resilience in mechanical engineering (i.e. the speed of rebound to a single equilibrium) and, consequently, is commonly termed "engineering resilience". Both definitions of resilience can give insights into the factors maintaining the stability of ecosystem functioning and service delivery; however only the ecological resilience model is capable of accounting for the potential for ecosystems to exist in multiple alternative configurations. Lakes are currently perhaps the best-studied ecosystem types regarding the existence of alternative equilibria (clear water and submerged aquatic vegetation versus turbid waters and algae; Scheffer, 1997), though regime shifts have been documented in other ecosystem types also (Angeler et al., 2013b; Dent et al., 2002; Heffernan, 2008; Scheffer et al., 2001). In many ecosystem types, including running waters, it is more common to observe a gradual erosion of key ecosystem parameters rather than rapid and extreme shifts in equilibria (e.g. Mancinelli and Mulder, 2015). However, recent palaeoecological research indicates that not all regime shifts are sudden, and that prolonged periods of instability prior to regime shifts may exceed human life spans (Angeler et al., 2015a; Spanbauer et al., 2014), emphasising the need for the development of tools and metrics that can identify when ecosystems and their ecosystem services are at risk of an impending regime shift.

A key linkage between B-EF and ecological resilience research is the *insurance effect hypothesis* (Naeem and Li, 1997; Yachi and Loreau, 1999), which states that greater biodiversity should enhance the stability of functioning in ecosystems under stress, due to the higher likelihood that a species-rich assemblage will include functionally redundant, tolerant species, able to compensate for those negatively affected (suppressed or exterminated) by the disturbance (Elmqvist et al., 2003; Loreau et al., 2002). Furthermore, functioning should be inherently more stable for species-rich systems, as the responses of extreme species are diluted over a more diverse assemblage (i.e. "statistical averaging"; Doak et al., 1998). Importantly, connectivity among habitat patches or ecosystem compartments can contribute to the maintenance of this stability, by extending the potential pool of functionally redundant, replacement species from those present in the local habitat patch. *The cross-scale model of resilience* (Peterson et al., 1998) builds on these concepts to predict that resilience of an ecological network is enhanced when:

- **(i)** Functional redundancy within and across ecological scales (e.g. among size classes in consumer guilds, or across habitat patches) is high;
- **(ii)** The diversity of rare species, which may contribute to future "insurance effects", is also high within and across ecological scales;
- **(iii)** Key functional effect traits are more strongly associated with tolerant rather than sensitive species;
- **(iv)** Connectivity among ecological compartments is high, facilitating maintenance of functional redundancy/key functional traits at local scales under stress (e.g. by dispersal-mediated "mass effects").

Significantly, all these properties can be estimated by applying a trait-based framework to taxonomic data, even in the absence of direct quantification of ecosystem processes (Angeler et al., 2013a). This potentially allows assessment of species traits and ecosystem resilience alongside more traditional biodiversity and ecological status classifications (Allen et al., 2005; Elmqvist et al., 2003). Here, "ecosystem compartments" could refer to food web compartments (e.g. defined based on body size), different habitat patches in space, or the same habitat patches occupied by different species or life stages, or supporting different ecosystem process, in time. Statistical approaches, such as discontinuity analysis, can be used to identify the level of scaling constraining the activities and movements of species, based, for example, on biomass spectra, or fluctuations in species abundances in space and time (Angeler et al., 2015a). Once the appropriate ecological scales have been identified, an assessment of distributions of species and their associated functional traits within and across the scales, and connectivity among ecological compartments, allows for an assessment of resilience, and in turn ecosystem vulnerability.

Identification of a loss of functional redundancy among ecological scales, or reduced connectivity between assemblages in an ecological compartment and their broader metacommunity, might be a warning sign of an impending regime state shift, or indicate that the gradual erosion of the existing regime is in danger of threatening key species, ecosystem processes, and ecosystem services. This is potentially a highly useful tool in ecosystem service management, since regime shifts, which often have highly uncertain outcomes, are generally associated with negative consequences regarding ecosystem functioning, ecosystem service provisioning, and thus human well-being and societal development (Rocha et al., 2015). For instance, once a clear-water lake tips into the stable alternative turbid-water state due to excessive nutrient loading, important recreational (swimming, boating) and provisioning services (fisheries) may be jeopardised because of the development of dense, often toxic, algal blooms, and the loss of top predators in the food web (piscivorous fish) (Pope et al., 2014). Early detection of an impending regime shift, based on declining functional redundancy and ecological connectivity, could be applied in such cases to the protection of these recreational and provisioning services (Angeler et al., 2014).

3.3 Incorporating Resilience Aspects into Portfolio Theory for Management of Ecosystem Services

Recently, a framework from the socio-economic sciences has been proposed to quantify the reliability of performance of ecosystem functioning and service provisioning (Griffiths et al., 2014). Portfolio theory has a long history in financial management, and also has applications in ecology (Schindler et al., 2015). The theory has been used to quantify links between the risk and reward of individual assets or commodities, like for instance bank deposits or timber, and the risk and return associated with a portfolio of financial resources. As economic conditions change, investments can be bought or sold, thereby maintaining a desired reward–risk balance and reliability of portfolio performance. Ecological management, particularly in the context of ecosystem service provisioning, has many parallels with financial portfolio management. Both aim to achieve high returns while minimising risk under uncertainty. Rapid environmental change outcomes are highly uncertain, yet sustained ecosystem service provisioning is one of humanity's priorities. Portfolio theory offers a quantitative way to evaluate management options so that the portfolio makes the preferred trade-off of reward versus risk, and provides a clear example of the growing convergence of ecology and economics within the ecosystem services paradigm (see also Mulder et al. (2015)).

Griffiths et al. (2014) used portfolio theory to demonstrate how ecosystem services can be quantified, and trade-offs among management alternatives assessed, to target conservation efforts. "Assets" in their study were Pacific salmon (*Oncorhynchus* spp.) runs from one or more populations within a catchment (see Durance et al., 2016 for further fisheries-based examples). They evaluated the variability of returns within and across individual populations to assess the performance and reliability of salmon fisheries ("portfolios") across catchments over time. They also studied how human impacts influence fishery portfolios. They found that salmon portfolios in near-pristine ecosystems were more reliable than where human impacts on catchments (e.g. dams and land-use change) were more pronounced. Based on the identification of impact sources to portfolio risk, Griffiths et al. (2014) concluded that specific management actions (habitat protection, specific harvest strategies, maintenance of a diverse disturbance regime) can be guided to improve restoration activities and maintain existing resilience. Thus, portfolio theory combined with an adaptive management approach allows for recalibration and adjustments to changing environmental conditions (Allen et al., 2011). This adaptive approach contrasts with traditional conservation planning, which uses optimisation algorithms for defining fixed sets of conservation priorities based on a static view of the distribution of biodiversity elements in relation to threats (Hoekstra, 2012), with little direct consideration given to ecosystem services.

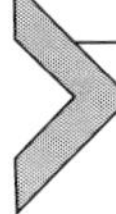

4. EXTENDING AND PARAMETERISING A TRAIT-BASED FRAMEWORK FOR PREDICTING FUNCTIONAL REDUNDANCY AND OUTCOMES FOR ECOSYSTEM FUNCTIONING AND SERVICES

The species trait concept is central in research on ecological resilience, and quantification of functional redundancy based on species traits has enormous potential in monitoring and management, for identifying ecosystems where functional capacity and hence ecosystem service delivery are at risk of being compromised. However, a key theme of this review is that expression of specific functional traits can depend strongly on environmental context, interactions among species and their traits, and trophic interactions and feedbacks. As such, traits are best regarded as tools for *predicting* or diagnosing impacts of stressors on communities, and outcomes of stressors and biodiversity change for functioning and the delivery of the associated services.

Earlier, Reiss et al. (2009) presented a conceptual framework linking individuals, species, and traits within B-EF research. Here, we extend this approach to focus more strongly on functional redundancy, species interactions, and both additive and non-additive outcomes for multiple ecosystem processes and ecosystem services (Box 1). These scenarios recognise that multiple species contribute to multiple services, and that changes in diversity

BOX 1 Integrating species interactions and ecosystem services into a trait-based framework

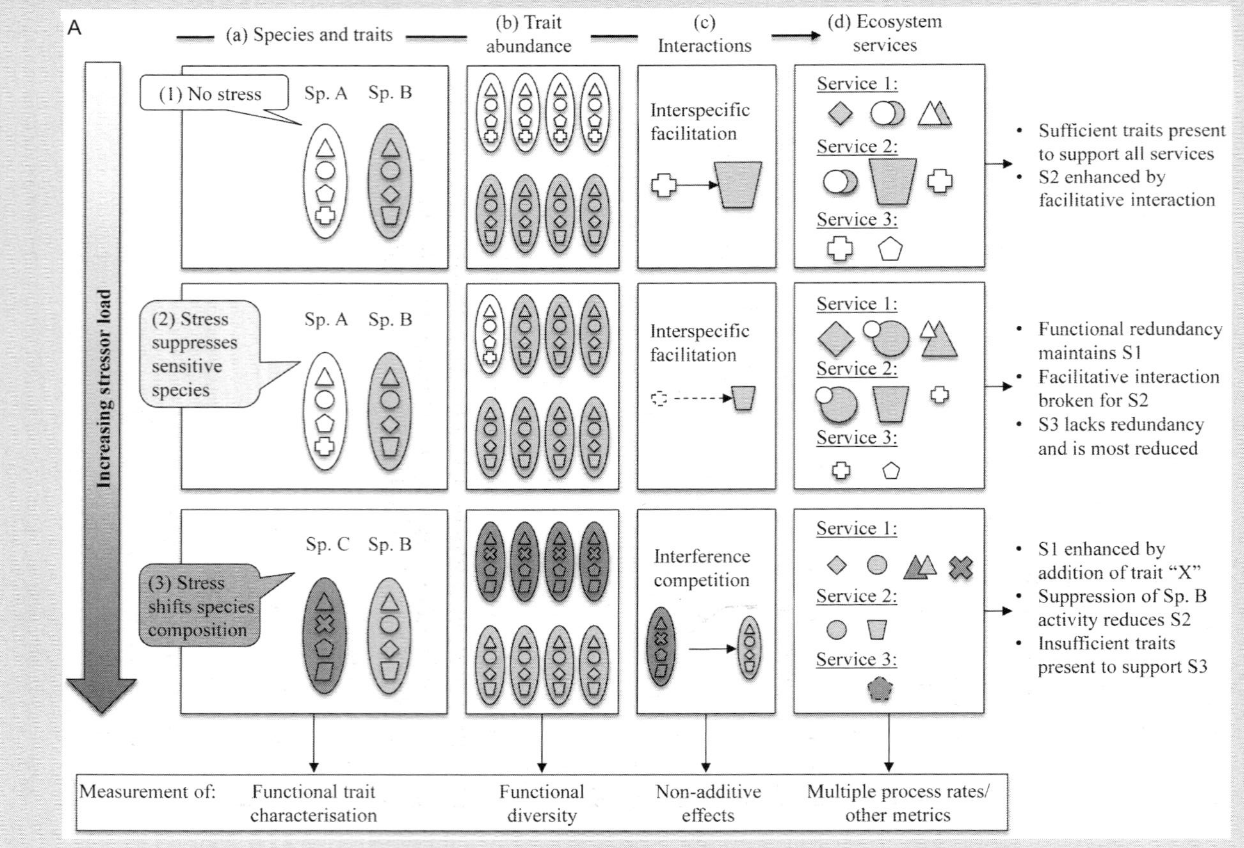

BOX 1 Integrating species interactions and ecosystem services into a trait-based framework—cont'd

(A) Shifts in species, traits, species interactions, and ecosystem services (ES) under an increasing stressor load. Species traits, represented by different symbols (e.g. triangle, circle, square, cross, etc.), are characteristics of different consumer species, represented by ovals (column a). Increases in stressor loadings affect species richness and evenness and hence trait diversity (column b), and the intensity of particular species interactions (column c). Together, the presence and relative abundance of species traits and the intensity of species interactions regulate efficiency of three different ecosystem services (column d, either quantified as appropriate ecosystem processes or some other ES metric, see below). Services 1–2 require the presence of at least three unique traits to occur, whilst Service 3 requires two traits.

In *scenario 1*, the species assemblage is not under anthropogenic stress, and species are present with equal abundances, but two traits (the triangle and circle) occur with twice the abundance of the others, due to functional redundancy among the species (box 1b). One trait of species A (the cross) facilitates the activities of species B (box 1c), increasing expression of another trait (the tetrahedron). Functional redundancy is greatest in traits driving service 1 and lowest for service 3, which is wholly dependent on species A (box 1d). The facilitative interaction enhances service 2 (box 1d).

In *scenario 2*, a stressor reduces the abundance of the sensitive (purple (white in the print version)) species A relative to the insensitive species B (blue (dark grey in the print version)), reducing the relative abundance of the unique traits contributed by species A, but not of those shared by both species (box 2b). Consequently, species A is no longer present at sufficient densities to support the interspecific facilitation of species B (box 2c). Outcomes for the three services contrast markedly (box 2d). Service 1 remains unaffected, due to the high degree of redundancy in the key traits required. The efficiency of Service 2 declines, reflecting the loss of the facilitative interaction that enhanced it under scenario 1, and the lack of functional redundancy in one of the impacted traits (the cross). Service 3 is impacted most of all due to the lack of redundancy in any of the key underpinning traits.

In *scenario 3*, further increases in stressor load drives a change in species composition, as the most sensitive species A goes extinct and a tolerant (red (dark grey in the print version)) species colonises (box 3a), bringing two novel traits, and two traits which overlap with those of species A and B (the hexagon and triangle, respectively) (box 3b). The activities of species C partly interfere with those of species B (box 3c). Key traits contributed by species B to Service 1 are consequently suppressed. However, this is partly compensated by (i) redundancy in one trait (the triangle) between species B and C, and (ii) introduction of a novel trait (X) with species C, which enhances service 1 (box 3d). Service 2 is also reduced, reflecting the loss of all underlying functional redundancy. Species

Continued

BOX 1 Integrating species interactions and ecosystem services into a trait-based framework—cont'd

C reintroduces one of the traits lost with species A (the hexagon), but this is insufficient to maintain Service 3, since the other key underpinning trait (the cross) has been lost.

Presently, solid methodologies and indices exist for allocating traits to species, quantifying functional diversity, and measuring supporting ecosystem processes as process rates, for at least some organism groups (e.g. freshwater macroinvertebrates, terrestrial plants) and some supporting processes (leaf decomposition, primary productivity). Ecosystem process rate measures may be useful proxies for ecosystem services, especially when a socio-economic value can be allocated to the process. Alternatively, other ES metrics (e.g. recreational fish standing stocks) might be used. Quantification of non-additive effects of species interactions on trait expression and ecosystem processes is most challenging, particularly as such interactions are often highly context dependent. Approaches to assessing non-additivity have nevertheless been developed in B-EF research, and development of a predictive trait-based framework might allow estimation of this component based on departures from the predicted effects of traits on functioning and services.

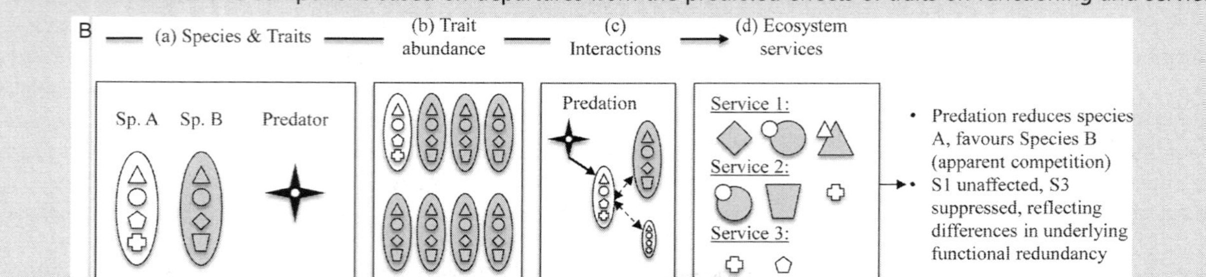

(B) Incorporation of further species interactions. In the above example, a predator preferentially consumes species A, in turn favouring species B in an "apparent competition" scenario. Outcomes for ecosystem processes are similar to those induced by the stressor in scenario 2 above, but the underlying driver (stressor vs. predator) is different.

After Reiss et al. (2009).

or trait expression that enhance some services will not necessarily enhance others. As in Reiss et al.'s (2009) original framework, the scheme presented in Box 1 is capable of dealing with intraspecific trait variation, since traits can be allocated to individuals within species. However, for simplicity we have not varied traits within species for the scenarios discussed here.

In general, it is expected that a greater redundancy of functional traits will allow maintenance of ecosystem services even when stressors reduce diversity, if unaffected species compensate for those lost (Box 1A). Conversely, increasing stressor loads may rapidly compromise supporting services that depend on unique traits possessed by sensitive species (Box 1A). Consequences of changes in biodiversity and species composition can also be predicted based on their species traits. However, non-additive effects associated with species interactions (facilitation, interference competition), or context-specific variation in trait expression, can modify these predictions (Box 1A). Within a trait-based framework, it is possible to explore outcomes of different types of species interactions on supporting ecosystem services, including quite complex cases. For example, apparent competition associated with selective predation may have similar outcomes for services as a stressor that selectively suppresses and favours different species, and distinguishing these underlying mechanisms has significance in ecosystem assessment and management (Box 1B).

The scheme presented in Box 1 requires quantification of four main types of parameters: (i) allocation of traits to species (and ideally individuals, in cases of high intraspecific variability); (ii) calculation of various metrics characterising the functional diversity and composition of species traits; (iii) quantification of supporting ecosystem processes, ideally as ecosystem process rates, and identifying linkages with final ecosystem services; and (iv) estimation of non-additivity arising from species interactions and context-specific variability in trait expression. Of these, robust methodologies already exist for allocating species traits and quantifying supporting ecosystem processes, for at least some organism groups and processes (e.g. aquatic macroinvertebrates, terrestrial plants, and litter decomposition and primary productivity; Fig. 2), and multiple metrics for characterising functional diversity and trait averages are available (Gamfeldt et al., 2008; Lefcheck et al., 2015; Maestre et al., 2012). Direct quantification of "non-additive" effects will often be very challenging logistically, outside of the highly controlled settings of B-EF experiments. However, departures from predictions based on the diversity and composition of species traits might potentially be used in estimating the size of the non-additive effect on functioning with, for example, computer-based simulations (Bohan et al., 2013; Tamaddoni-Nezhad et al., 2013). We suggest that Reiss et al.'s (2009) original visualisation of individuals, species, and traits can have further value as a heuristic and possibly predictive tool linking species, interactions, and functioning, and for testing the assumptions underlying the use of species traits within the species resilience framework.

5. CONCLUDING REMARKS

Each step in this compilation of theory and literature on the drivers of functioning and outcomes for ecosystem services has entailed the addition of extra layers of complexity—from abiotic and biotic drivers through extended trophic and spatial dimensions—for a more complete understanding. However, the management of ecosystem services might not always need to account for all this detail, and a key challenge for future research is to identify when a simpler versus more complex approach to services can be taken. For example, in assessing the effects of stream networks on nutrient sequestration in the context of landscape-level nutrient budgets, it may often be sufficient to primarily characterise only the abiotic drivers and proxies for the processes underlying this ecosystem service (e.g. nutrient inputs and outputs, hydrology, temperature), and to regard the biotic components of the ecosystem as a "black box" (as standard in most current biogeochemical modelling—e.g. Alexander et al., 2000; Seitzinger et al., 2006). However, in some cases, variability in community composition may have consequences for nutrient dynamics which need to be accounted for even at whole-catchment scale, as when invasive species achieve very high biomasses over an extended spatial distribution (Hall et al., 2003). Similarly, substantial variability in the process of leaf decomposition might often be explained based on a few abiotic variables (nutrients, temperature) and a single biotic variable: the species composition of litter entering the stream, since decomposition rates are regulated strongly by the characteristics (nutrient content, refractory compounds) of the litter itself (Meentemeyer, 1978). However, two decades of B-EF research have highlighted the potential for non-additive effects of interactions among microbes, detritivores, and across trophic levels to alter decomposition rates away from predictions based solely on litter characteristics, even if this variability has not always been systematically associated with changes in diversity *per se* (Gessner et al., 2010; Handa et al., 2014). Finally, the trait-based methodology of the cross-scale resilience model offers a powerful approach for detecting losses of functional diversity and redundancy, and predicting outcomes for ecosystem functioning and service provision. However, the high potential for non-additive effects of species interactions on ecosystem processes generates much uncertainty within this framework.

Accordingly, a number of challenges remain for future research, particularly in quantifying the level of *uncertainty* in ecosystem processes and the services supported by those processes, associated with variability in biodiversity, species interactions, and species trait expression from local through landscape and larger scales. These include:

(i) The range of potentially complex synergistic and antagonistic interactions between multiple abiotic stressors, biodiversity change, and shifts in trophic structure implies large uncertainties when predicting outcomes for communities, ecosystem functioning, and ecosystem services. Research should focus on not only documenting these non-additive interactions but also understanding the environmental, biological, and spatial factors regulating *when and where* such interactions are most likely to occur, and when their outcomes are *greatest/most important* for communities, functioning, and services. This would facilitate incorporation of non-additive stressor–biodiversity interactions into a cross-scale predictive framework based on meta-ecosystem and resilience theory.

(ii) Species traits offer a potentially powerful tool for predicting effects of environmental change on community structure and ecosystem functioning and are a key component in the further development and application of the meta-ecosystem and resilience frameworks in the context of ecosystem services. However, to date robust species trait databases, mostly comprising response rather than true functional effect traits, have been compiled for very few organism groups. Furthermore, most of these databases are static, with limited coverage of intraspecific variation within populations or across a species' range. Developing objective trait classifications across a broad range of groups requires a great research effort which will generally be beyond the resources of any one research group, highlighting the need for collaborative approaches.

(iii) Compilation of databases recording true functional effect traits (e.g. species-specific growth and consumption rates) for entire biological groups and with broad biogeographic scope is even more challenging, particularly since trait expression in terms of functional processing rates is likely to be highly context specific (i.e. depending *inter alia* on environmental characteristics, individual fitness, and interactions with other organisms). Partly, the lack of good functional effect trait information can be addressed by more explicitly tying key response traits to particular ecosystem processes, and identifying traits of particular importance in community assembly to predict variations in ecosystem functioning and ultimately ecosystem services. In practice, this is currently how most research linking species response traits with functioning proceeds. Quantifying the level of uncertainty associated with the potential mismatch between functional effect trait presence (quantified either as a response trait proxy or effect trait mean value) and functional effect trait expression (i.e. depending on local environmental context), across a broad range of environmental conditions, would assist in a more robust development of trait-based approaches in B-EF, meta-ecosystem and resilience research, and in management of ecosystem services.

(iv) Several ecosystem processes are easily characterised as supporting the delivery of ecosystem services (e.g. leaf decomposition rates, primary productivity, biofilm-mediated nutrient uptake rates), and in some cases, where clear socio-economic values can be assigned, these processes may be understood as provisioning or regulating services in their own right. However, linking specific processes with final ecosystem services remains generally challenging, given that final services are often underpinned by several supporting processes simultaneously (e.g. nutrient and carbon sequestration as final services delivered by streams depends on a complex interplay between primary productivity, organic matter breakdown, nutrient uptake rates, and so on). These processes do not necessarily respond to stressors or losses of biodiversity in the same way. Research should focus on which processes/supporting service measures, and possibly ecosystem functioning "multimetrics", work best as proxies for final ecosystem services, across a range of environmental and socio-economic contexts (and hence this will require interdisciplinary collaboration).

(v) Broader application of concepts from the meta-ecosystem framework, cross-scale resilience model, portfolio theory, and B-EF and B-ES research all seem warranted in ecology, particularly in management and conservation of ecosystem functioning and ecosystem services. However, the potential for doing so is currently limited by the lack of extensive, long-term data linking communities with supporting ecosystem processes, and final ecosystem services, all within an explicit spatial context allowing assessment of ecological connectivity. Such data sets might arise as functional-based approaches are incorporated more broadly into routine bioassessment at regional and national scales. However, even compilation of a few, comprehensive databases would allow an assessment of the robustness of resilience methodologies that rely on static functional traits (e.g. within the framework presented in Box 1), with associated uncertainties quantified, or how spatial connectivity among habitat patches at regional scales affects local-scale ecosystem functioning and service delivery. Compilation and analysis of such data in collaboration with social scientists, policy makers, and stakeholders would allow scientists to develop new avenues of research and assist in designing monitoring programmes (Durance et al., 2016; Palomo et al., 2016).

Through this review, we have argued that unification of trait-based ecology, the meta-ecosystem concept, and B-EF and B-ES research within an ecological resilience framework will improve capacities of scientists and managers for predicting, explaining, and ultimately managing the effects of stressors, biodiversity loss and altered ecological connectivity on multiple ecosystem attributes, and over multiple spatiotemporal scales. These attributes include the stability of food webs, ecosystem functioning, and ecosystem services. Despite the challenges identified here, we suggest that the necessary theory (including B-EF, metacommunity, resilience, and trait-driver theories) and tools (species traits, methods for quantifying functioning, statistical techniques) already exist for at least some ecosystems and organism groups to begin developing such a unified approach (e.g. Angeler et al., 2014, 2015b; Nash et al., 2014; Wardle and Jonsson, 2014) and to test its application in ecosystem assessment and management, including of ecosystem services.

ACKNOWLEDGEMENTS

The authors are grateful to Guy Woodward and Isabelle Durance for their insightful comments on an earlier version of our manuscript, which greatly improved its quality. We thank the following colleagues for providing photographs used in Fig. 2: Dr. Jon Benstead (the chamber picture), Dr. Frauke Ecke (aerial view of macrophytes in a lake), and Dr. Steffi Gottschalk (the diatom). This work was supported by the project WATERS (to R.K.J. and B.G.M.), funded by the Swedish Environmental Protection Agency (Dnr 10/179) and the Swedish Agency for Marine and Water Management.

REFERENCES

Alexander, R.B., Smith, R.A., Schwarz, G.E., 2000. Effect of stream channel size on the delivery of nitrogen to the Gulf of Mexico. Nature 403, 758–761.

Allen, C.R., Gunderson, L., Johnson, A.R., 2005. The use of discontinuities and functional groups to assess relative resilience in complex systems. Ecosystems 8, 958–966.

Allen, C.R., Fontaine, J.J., Pope, K.L., Garmestani, A.S., 2011. Adaptive management for a turbulent future. J. Environ. Manag. 92, 1339–1345.

Angeler, D.G., Allen, C.R., Johnson, R.K., 2013a. Measuring the relative resilience of subarctic lakes to global change: redundancies of functions within and across temporal scales. J. Appl. Ecol. 50, 572–584.

Angeler, D.G., Allen, C.R., Rojo, C., Alvarez-Cobelas, M., Rodrigo, M.A., Sánchez-Carrillo, S., 2013b. Inferring the relative resilience of alternative states. PLoS One 8, e77338c.

Angeler, D.G., Allen, C.R., Birgé, H.E., Drakare, S., McKie, B.G., Johnson, R.K., 2014. Assessing and managing freshwater ecosystems vulnerable to environmental change. AMBIO 43, 113–125.

Angeler, D.G., Allen, C.R., Barichievy, C., Eason, T., Garmestani, A.S., Graham, N.A.J., Granholm, D., Gunderson, L., Knutson, M., Nash, K.L., Nelson, R.J., Nyström, M., Spanbauer, T.E., Stow, C.A., Sundstrom, S.M., 2015a. Management applications of discontinuity theory. J. Appl. Ecol. http://dx.doi.org/10.1111/1365-2664.12494.

Angeler, D.G., Baho, D.L., Allen, C.R., Johnson, R.K., 2015b. Linking degradation status with ecosystem vulnerability to environmental change. Oecologia 178, 899–913.

Astegiano, J., Guimarães Jr., P.R., Cheptou, P.-O., Vidal, M.M., Mandai, C.Y., Ashworth, L., Massol, F., 2015. Persistence of plants and pollinators in the face of habitat loss: insights from trait-based metacommunity models. Adv. Ecol. Res. 53, 201–257.

Balvanera, P., Pfisterer, A.B., Buchmann, N., He, J.S., Nakashizuka, T., Raffaelli, D., Schmid, B., 2006. Quantifying the evidence for biodiversity effects on ecosystem functioning and services. Ecol. Lett. 9, 1146–1156.

Bastian, M., Pearson, R.G., Boyero, L.U.Z., 2008. Effects of diversity loss on ecosystem function across trophic levels and ecosystems: a test in a detritus-based tropical food web. Austral Ecol. 33, 301–306.

Bastian, O., Grunewald, K., Syrbe, R.-U., 2015. Classification of ecosystem services. In: Grunewald, K., Bastian, O. (Eds.), Ecosystem Services—Concept, Methods and Case Studies. Springer, Berlin.

Baulch, H.M., Stanley, E.H., Bernhardt, E.S., 2011. Can algal uptake stop NO_3^- pollution? Nature 477, 86–89.

Bohan, D.A., Raybould, A., Mulder, C.P., Woodward, G., Tamaddoni-Nezhad, A., Bluthgen, N., Pocock, M.J.O., Muggleton, S., Evans, D.M., Astegiano, J., Massol, F., Loeuille, N., Petit, S., Macfadyen, S., 2013. Networking agroecology: integrating the diversity of agroecosystem interactions. In: Woodward, G., Bohan, D.A. (Eds.), Ecological Networks in an Agricultural World. In: Advances in Ecological Research, Academic Press, San Diego.

Bommarco, R., Biesmeijer, J.C., Meyer, B., Potts, S.G., Pöyry, J., Roberts, S.P.M., Steffan-Dewenter, I., Öckinger, E., 2010. Dispersal capacity and diet breadth modify the response of wild bees to habitat loss. Proc. R. Soc. Lond. B Biol. Sci. 277, 2075–2082.

Bonan, G.B., Oleson, K.W., Fisher, R.A., Lasslop, G., Reichstein, M., 2012. Reconciling leaf physiological traits and canopy flux data: use of the TRY and FLUXNET databases in the community land model version 4. J. Geophys. Res. Biogeosci. 117, G02026. http://dx.doi.org/10.1029/2011JG001913.

Bott, T.L., 2006. Primary productivity and community respiration. In: Hauer, F.R., Lamberti, G.A. (Eds.), Methods in Stream Ecology, second ed. Elsevier, Amsterdam.

Boyd, J., Banzhaf, S., 2007. What are ecosystem services? The need for standardized environmental accounting units. Ecol. Econ. 63, 616–626.

Brown, J.H., Gillooly, J.F., Allen, A.P., Savage, V.M., West, G.B., 2004. Toward a metabolic theory of ecology. Ecology 85, 1771–1789.

Bunn, S.E., Davies, P.M., 2000. Biological processes in running waters and their implications for the assessment of ecological integrity. Hydrobiologia 422/423, 61–70.

Burrell, T.K., O'Brien, J.M., Graham, S.E., Simon, K.S., Jon, S.H., McIntosh, A.R., 2014. Riparian shading mitigates stream eutrophication in agricultural catchments. Freshw. Sci. 33, 73–84.

Burrows, R.M., Hotchkiss, E.R., Jonsson, M., Laudon, H., McKie, B.G., Sponseller, R.A., 2015. Nitrogen limitation of heterotrophic biofilms in boreal streams. Freshw. Biol. 60, 1237–1251.

Caissie, D., 2006. The thermal regime of rivers: a review. Freshw. Biol. 51, 1389–1406.

Calcagno, V., Massol, F., Mouquet, N., Jarne, P., David, P., 2011. Constraints on food chain length arising from regional metacommunity dynamics. Proc. R. Soc. Lond. B Biol. Sci. 278, 3042–3049.

Cardinale, B.J., 2011. Biodiversity improves water quality through niche partitioning. Nature 472, 86–89.

Cardinale, B.J., Palmer, M.A., Collins, S.L., 2002. Species diversity enhances ecosystem functioning through interspecific facilitation. Nature 415, 426–429.

Cardinale, B.J., Ives, A.R., Inchausti, P., 2004. Effects of species diversity on the primary productivity of ecosystems: extending our spatial and temporal scales of inference. Oikos 104, 437–450.

Cardinale, B.J., Srivastava, D.S., Duffy, J.E., Wright, J.P., Downing, A.L., Sankaran, M., Jouseau, C., 2006. Effects of biodiversity on the functioning of trophic groups and ecosystems. Nature 443, 989–992.

Cardinale, B.J., Wright, J.P., Cadotte, M.W., Carroll, I.T., Hector, A., Srivastava, D.S., Loreau, M., Weis, J.J., 2007. Impacts of plant diversity on biomass production increase through time because of species complementarity. Proc. Natl. Acad. Sci. U.S.A. 104, 18123–18128.

Cardinale, B.J., Duffy, J.E., Gonzalez, A., Hooper, D.U., Perrings, C., Venail, P., Narwani, A., Mace, G.M., Tilman, D., Wardle, D.A., Kinzig, A.P., Daily, G.C., Loreau, M., Grace, J.B., Larigauderie, A., Srivastava, D.S., Naeem, S., 2012. Biodiversity loss and its impact on humanity. Nature 486, 59–67.

Carpenter, S.R., Kitchell, J.F., Hodgson, J.R., 1985. Cascading trophic interactions and lake productivity. BioScience 35, 634–639.

Carpenter, S.R., Kitchell, J.F., Hodgson, J.R., Cochran, P.A., Elser, J.J., Elser, M.M., Lodge, D.M., Kretchmer, D., He, X., Von Ende, C.N., 1987. Regulation of lake primary productivity by food web structure. Ecology 68, 1863–1876.

Comas, L.H., Becker, S.R., Von Mark, V.C., Byrne, P.F., Dierig, D.A., 2013. Root traits contributing to plant productivity under drought. Front. Plant Sci. 4, 1–16.

Creed, R.P., Cherry, R.P., Pflaum, J.R., Wood, C.J., 2009. Dominant species can produce a negative relationship between species diversity and ecosystem function. Oikos 118, 723–732.

Cummins, K.W., 1974. Structure and function of stream ecosystems. BioScience 24, 631–641.

Daily, G.C., 1997. Nature's Services: Societal Dependence on Natural Ecosystems. Island Press, Washington, D.C.

Daily, G.C., Söderqvist, T., Aniyar, S., Arrow, K., Dasgupta, P., Ehrlich, P.E., Folke, C., Jansson, A., Jansson, B.O., Kautsky, N., Levin, S., Lubchenco, J., Mäler, K.G.,

Simpson, D., Starrett, D., Tilman, D., Walker, B., 2000. The value of nature and the nature of value. Science 289, 395–396.

Dangles, O., Malmqvist, B., 2004. Species richness-decomposition relationships depend on species dominance. Ecol. Lett. 7, 395–402.

Dangles, O., Crespo-Pérez, V., Andino, P., Espinosa, R., Calvez, R., Jacobsen, D., 2011. Predicting richness effects on ecosystem function in natural communities: insights from high-elevation streams. Ecology 92, 733–743.

Darwin, C., 1859. The Origin of Species by Means of Natural Selection. Penguin Classics, London.

Dent, C.L., Cumming, G.S., Carpenter, S.R., 2002. Multiple states in river and lake ecosystems. Philos. Trans. R. Soc. Lond. B Biol. Sci. 357, 635–645.

Díaz, S., Cabido, M., 2001. Vive la différence: plant functional diversity matters to ecosystem processes. Trends Ecol. Evol. 16, 646–655.

Dirzo, R., Young, H.S., Galetti, M., Ceballos, G., Isaac, N.J., Collen, B., 2014. Defaunation in the anthropocene. Science 345, 401–406.

Doak, D.F., Bigger, D., Harding, E.K., Marvier, M.A., O'Malley, R.E., Thomson, D., 1998. The statistical inevitability of stability-diversity relationships in community ecology. Am. Nat. 151, 265–276.

Downing, A.L., Leibold, M.A., 2002. Ecosystem consequences of species richness and composition in pond food webs. Nature 416, 837–841.

Duffy, J.E., 2009. Why biodiversity is important to the functioning of real-world ecosystems. Front. Ecol. Environ. 7, 437–444.

Duffy, J.E., Cardinale, B.J., France, K.E., McIntyre, P.B., Thebault, E., Loreau, M., 2007. The functional role of biodiversity in ecosystems: incorporating trophic complexity. Ecol. Lett. 10, 522–538.

Durance, I., Bruford, M.W., Chalmers, R., Chappell, N.A., Christie, M., Jack Cosby, B., Noble, D., Ormerod, S.J., Prosser, H., Weightman, A., Woodward, G., 2016. The challenges of linking ecosystem services to biodiversity: lessons from a large-scale freshwater study. Adv. Ecol. Res. 54, in press.

Ellison, A.M., Bank, M.S., Clinton, B.D., Colburn, E.A., Elliott, K., Ford, C.R., Foster, D.R., Kloeppel, B.D., Knoepp, J.D., Lovett, G.M., Mohan, J., Orwig, D.A., Rodenhouse, N.L., Sobczak, W.V., Stinson, K.A., Stone, J.K., Swan, C.M., Thompson, J., Von Holle, B., Webster, J.R., 2005. Loss of foundation species: consequences for the structure and dynamics of forested ecosystems. Front. Ecol. Environ. 3, 479–486.

Elmqvist, T., Folke, C., Nystrom, M., Peterson, G., Bengtsson, J., Walker, B., Norberg, J., 2003. Response diversity, ecosystem change, and resilience. Front. Ecol. Environ. 1, 488–494.

Enquist, B.J., Norberg, J., Bonser, S.P., Violle, C., Webb, C.T., Henderson, A., Sloat, L.L., Savage, V.M., 2015. Scaling from traits to ecosystems: developing a general trait driver theory via integrating trait-based and metabolic scaling theories. In: Pawar, S., Woodward, G., Dell, A.I. (Eds.), Trait-Based Ecology—From Structure to Function. In: Advances in Ecological Research, Academic Press, San Diego.

Ewers, R.M., Didham, R.K., 2006. Confounding factors in the detection of species responses to habitat fragmentation. Biol. Rev. 81, 117–142.

Ferreira, V., Chauvet, E., 2011. Synergistic effects of water temperature and dissolved nutrients on litter decomposition and associated fungi. Glob. Chang. Biol. 17, 551–564.

Frainer, A., McKie, B.G., 2015. Shifts in the diversity and composition of consumer traits constrain the effects of land use on stream ecosystem functioning. In: Pawar, S., Woodward, G., Dell, A.I. (Eds.), Trait-Based Ecology—From Structure to Function. In: Advances in Ecological Research, Academic Press, San Diego.

Frainer, A., McKie, B.G., Malmqvist, B., 2014. When does diversity matter? Species functional diversity and ecosystem functioning across habitats and seasons in a field experiment. J. Anim. Ecol. 83, 460–469.

France, K.E., Duffy, J.E., 2006. Diversity and dispersal interactively affect predictability of ecosystem function. Nature 441, 1139–1143.

Gamfeldt, L., Hillebrand, H., Jonsson, P.R., 2008. Multiple functions increase the importance of biodiversity for overall ecosystem functioning. Ecology 89, 1223–1231.

Gamfeldt, L., Snall, T., Bagchi, R., Jonsson, M., Gustafsson, L., Kjellander, P., Ruiz-Jaen, M.C., Froberg, M., Stendahl, J., Philipson, C.D., Mikusinski, G., Andersson, E., Westerlund, B., Andren, H., Moberg, F., Moen, J., Bengtsson, J., 2013. Higher levels of multiple ecosystem services are found in forests with more tree species. Nat. Commun. 4, 1340–1347.

Gessner, M.O., Chauvet, E., 2002. A case for using litter breakdown to assess functional stream integrity. Ecol. Appl. 12, 498–510.

Gessner, M.O., Swan, C.M., Dang, C.K., McKie, B.G., Bardgett, R.D., Wall, D.H., Hättenschwiler, S., 2010. Diversity meets decomposition. Trends Ecol. Evol. 25, 372–380.

Gilbert, A.J., 2009. Connectance indicates the robustness of food webs when subjected to species loss. Ecol. Indic. 9, 72–80.

Gill, R.J., Baldock, K.C.R., Brown, M.J.F., Cresswell, J.E., Dicks, L.V., Fountain, M.T., Garratt, M.P.D., Gough, L.A., Heard, M.S., Holland, J.M., Ollerton, J., Stone, G.N., Tang, C.Q., Vanbergen, A.J., Vogler, A.P, Woodward, G., Arce, A.N., Boatman, N.D., Brand-Hardy, R., Breeze, T.D., Green, M., Hartfield, C.M., O'Connor, R.S., Osborne, J.L., Phillips, J., Sutton, P.B., Potts, S.G., 2016. Protecting an ecosystem service: approaches to understanding and mitigating threats to wild insect pollinators. Adv. Ecol. Res. 54, in press.

Gravel, D., Mouquet, N., Loreau, M., Guichard, F., 2010. Patch dynamics, persistence, and species coexistence in metaecosystems. Am. Nat. 176, 289–302.

Gravel, D., Canard, E., Guichard, F., Mouquet, N., 2011a. Persistence increases with diversity and connectance in trophic metacommunities. PLoS One 6, e19374.

Gravel, D., Massol, F., Canard, E., Mouillot, D., Mouquet, N., 2011b. Trophic theory of island biogeography. Ecol. Lett. 14, 1010–1016.

Griffiths, J.R., Schindler, D.E., Armstrong, J.B., Scheuerell, M.D., Whited, D.C., Clark, R.A., Hilborn, R., Holt, C.A., Lindley, S.T., Stanford, J.A., Volk, E.C., 2014. Performance of salmon fishery portfolios across western North America. J. Appl. Ecol. 51, 1554–1563.

Grime, J.P., 1998. Benefits of plant diversity to ecosystems: immediate, filter and founder effects. J. Ecol. 86, 902–910.

Gunderson, L.H., 2000. Ecological resilience—in theory and application. Annu. Rev. Ecol. Syst. 31, 425–439.

Hagen, M., Kissling, W.D., Rasmussen, C., De Aguiar, M.A.M., Brown, L.E., Carstensen, D.W., Alves-Dos-Santos, I., Dupont, Y.L., Edwards, F.K., Genini, J., Guimarães Jr., P.R., Jenkins, G.B., Jordano, P., Kaiser-Bunbury, C.N., Ledger, M.E., Maia, K.P., Marquitti, F.M.D., McLaughlin, Ó., Morellato, L.P.C., O'Gorman, E.J., Trøjelsgaard, K., Tylianakis, J.M., Vidal, M.M., Woodward, G., Olesen, J.M., 2012. Biodiversity, species interactions and ecological networks in a fragmented world. In: Jacob, U., Woodward, G. (Eds.), Global Change in Multispecies Systems Part 1. In: Advances in Ecological Research, Academic Press, San Diego.

Haines-Young, R., Potschin, M., 2010. The links between biodiversity, ecosystem services and human well-being. In: Raffaelli, D.G., Frid, C.L.J. (Eds.), Ecosystem Ecology: A New Synthesis. Cambridge University Press, New York.

Hall, R.O., Tank, J.L., Dybdahl, M.F., 2003. Exotic snails dominate nitrogen and carbon cycling in a highly productive stream. Front. Ecol. Environ. 1, 407–411.

Handa, I.T., Aerts, R., Berendse, F., Berg, M.P., Bruder, A., Butenschoen, O., Chauvet, E., Gessner, M.O., Jabiol, J., Makkonen, M., McKie, B.G., Malmqvist, B., Peeters, E.T., Scheu, S., Schmid, B., Van Ruijven, J., Vos, V.C., Hattenschwiler, S., 2014. Consequences of biodiversity loss for litter decomposition across biomes. Nature 509, 218–221.

Hauer, F.R., Hill, W.R., 2006. Temperature, light and oxygen. In: Hauer, F.R., Lamberti, G.A. (Eds.), Methods in Stream Ecology, second ed. Elsevier Ltd.

Heemsbergen, D.A., Berg, M.P., Loreau, M., Van Hal, J.R., Faber, J.H., Verhoef, H.A., 2004. Biodiversity effects on soil processes explained by interspecific functional dissimilarity. Science 306, 1019–1020.

Heffernan, J.B., 2008. Wetlands as an alternative stable state in desert streams. Ecology 89, 1261–1271.

Hillebrand, H., Matthiessen, B., 2009. Biodiversity in a complex world: consolidation and progress in functional biodiversity research. Ecol. Lett. 12, 1405–1419.

Hines, J., van der Putten, W.H., De Deyn, G.B., Wagg, C., Voigt, W., Mulder, C., Weisser, W.W., Engel, J., Melian, C., Scheu, S., Birkhofer, K., Ebeling, A., Scherber, C., Eisenhauer, N., 2015. Towards an integration of biodiversity-ecosystem functioning and food web theory to evaluate relationships between multiple ecosystem services. Adv. Ecol. Res. 53, 161–199.

Hladyz, S., Åbjörnsson, K., Chauvet, E., Dobson, M., Elosegi, A., Ferreira, V., Fleituch, T., Gessner, M.O., Giller, P.S., Gulis, V., Hutton, S.A., Lacoursière, J.O., Lamothe, S., Lecerf, A., Malmqvist, B., McKie, B.G., Nistorescu, M., Preda, E., Riipinen, M.P., Rîşnoveanu, G., Schindler, M., Tiegs, S.D., Vought, L.B.M., Woodward, G., 2011. Stream ecosystem functioning in an agricultural landscape: the importance of terrestrial–aquatic linkages. In: Woodward, G. (Ed.), Advances in Ecological Research. Academic Press, San Diego.

Hlúbiková, D., Novais, M.H., Dohet, A., Hoffmann, L., Ector, L., 2014. Effect of riparian vegetation on diatom assemblages in headwater streams under different land uses. Sci. Total Environ. 475, 234–247.

Hoekstra, J., 2012. Improving biodiversity conservation through modern portfolio theory. Proc. Natl. Acad. Sci. U.S.A. 109, 6360–6361.

Holling, C.S., 1973. Resilience and stability of ecological systems. Annu. Rev. Ecol. Syst. 4, 1–23.

Hooper, D.U., Solan, M., Symstad, A.J., Díaz, S., Gessner, M.O., Buchmann, N., Degrange, V., Grime, P., Hulot, F., Mermillod-Blondin, F., Roy, J., Spehn, E., Van Peer, L., 2002. Species diversity, functional diversity, and ecosystem functioning. In: Loreau, M., Naeem, S., Inchausti, P. (Eds.), Biodiversity and Ecosystem Functioning, Synthesis and perspectives. Oxford University Press, New York.

Howeth, J.G., Leibold, M.A., 2010. Species dispersal rates alter diversity and ecosystem stability in pond metacommunities. Ecology 91, 2727–2741.

Huston, M.A., 1997. Hidden treatments in ecological experiments: re-evaluating the ecosystem function of biodiversity. Oecologia 110, 449–460.

Jabiol, J., McKie, B.G., Bruder, A., Bernadet, C., Gessner, M.O., Chauvet, E., 2013. Trophic complexity enhances ecosystem functioning in an aquatic detritus-based model system. J. Anim. Ecol. 82, 1042–1051.

Johnson, K.H., Vogt, K.A., Clark, H.J., Schmitz, O.J., Vogt, D.J., 1996. Biodiveristy and the productivity and stability of ecosystems. Trends Ecol. Evol. 11, 372–377.

Jones, C.G., Lawton, J.H., Shachak, M., 1997. Positive and negative effects of organisms as physical ecosystem engineers. Ecology 78, 946–1957.

Jonsson, M., 2006. Species richness effects on ecosystem functioning increase with time in an ephemeral resource system. Acta Oecol. 29, 72–77.

Jonsson, M., Malmqvist, B., 2003. Mechanisms behind positive diversity effects on ecosystem functioning: testing the facilitation and interference hypotheses. Oecologia 134, 554–559.

Jonsson, M., Hedström, P., Stenroth, K., Hotchkiss, E.R., Vasconcelos, F.R., Karlsson, J., Byström, P., 2015a. Climate change modifies the size structure of assemblages of emerging aquatic insects. Freshw. Biol. 60, 78–88.

Jonsson, M., Kardol, P., Gundale, M.J., Bansal, S., Nilsson, M.C., Metcalfe, D.B., Wardle, D.A., 2015b. Direct and indirect drivers of moss community structure, function, and associated microfauna across a successional gradient. Ecosystems 18, 154–169.

Kattge, J., Díaz, S., Lavorel, S., Prentice, I.C., Leadley, P., Bönisch, G., Garnier, E., Westoby, M., Reich, P.B., Wright, I.J., Cornelissen, J.H.C., Violle, C., Harrison, S.P., Van Bodegom, P.M., Reichstein, M., Enquist, B.J., Soudzilovskaia, N.A., Ackerly, D.D., Anand, M., Atkin, O., Bahn, M., Baker, T.R., Baldocchi, D., Bekker, R., Blanco, C.C., Blonder, B., Bond, W.J., Bradstock, R., Bunker, D.E., Casanoves, F., Cavender-Bares, J., Chambers, J.Q., Chapin III, F.S., Chave, J., Coomes, D., Cornwell, W.K., Craine, J.M., Dobrin, B.H., Duarte, L., Durka, W., Elser, J., Esser, G., Estiarte, M., Fagan, W.F., Fang, J., Fernández-Méndez, F., Fidelis, A., Finegan, B., Flores, O., Ford, H., Frank, D., Freschet, G.T., Fyllas, N.M., Gallagher, R.V., Green, W.A., Gutierrez, A.G., Hickler, T., Higgins, S.I., Hodgson, J.G., Jalili, A., Jansen, S., Joly, C.A., Kerkhoff, A.J., Kirkup, D., Kitajima, K., Kleyer, M., Klotz, S., Knops, J.M.H., Kramer, K., Kühn, I., Kurokawa, H., Laughlin, D., Lee, T.D., Leishman, M., Lens, F., Lenz, T., Lewis, S.L., Lloyd, J., Llusià, J., Louault, F., Ma, S., Mahecha, M.D., Manning, P., Massad, T., Medlyn, B.E., Messier, J., Moles, A.T., Müller, S.C., Nadrowski, K., Naeem, S., Niinemets, Ü., Nöllert, S., Nüske, A., Ogaya, R., Oleksyn, J., Onipchenko, V.G., Onoda, Y., Ordoñez, J., Overbeck, G., Ozinga, W.A., et al., 2011. TRY—a global database of plant traits. Glob. Chang. Biol. 17, 2905–2935.

Lagrue, C., Kominoski, J.S., Danger, M., Baudoin, J.M., Lamothe, S., Lambrigot, D., Lecerf, A., 2011. Experimental shading alters leaf litter breakdown in streams of contrasting riparian canopy cover. Freshw. Biol. 56, 2059–2069.

Lavorel, S., Garnier, E., 2002. Predicting changes in community composition and ecosystem functioning from plant traits: revisiting the Holy Grail. Funct. Ecol. 16, 545–556.

Lefcheck, J.S., Byrnes, J.E.K., Isbell, F., Gamfeldt, L., Griffin, J.N., Eisenhauer, N., Hensel, M.J.S., Hector, A., Cardinale, B.J., Duffy, J.E., 2015. Biodiversity enhances ecosystem multifunctionality across trophic levels and habitats. Nat. Commun. 6.

Leibold, M.A., Holyoak, M., Mouquet, N., Amarasekare, P., Chase, J.M., Hoopes, M.F., Holt, R.D., Shurin, J.B., Law, R., Tilman, D., Loreau, M., Gonzalez, A., 2004. The metacommunity concept: a framework for multi-scale community ecology. Ecol. Lett. 7, 601–613.

Loreau, M., 2000. Biodiversity and ecosystem functioning: recent theoretical advances. Oikos 91, 3–17.

Loreau, M., Hector, A., 2001. Partitioning selection and complementarity in biodiversity experiments. Nature 412, 72–76.

Loreau, M., Downing, A., Emmerson, M., Gonzalez, A., Hughes, J., Inchausti, P., Joshi, J., Norberg, J., Sala, O., 2002. A new look at the relationship between diversity and stability. In: Loreau, M., Naeem, S., Inchausti, P. (Eds.), Biodiversity and Ecosystem Functioning: Synthesis and Perspectives. Oxford University Press, Oxford.

Loreau, M., Mouquet, N., Holt, R.D., 2003. Meta-ecosystems: a theoretical framework for a spatial ecosystem ecology. Ecol. Lett. 6, 673–679.

Mace, G.M., Norris, K., Fitter, A.H., 2012. Biodiversity and ecosystem services: a multilayered relationship. Trends Ecol. Evol. 27, 19–26.

Maestre, F.T., Escudero, A., Martinez, I., Guerrero, C., Rubio, A., 2005. Does spatial pattern matter to ecosystem functioning? Insights from biological soil crusts. Funct. Ecol. 19, 566–573.

Maestre, F.T., Quero, J.L., Gotelli, N.J., Escudero, A., Ochoa, V., Delgado-Baquerizo, M., García-Gómez, M., Bowker, M.A., Soliveres, S., Escolar, C., García-Palacios, P., Berdugo, M., Valencia, E., Gozalo, B., Gallardo, A., Aguilera, L., Arredondo, T., Blones, J., Boeken, B., Bran, D., Conceição, A.A., Cabrera, O., Chaieb, M., Derak, M., Eldridge, D.J., Espinosa, C.I., Florentino, A., Gaitán, J., Gatica, M.G., Ghiloufi, W., Gómez-González, S., Gutiérrez, J.R., Hernández, R.M., Huang, X., Huber-Sannwald, E., Jankju, M., Miriti, M., Monerris, J., Mau, R.L., Morici, E., Naseri, K., Ospina, A., Polo, V., Prina, A., Pucheta, E., Ramírez-Collantes, D.A., Romão, R., Tighe, M., Torres-Díaz, C., Val, J., Veiga, J.P., Wang, D., Zaady, E., 2012. Plant species richness and ecosystem multifunctionality in global drylands. Science 335, 214–218.

Malmqvist, B., Rundle, S., 2002. Threats to the running water ecosystems of the world. Environ. Conserv. 29, 134–153.

Mancinelli, G., Mulder, C., 2015. Detrital dynamics and cascading effects on supporting ecosystem services. Adv. Ecol. Res. 53, 97–160.

Marbuah, G., Gren, I.-M., McKie, B.G., 2014. Economics of harmful invasive species: a review. Diversity 6, 500–523.

Massol, F., Petit, S., 2013. Interaction networks in agricultural landscape mosaics. In: Woodward, G., Bohan, D.A. (Eds.), Ecological Networks in an Agricultural World. In: Advances in Ecological Research, Academic Press, San Diego.

Matthiessen, B., Hillebrand, H., 2006. Dispersal frequency affects local biomass production by controlling local diversity. Ecol. Lett. 9, 652–662.

Matthiessen, B., Gamfeldt, L., Jonsson, P.R., Hillebrand, H., 2007. Effects of grazer richness and composition on algal biomass in a closed and open marine system. Ecology 88, 178–187.

McKie, B.G., Cranston, P.S., 2001. Colonisation of experimentally immersed wood in south eastern Australia: responses of feeding groups to changes in riparian vegetation. Hydrobiologia 452, 1–14.

McKie, B.G., Pearson, R.G., 2006. Environmental variation and the predator-specific responses of tropical stream insects: effects of temperature and predation on survival and development of Australian Chironomidae (Diptera). Oecologia 149, 328–339.

McKie, B.G., Malmqvist, B., 2009. Assessing ecosystem functioning in streams affected by forest management: increased leaf decomposition occurs without changes to the composition of benthic assemblages. Freshw. Biol. 54, 2086–2100.

McKie, B.G., Woodward, G., Hladyz, S., Nistorescu, M., Preda, E., Popescu, C., Giller, P.S., Malmqvist, B., 2008. Ecosystem functioning in stream assemblages from different regions: contrasting responses to variation in detritivore richness, evenness and density. J. Anim. Ecol. 77, 495–504.

McKie, B.G., Schindler, M., Gessner, M.O., Malmqvist, B., 2009. Placing biodiversity and ecosystem functioning in context: environmental perturbations and the effects of species richness in a stream field experiment. Oecologia 160, 757–770.

Meentemeyer, V., 1978. Macroclimate the lignin control of litter decomposition rates. Ecology 59, 465–472.

Montoya, J.M., Rodríguez, M.A., Hawkins, B.A., 2003. Food web complexity and higher-level ecosystem services. Ecol. Lett. 6, 587–593.

Montoya, D., Zavala, M.A., Rodríguez, M.A., Purves, D.W., 2008. Animal versus wind dispersal and the robustness of tree species to deforestation. Science 320, 1502–1504.

Mouchet, M., Guilhaumon, F., Vlleger, S., Mason, N.W.H., Tomasini, J.A., Mouillot, D., 2008. Towards a consensus for calculating dendrogram-based functional diversity indices. Oikos 117, 794–800.

Mouillot, D., Mason, W.H., Dumay, O., Wilson, J.B., 2005. Functional regularity: a neglected aspect of functional diversity. Oecologia 142, 353–359.

Mouillot, D., Bellwood, D.R., Baraloto, C., Chave, J., Galzin, R., Harmelin-Vivien, M., Kulbicki, M., Lavergne, S., Lavorel, S., Mouquet, N., Paine, C.E.T., Renaud, J., Thuiller, W., 2013. Rare species support vulnerable functions in high-diversity ecosystems. PLoS Biol. 11, e1001569.

Mouquet, N., Loreau, M., 2003. Community patterns in source-sink metacommunities. Am. Nat. 162, 544–557.

Mulder, C.P., Uliassi, D.D., Doak, D.F., 2001. Physical stress and diversity-productivity relationships: the role of positive interactions. Proc. Natl. Acad. Sci. U.S.A. 98, 6704–6708.

Mulder, C., Bennett, E.M., Bohan, D.A., Bonkowski, M., Carpenter, S.R., Chalmers, R., Cramer, W., Durance, I., Eisenhauer, N., Fontaine, C., Haughton, A.J., Hettelingh, J.-P., Hines, J., Ibanez, S., Jeppesen, E., Krumins, J.A., Ma, A., Mancinelli, G., Massol, F., McLaughlin, Ó., Naeem, S., Pascual, U., Peñuelas, J., Pettorelli, N., Pocock, M.J.O., Raffaelli, D., Rasmussen, J.J., Rusch, G.M., Scherber, C., Setälä, H., Sutherland, W.J., Vacher, C., Voigt, W., Vonk, J.A., Wood, S.A., Woodward, G., 2015. 10 Years later: revisiting priorities for science and society a decade after the millennium ecosystem assessment. Adv. Ecol. Res. 53, 1–53.

Myers, R., 2003. The Basics of Chemistry. Greenwood Press, Westport.

Naeem, S., Li, S., 1997. Biodiversity enhances ecosystem reliability. Nature 390, 507–509.

Naeem, S., Wright, J.P., 2003. Disentangling biodiversity effects on ecosystem functioning: deriving solutions to a seemingly insurmountable problem. Ecol. Lett. 6, 567–579.

Naeem, S., Loreau, M., Inchausti, P., 2002. Biodiversity and ecosystem functioning: the emergence of a synthetic ecological framework. In: Loreau, M., Naeem, S., Inchausti, P. (Eds.), Biodiversity and Ecosystem Functioning. Synthesis and Perspectives. Oxford University Press, New York.

Nash, K.L., Allen, C.R., Angeler, D.G., Barichievy, C., Eason, T., Garmestani, A.S., Graham, N.A.J., Granholm, D., Knutson, M., Nelson, R.J., Nyström, M., Stow, C.A., Sundstrom, S.M., 2014. Discontinuities, cross-scale patterns, and the organization of ecosystems. Ecology 95, 654–667.

Öckinger, E., Smith, H.G., 2007. Semi-natural grasslands as population sources for pollinating insects in agricultural landscapes. J. Appl. Ecol. 44, 50–59.

Öckinger, E., Schweiger, O., Crist, T.O., Debinski, D.M., Krauss, J., Kuussaari, M., Petersen, J.D., Pöyry, J., Settele, J., Summerville, K.S., Bommarco, R., 2010. Life-history traits predict species responses to habitat area and isolation: a cross-continental synthesis. Ecol. Lett. 13, 969–979.

O'Connor, N.E., Donohue, I., 2013. Environmental context determines multi-trophic effects of consumer species loss. Glob. Chang. Biol. 19, 431–440.

Paine, R.T., 1966. Food web complexity and species diversity. Am. Nat. 100, 65–75.

Palomo, I., Felipe-Lucia, M.R., Bennett, E.M., Martín-López, B., Pascual, U., 2016. Disentangling the pathways and effects of ecosystem service co-production. Adv. Ecol. Res. 54, in press.

Peierls, B.L., Caraco, N.F., Pace, M.L., Cole, J.J., 1991. Human influence on river nitrogen. Nature 350, 386–387.

Perkins, D.M., Bailey, R.A., Dossena, M., Gamfeldt, L., Reiss, J., Trimmer, M., Woodward, G., 2015. Higher biodiversity is required to sustain multiple ecosystem processes across temperature regimes. Glob. Chang. Biol. 21, 396–406.

Petchey, O.L., Gaston, K.J., 2006. Functional diversity: back to basics and looking forward. Ecol. Lett. 9, 741–758.

Petchey, O.L., Gaston, K.J., 2007. Dendrograms and measuring functional diversity. Oikos 116, 1422–1426.

Petchey, O.L., McPhearson, P.T., Casey, T.M., Morin, P.J., 1999. Environmental warming alters food-web structure and ecosystem function. Nature 402, 69–72.

Peterson, G., Allen, C.R., Holling, C.S., 1998. Ecological resilience, biodiversity, and scale. Ecosystems 1, 6–18.

Piggott, J.J., Townsend, C.R., Matthaei, C.D., 2015. Climate warming and agricultural stressors interact to determine stream macroinvertebrate community dynamics. Glob. Chang. Biol. 21, 1887–1906.

Pillai, P., Gonzalez, A., Loreau, M., 2011. Metacommunity theory explains the emergence of food web complexity. Proc. Natl. Acad. Sci. U.S.A. 108, 19293–19298.

Pimm, S.L., 1991. The Balance of Nature? Ecological Issues in the Conservation of Species and Communities. University of Chicago Press, Chicago.

Poff, N.L., Olden, J.D., Vieira, N.K.M., Finn, D.S., Simmons, M.P., Kondratieff, B.C., 2006. Functional trait niches of North American lotic insects: traits-based ecological applications in light of phylogenetic relationships. J. N. Am. Benthol. Soc. 25, 730–755.

Pope, K.L., Allen, C.R., Angeler, D.G., 2014. Fishing for resilience. Trans. Am. Fish. Soc. 143, 467–478.

Power, M.E., Tilman, D., Estes, J.A., Menge, B.A., Bond, W.J., Mills, L.S., Daily, G., Castilla, J.C., Lubchenco, J., Paine, R.T., 1996. Challenges in the quest for keystones. BioScience 46, 609–620.

Raffaelli, D., Van Der Putten, W.H., Persson, L., Wardle, D.A., Petchey, O.L., Koricheva, J., Van Der Heijden, M., Mikola, J., Kennedy, T., 2002. Multi-trophic dynamics and ecosystem processes. In: Loreau, M., Naeem, S., Inchausti, P. (Eds.), Biodiversity and Ecosystem Functioning, Synthesis and Perspectives. Oxford University Press, New York.

Reiss, J., Bridle, J.R., Montoya, J.M., Woodward, G., 2009. Emerging horizons in biodiversity and ecosystem functioning research. Trends Ecol. Evol. 24, 505–514.

Rocha, J., Yletyinen, J., Biggs, R., Blenckner, T., Peterson, G., 2015. Marine regime shifts: drivers and impacts on ecosystems services. Philos. Trans. R. Soc. B 370, 20130273.

Roscher, C., Schumacher, J., Gubsch, M., Lipowsky, A., Weigelt, A., Buchmann, N., Schmid, B., Schulze, E.D., 2012. Using plant functional traits to explain diversity–productivity relationships. PLoS One 7, e36760.

Rosemond, A.D., Benstead, J.P., Bumpers, P.M., Gulis, V., Kominoski, J.S., Manning, D.W.P., Suberkropp, K., Wallace, J.B., 2015. Experimental nutrient additions accelerate terrestrial carbon loss from stream ecosystems. Science 347, 1142–1145.

Sabater, S., Butturini, A., Muños, I., Romaní, A., Wray, J., Sabater, F., 1998. Effects of removal of riparian vegetation on algae and heterotrophs in a Mediterranean stream. J. Aquat. Ecosyst. Stress Recovery 6, 129–140.

Scheffer, M., 1997. Ecology of Shallow Lakes. Chapman and Hall, London, UK.

Scheffer, M., Carpenter, S., Foley, J.A., Folke, C., Walker, B., 2001. Catastrophic shifts in ecosystems. Nature 413, 591–596.

Schindler, D.E., Armstrong, J.B., Reed, T.E., 2015. The portfolio concept in ecology and evolution. Front. Ecol. Environ. 13, 257–263.

Schmidt-Kloiber, A., Hering, D., 2015. www.freshwaterecology.info—an online tool that unifies, standardises and codifies more than 20,000 European freshwater organisms and their ecological preferences. Ecol. Indic. 53, 271–282.

Seitzinger, S., Harrison, J.A., Böhlke, J.K., Bouwman, A.F., Lowrance, R., Peterson, B., Tobias, C., Drecht, G.V., 2006. Denitrification across landscapes and waterscapes: a synthesis. Ecol. Appl. 16, 2064–2090.

Smith, P., Ashmore, M.R., Black, H.I.J., Burgess, P.J., Evans, C.D., Quine, T.A., Thomson, A.M., Hicks, K., Orr, H.G., 2013. The role of ecosystems and their management in regulating climate, and soil, water and air quality. J. Appl. Ecol. 50, 812–829.

Spanbauer, T.L., Allen, C.R., Angeler, D.G., Eason, T., Fritz, S.C., Garmestani, A.S., Nash, K.L., Stone, J.R., 2014. Prolonged instability prior to a regime shift. PLoS One 9, e108936.

Srivastava, D.S., 2002. The role of conservation in expanding biodiversity research. Oikos 98, 351–360.

Srivastava, D.S., Vellend, M., 2005. Biodiversity-ecosystem function research: is it relevant to conservation? Annu. Rev. Ecol. Evol. Syst. 36, 267–294.

Stachowicz, J.J., Graham, M.H., Bracken, M.E.S., Szoboszlai, A.I., 2008. Diversity enhances cover and stability of seaweed assemblages: the role of heterogeneity and time. Ecology 89, 3008–3019.

Stanley, E.H., Powers, S.M., Lottig, N.R., 2010. The evolving legacy of disturbance in stream ecology: concepts, contributions, and coming challenges. J. N. Am. Benthol. Soc. 29, 67–83.

Sundqvist, M.K., Sanders, N.J., Wardle, D.A., 2013. Community and ecosystem responses to elevational gradients: processes, mechanisms, and insights for global change. Annu. Rev. Ecol. Evol. Syst. 44, 261–280.

Tamaddoni-Nezhad, A., Milani, G.A., Raybould, A., Muggleton, S., Bohan, D.A., 2013. Construction and validation of food webs using logic-based machine learning and text mining. In: Woodward, G., Bohan, D.A. (Eds.), Ecological Networks in an Agricultural World. In: Advances in Ecological Research, Academic Press, San Diego.

Thébault, E., Loreau, M., 2003. Food-web constraints on biodiversity-ecosystem functioning relationships. Proc. Natl. Acad. Sci. U.S.A. 100, 14949–14954.

Thébault, E., Loreau, M., 2006. The relationship between biodiversity and ecosystem functioning in food webs. Ecol. Res. 21, 17–25.

Thompson, R., Starzomski, B.M., 2007. What does biodiversity actually do? A review for managers and policy makers. Biodivers. Conserv. 16, 1359–1378.

Thompson, M.S.A., Bankier, C., Bell, T., Dumbrell, A.J., Gray, C., Ledger, M.E., Lehmann, K., McKew, B.A., Sayer, C.D., Shelley, F., Trimmer, M., Warren, S.L., Woodward, G., in press. Gene-to-ecosystem impacts of a catastrophic pesticide spill: testing a multilevel bioassessment approach in a river ecosystem. Freshw. Biol. http://dx.doi.org/10.1111/fwb.12676

Tilman, D., 2001. Functional diversity. In: Levin, S.A. (Ed.), Encyclopedia of Biodiversity. Academic Press, San Diego, CA.

Tilman, D., Downing, J.A., 1994. Biodiversity and stability in grassland. Nature 367, 363–365.

Tilman, D., Reich, P.B., Knops, J.M.H., Wedin, D., Mielke, T., Lehman, C.L., 2001. Diversity and productivity in a long-term grassland experiment. Science 294, 843–845.

Tiunov, A.V., Scheu, S., 2005. Facilitative interactions rather than resource partitioning drive diversity-functioning relationships in laboratory fungal communities. Ecol. Lett. 8, 618–625.

Tolkkinen, M., Mykrä, H., Markkola, A.M., Aisala, H., Vuori, K.M., Lumme, J., Pirttilä, A.M., Muotka, T., 2013. Decomposer communities in human-impacted streams: species dominance rather than richness affects leaf decomposition. J. Appl. Ecol. 50, 1142–1151.

Törnroos, A., Nordström, M.C., Aarnio, K., Bonsdorff, E., 2015. Environmental context and trophic trait plasticity in a key species, the tellinid clam Macoma balthica L. J. Exp. Mar. Biol. Ecol. 472, 32–40.

Townsend, C.R., Hildrew, A.G., 1994. Species traits in relation to a habitat templet for river systems. Freshw. Biol. 31, 265–275.

Townsend, C.R., Uglmann, S.S., Matthaei, C.D., 2008. Individual and combined responses of stream ecosystems to multiple stressors. J. Appl. Ecol. 45, 1810–1819.
Tylianakis, J.M., Tscharntke, T., Lewis, O.T., 2007. Habitat modification alters the structure of tropical host-parasitoid food webs. Nature 445, 202–205.
UNEP, 2005. The Millennium Ecosystem Assessment Programme. Ecosystems and human well-being: Synthesis. Island Press, Washington, DC.
Valiente-Banuet, A., Aizen, M.A., Alcántara, J.M., Arroyo, J., Cocucci, A., Galetti, M., García, M.B., García, D., Gómez, J.M., Jordano, P., Medel, R., Navarro, L., Obeso, J.R., Oviedo, R., Ramírez, N., Rey, P.J., Traveset, A., Verdú, M., Zamora, R., 2015. Beyond species loss: the extinction of ecological interactions in a changing world. Funct. Ecol. 29, 299–307.
Vaughn, C.C., 2010. Biodiversity losses and ecosystem function in freshwaters: emerging conclusions and research directions. BioScience 60, 25–35.
Venail, P.A., Maclean, R.C., Meynard, C.N., Mouquet, N., 2010. Dispersal scales up the biodiversity–productivity relationship in an experimental source-sink metacommunity. Proc. R. Soc. Lond. B Biol. Sci. 277, 2339–2345.
Verberk, W.C.E.P., Van Noordwijk, C.G.E., Hildrew, A.G., 2013. Delivering on a promise: integrating species traits to transform descriptive community ecology into a predictive science. Freshw. Sci. 32, 531–547.
Villeger, S., Mason, N.W.H., Mouillot, D., 2008. New multidimensional functional diversity indices for a multifaceted framework in functional ecology. Ecology 89, 2290–2301.
Violle, C., Navas, M.-L., Vile, D., Kazakou, E., Fortunel, C., Hummel, I., Garnier, E., 2007. Let the concept of trait be functional!. Oikos 116, 882–892.
Wardle, D.A., Jonsson, M., 2014. Long-term resilience of above- and belowground ecosystem components among contrasting ecosystems. Ecology 95, 1836–1849.
Wardle, D.A., Hyodo, F., Bardgett, R.D., Yeates, G.W., Nilsson, M.C., 2011. Long-term aboveground and belowground consequences of red wood ant exclusion in boreal forest. Ecology 92, 645–656.
Whittaker, R.H., 1960. Vegetation of the Siskiyou Mountains, Oregon and California. Ecol. Monogr. 30, 279–338.
Whittaker, R.H., 1972. Evolution and measurement of species diversity. Taxon 21, 213–251.
Wilson, E.O., 1992. The Diversity of Life. Harvard University Press, Massachusetts.
Woodward, G., Gessner, M.O., Giller, P.S., Gulis, V., Hladyz, S., Lecerf, A., Malmqvist, B., McKie, B.G., Tiegs, S.D., Cariss, H., Dobson, M., Elosegi, A., Ferreira, V., Graca, M.A.S., Fleituch, T., Lacoursiere, J.O., Nistorescu, M., Pozo, J., Risnoveanu, G., Schindler, M., Vadineanu, A., Vought, L.B.M., Chauvet, E., 2012. Continental-scale effects of nutrient pollution on stream ecosystem functioning. Science 336, 1438–1440.
Yachi, S., Loreau, M., 1999. Biodiversity and ecosystem productivity in a fluctuating environment: the insurance hypothesis. Proc. Natl. Acad. Sci. U.S.A. 96, 1463–1468.
Young, R.G., Huryn, A.D., 1999. Effects of land use on stream metabolism and organic matter turnover. Ecol. Appl. 9, 1359–1376.

CHAPTER THREE

Detrital Dynamics and Cascading Effects on Supporting Ecosystem Services

Giorgio Mancinelli*[,1], Christian Mulder[†,1]

*Department of Biological and Environmental Sciences and Technologies, Centro Ecotekne, University of Salento, Lecce, Italy

[†]Department of Environmental Effects and Ecosystems, Centre for Sustainability, Environment and Health, National Institute for Public Health and the Environment (RIVM), Bilthoven, The Netherlands

[1]Corresponding authors: e-mail address: giorgio.mancinelli@unisalento.it; christian.mulder@rivm.nl

Contents

1. Introduction 98
2. Data Analysis 105
   2.1 General Characteristics of the Data Set 105
   2.2 Effects of Categorical Variables 112
3. Discussion 121
4. Future Ecosystem Services Research 124
Acknowledgements 127
Appendix 127
   A.1 Data Sources and Selection Criteria 127
   A.2 Quantifying Predator Effects 128
   A.3 Meta-Analytic Procedures 129
   A.4 Publication and Experimental Biases 134
References 148

Abstract

Top-down trophic cascades are well known in many autotrophic systems, yet their role in heterotrophic food webs is less clear. We collated data from 78 investigations and applied meta-analysis to evaluate the strength of detrital trophic cascades in freshwater and terrestrial food webs. Predators exerted significant, indirect controls on detrital resources, in line with theoretical predictions, whereas this was not the case for omnivores, suggesting that detritivory prevailed over predation and disrupted trophic cascades. Significant relationships were observed for both types of consumer in terms of their responses to detrital quality: specifically, unimodal curves across C:N and N:P gradients were the best fits for predators, whilst cascade strength responses to detrital quality were saddle shaped. These insights suggest that while predatory strategy is determining cascades within detrital-based systems, resource quality has bottom-up role effects on predators and on preferential consumption by omnivores. As such, these

Advances in Ecological Research, Volume 53
ISSN 0065-2504
http://dx.doi.org/10.1016/bs.aecr.2015.10.001

environmental responses seem to mirror some provisioning and supporting services; our findings are discussed within conceptual frameworks related to ecological stoichiometry and ecosystem services.

1. INTRODUCTION

Vegetation governs the quality and quantity of plant litter produced within an ecosystem which, in turn, influences its environmental quality: as soil organic matter in terrestrial systems and sediment organic matter in aquatic ecosystems have intrinsically low qualities, decomposition is stimulated by the supply of fresh organic matter like leaf litter (Mulder, 2006; Vannote et al., 1980; Woodward et al., 2008). Water-soluble organic matter containing energy-rich carbon compounds is considered a labile organic fraction that is easily degradable by microbial *r*-strategists, like most soil bacteria and some aquatic microfungi (Fontaine and Barot, 2005; Fontaine et al., 2004). This leads to the conventional perspective of 'top-down' forces determining ecosystem functioning, and subsequently influencing the resultant ecosystem services. However, we have to take into account the possibility that without considering bottom-up constraints related to the environmental quality of the available resources, the picture of ecosystem functioning and the related supporting ecosystem services is incomplete. Understanding how services such as carbon sequestration, water purification and fish biomass production, for instance, requires insight into both bottom-up and top-down processes within the food web, so it is critically important that we consider both directions in these trophic feedbacks, rather than simply continue to focus on each of the two individually. To do this, we need to review how the top-down view came to predominate and to re-evaluate it in the light of the more complete understanding that is emerging from recent research in trophic ecology.

Historically, the influence of predators on community structure has received considerable attention in terms of both its direct and indirect effects. Trophic cascades are the most familiar example of the latter (Hairston et al., 1960), and these common phenomena are manifested as the propagation of indirect mutualism between non-adjacent trophic levels (*sensu* Oksanen et al., 1981; Paine, 1980; Carpenter et al., 1985). Top-down cascades were once assumed to be primarily restricted to relatively simple aquatic systems, yet we now know they are far more ubiquitous and have been reported in tropical rainforests, soil food webs and an increasingly diverse set of ecosystems. Their potential to shape the delivery of ecosystem services is huge, yet has been largely ignored to date, with the notable exception of commercial marine fisheries where they can dictate not just the yield of fisheries caught but also the species composition of the catch by inducing powerful regime shifts within the food web, especially when allied to other anthropogenic stressors (Branch et al., 2010; Casini et al., 2009; Jennings et al., 2014; Pikitch et al., 2014).

The progressive identification of factors limiting the strength of trophic cascades over the last few decades (reviewed in Persson, 1999; Polis et al., 2000) has developed in parallel to a long-running dispute about the importance of a consumer-driven versus a resource-driven regulation. Increasingly, however, it is becoming clear that both processes may be operating in concert within an ecosystem, albeit with different relative strengths depending on the environmental and biotic context (Borer et al., 2006; Faithfull et al., 2011; Frank et al., 2007; Hunter and Price, 1992; McQueen et al., 1989; Menge, 2000; Power, 1992). There is considerable interest in understanding the role of trophic cascades within the framework of global biodiversity loss and overexploitation, and this has been growing over the past couple of decades in particular (Pauly et al., 1998; Terborgh and Estes, 2010). When cascades are especially strong they can trigger pervasive regime shifts, as those observed in relation with the loss or the introduction of predatory species, often with unexpected consequences on ecological processes and ecosystem services as diverse as carbon sequestration and other biogeochemical cycles, the spread of invasive species and pathogens, and the loss of water quality and amenity value in freshwaters (Carpenter and Kitchell, 1988; Estes et al., 2011; Folke et al., 2004; Schmitz et al., 2010).

Empirical evidence from both aquatic and terrestrial ecosystems strongly supports a general donor-control model where supporting services, like organic matter decomposition, and provisioning and regulating ecosystem services, like the chemical quality of detritus, drive ecosystem functioning (Elser et al., 1998; Moore et al., 2004; Mulder et al., 2013; Sterner and Elser, 2002; Wolkovich et al., 2014). In freshwaters, for instance, most of the carbon fixed by plants enters the food web via detrital pathways, and this, when combined with allochthonous terrestrial input represents a pivotal energy source (Allan and Castillo, 2007; Cebrian, 1999; Cebrian and Lartigue, 2004; Premke et al., 2010; Tank et al., 2010; Vannote et al., 1980; Wolkovich et al., 2014). Its importance, particularly as an external subsidy, is especially pronounced in aquatic-terrestrial ecotones, such as shorelines and riparian zones, where detritus fuels the base of the food web (Lugo and Snedaker, 1974; Polis et al., 2000; Richardson et al., 2009; Vannote et al., 1980; Woodward and Hildrew, 2002). These critically important thin strips between aquatic and terrestrial ecosystems have long been overlooked by the mainstream of freshwater and terrestrial ecology, despite they span both disciplines, even though these habitats are often close to intensive land-use management practises. As such, they are especially vulnerable to a cocktail of anthropogenic stressors from urban areas or agroecosystems, which can lead to altered nutrient cycling, impaired ecosystem functioning and unstable detrital food webs (DFWs hereafter) (Mulder et al., 2015a; Thompson et al., 2016).

Of the 89 publications on aquatic ecosystems included in a previous meta-analysis by Borer et al. (2005), 45 were from freshwater benthic systems, in which predator-induced effects on primary producers were second in magnitude only to those seen in openwater marine systems, and as pronounced as in the pelagic zone of lakes, which are generally considered the two classic cases of trophic cascades in aquatic habitats (Carpenter and Kitchell, 1993; Shurin et al., 2006; Sommer, 2008; Strong, 1992). While studies on grazing 'green world' cascades have continued to accumulate, for 'brown world' DFWs such studies have received far less attention (but see Krumins et al., 2013; Moore et al., 2004; Wolkovich et al., 2014). This at least partly reflects both conceptual and semantic issues, as widely accepted definitions of cascades (Pace et al., 1999; Persson, 1999) emphasize the importance of reciprocal predator–prey interactions in determining inverse biomass patterns across trophic levels. DFWs, though, are considered donor-controlled (*sensu* Pimm, 1982), with the detrital resource regulating the abundance of primary consumers (e.g. Tiegs et al., 2008; Wallace et al., 1999) but not vice versa, i.e., they appear incompatible with cascade propagation down to the basal trophic level.

This view has long been challenged for terrestrial DFWs (Hawlena et al., 2012; Hines and Gessner, 2012; Lensing and Wise, 2006; Mikola and Setälä, 1998; Prather et al., 2012; Santos et al., 1981; Wardle, 2010; Wise et al., 1999) because (i) trophic relationships between detritivores and the detrital resource are influenced by the heterotrophic microflora and (ii) primary consumers can control the standing stock of producers and decomposers, their assimilation rate into the system, as well as the key ecosystem functions and service they deliver (Table 1). Increasingly, detrital and grazing pathways are being considered holistically as parts of a wider network of interacting species (Fig. 1). This perspective is even

Table 1 Ecosystem Services and Processes Provided by Different Trophic Levels in Aquatic and Terrestrial Detrital Food Webs (DFWs)

Service Types	Goods	Ecosystem Process	Habitat	Microflora	Intermediate Consumers	(Hyper) Predators	Aquatic DFWs	Terrestrial DFWs
Support and provisioning	*Nutrient cycling*	Decomposition/humification/mobilization	X	X	X	Controversial	O	O
		Pedogenesis (bioturbation, burrowing, particle binding)	X	X	X		O	O
		Regulation of nutrient loss by leaching/denitrification	X	X	Controversial		O	O
	Primary production	Carbon sequestration and storage in soils/sediments		X	X		O	O
		(Induced) Defense against herbivory	X	X		X	O	O
		(Genetic) Defense against pathogens		X	X	X	O	O
	Secondary production	Biomodification of pollutants, mitigation of atmospheric deposition		X			O	O
		Managed top-down control in fisheries			X	Controversial	O	
		Plant–plant interactions (vegetation structure)	X	X			O	O

Continued

Table 1 Ecosystem Services and Processes Provided by Different Trophic Levels in Aquatic and Terrestrial Detrital Food Webs (DFWs)—cont'd

Service Types	Goods	Ecosystem Process	Habitat	Microflora	Intermediate Consumers	(Hyper) Predators	Aquatic DFWs	Terrestrial DFWs
Regulation	*Climate regulation*	Infiltration and storage of water and oxygen in soils/sediments	X	X			O	O
		Stimulation/translocation of symbiotic activity in soils		X	X			O
	Water regulation	Production/consumption of greenhouse gases		X	X		O	O
		Water infiltration/sediment advection	X	X	X		O	O

Given the localization of many investigated DFWs, generally linked to 'lower' landscapes, identification of ecological processes and assignment to ecosystem services assumed a benthic characterization for freshwater systems. Worldwide, the majority of aquatic and terrestrial ecosystems are directed to serve human needs (Naeem, 2013). This makes the investigation of the processes behind services mandatory. In this meta-analysis, we will focus on the 'controversial' support of detritivores and some predator species in supporting and provisioning ecosystem services.

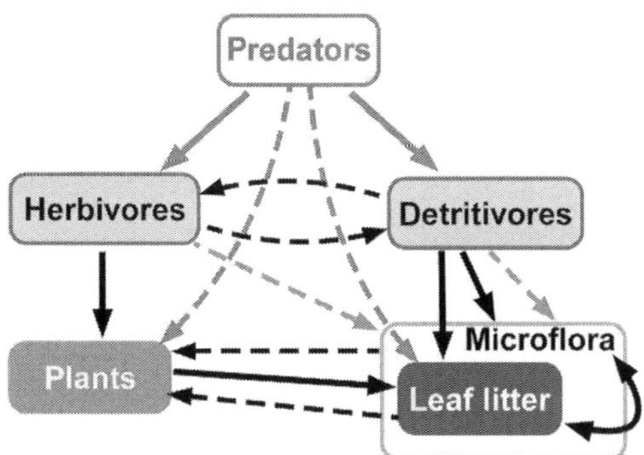

Figure 1 Conceptual model of direct (solid lines) and indirect (dashed lines) possible interactions in a detrital food web (DFW) where generalist predators feed on prey assemblages composed of both herbivore and detritivore consumers. The interaction litter–microflora (weakening trophic cascades) is explicitly shown and discussed in the text. Framework inspired by Hines and Gessner (2012); for ecosystem functions and related services provided by plants, leaf litter and microflora, see in Table 1. Grey dashed arrows suggest indirect fertilization effects on the heterotrophic microflora due to nutrient excretion.

now being extended to managed systems, as suggested by Mulder (2006) for agroecosystems, and subsequently corroborated by a number of theoretical and experimental studies (e.g. Leroux and Loreau, 2008, 2012; Schmitz et al., 2010; Ward et al., 2015; Wolkovich et al., 2014). Hence, even indirect biotic interactions involving detritivores can affect the basal trophic level by influencing the detritus processing rate and thus supporting services that underpin food provision, regulating services beyond carbon sequestration, waste processing and a host of other ecosystem services that are essential to human well-being (Fig. 1).

Adaptive responses and mutual independence between limiting factors are widespread (Cohen, 1995), and cascading effects are perfect examples of the often hidden indirect relationships embedded within complex food webs (Thompson et al., 2016; Woodward et al., 2008). In principle, trophic cascades may be even more prevalent and powerful in detrital versus grazing food webs, where consumer-altered renewal rates of resources can balance the consumption of primary producers (Rosemond et al., 2001; Srivastava et al., 2009). However, microbes may disrupt these processes by acting as (i) prey, enhancing the palatability and nutritional value of the detrital resource to detritivores (e.g. Bärlocher, 1985; Chung and Suberkropp, 2009; Dighton, 2003; Graça, 2001) and (ii) decomposers, mediating the environmental effects on key supporting ecosystem services like nutrient cycling (Allison, 2005; Bärlocher and Corkum, 2003; Dang et al., 2009; Dighton, 2003; Gessner et al., 2010; Kaspari and Yanoviak, 2008; Killham, 1994; Suberkropp et al., 2010). Dynamical responses of microbial biomass to alterations in consumers abundance have long been known to hamper cascading effects in terrestrial detrital systems (e.g. Mikola and Setälä, 1998; Taylor et al., 2010), although the opposite can be true for aquatic habitats (e.g. Mancinelli et al., 2002, 2009; Majdi et al., 2014). Furthermore, the recognition that microbial diversity *per se* may affect leaf-litter decomposition rate (Dang et al., 2005; Duarte et al., 2006) highlights how multiple biodiversity controls at the lower trophic levels can underpin the functioning of heterotrophic systems (Gessner et al., 2010; Kominoski et al., 2010; Sechi et al., 2015; Srivastava et al., 2009). These controls entail, beyond the microbial level, the resource (Kominoski et al., 2007, 2009; Swan and Palmer, 2004) and intermediate consumer levels (Dangles and Malmqvist, 2004; McKie et al., 2008).

Bacteria, fungi and plants interact to affect nutrient uptake and other processes (Laakso et al., 2000; Setälä, 2002) that operate among the basal trophic levels, and which generate supporting ecosystem services (Mulder et al., 2011, 2015a). Ecological stoichiometry and differences in C:N:P ratios of resources and the requirements of different consumers alter microbial communities (Högberg et al., 2007; Swift et al., 1979) and associated nutrient fluxes and decomposition rates. The efficiency of nutrient uptake and biomass production by different decomposers varies across environmental (or resource) gradients of C:N (Hodge et al., 2000; Killham, 1994) and potentially also P:N ratios (Güsewell and Gessner, 2009). Nutrient ratios can be used for gauging resource quality and realized versus potential process rates within a given ecosystem (Kaspari and Yanoviak, 2008; Sterner and Elser, 2002; Woodward and Hildrew, 2002), and this could provide a powerful tool for linking environmental change to key ecosystem services in a relatively direct and quantifiable manner.

This elemental mirror between abiotic and biotic components has been extended by Enquist et al. (2015) to develop a trait-driven theory of ecological stoichiometry, which assumes that traits can and will shift under a changing environment. At local scales, for instance, tissues of slow-growing tree species have a higher P content than fast-growing trees (Mulder et al., 2013) and the resulting differences in their elemental composition (Sardans and Peñuelas, 2015) could account for commonly observed shifts in resource quality, as seen in riverine ecosystems (*sensu* Vannote et al., 1980). Because foliar and litter C:N:P ratios are related to both (allochthonous) riparian trees and (autochthonous) emergent macrophytes, these multiple inputs entering the detrital pathway will influence DFWs, generating strong trait-driven shifts among the decomposers and ultimately triggering cascading effects. Functional responses will be manifested across several organizational levels and hence the mass loss determined by consumers (predators) could be considered a community trait in itself. At present, despite the potential for these 'brown world' cascades to interlink with those in the 'green world' in ways that fundamentally alter ecosystem functioning and services, they have been largely ignored.

Our primary goal here is to provide a new global quantitative assessment of the importance of trophic cascades in DFWs, the kind of nutrient cycling involved, and the likely consequences for ecosystem services. To this end, the results of published investigations testing whether predators influence leaf-litter processing were analyzed. Across these studies a widespread generic methodology—mass loss of litterbags in response to experimental treatments (e.g. Bärlocher, 2005; Kampichler and Bruckner, 2009)—was applied over the global range of freshwater and terrestrial ecosystems, climatic conditions, latitudes, and the large variety of predators, detritivorous prey and leaf-litter types found in these systems. Given the relative paucity of investigations performed in terrestrial habitats, we also included studies on ephemeral and patchy resources like dung-pats (Finn, 2001), whose characteristics in terms of patterns of decomposition and colonization are broadly comparable to those in litterbags (Mancinelli, 2009). Additionally, from a practical point of view the estimation of treatment effects—including predator occurrence—can be assessed by an identical comparison of remaining masses or, alternatively, of percent mass loss. Meta-analysis was conducted to detect potential publication artefacts, to synthesize overall estimates of effect size, and to evaluate the significance of the *a priori* factors for explaining patterns of inter-study variability in cascade strength.

2. DATA ANALYSIS

2.1 General Characteristics of the Data Set

Full details of the literature used to perform the meta-analyses are given in Appendix. Data were compiled from investigations that assessed the influence of predators on detrital (litter or dung) mass loss and either the mass or abundance of invertebrate prey; effects were quantified as the natural log ratio [$\mathrm{Ln}(p_{+}/p_{-})$] of resource and prey data in either presence (p_{+}) or absence (p_{-}) of predators (*sensu* Berlow et al., 1999). Furthermore, the dependence of predator effects on *a priori*-identified factors was investigated: from each publication, information was extracted and classified into categorical and continuous variables assumed to influence cascade strength (Table 2). Methodological information included the type of the caging devices used to manipulate predators and the initial mass of litter. The *a priori* factors were classified into categorical (booleans) and continuous variables expected to potentially bias effect-size estimations (reported in Table A2 together with a brief summary of their underlying implications). Subsequently, categorical and continuous variables were identified using Polis et al. (2000) as our main reference and considering current theories in freshwater and terrestrial ecology (Allan and Castillo, 2007; Terborgh and Estes, 2010). In addition, geographical and biological data were categorized into (i) habitat; (ii) climate, latitude; (iii) predator trophic strategy (trophic type); and (iv) resource quality (Table 2). Statistical power represented an additional criterion of choice; to this end, categorical factors had to be characterized by a minimum of 10 observations per category (Borenstein et al., 2009).

Our stringent literature search produced 94 candidate studies, of which 58.5% ($n=55$) examined the effects on detritus and detritivores of predatory species in freshwater DFWs (streams, lakes, microcosms and mesocosms, including artificial rain-forest container plants mimicking natural phytotelmata) and the remainder ($n=39$) examined the same effects in terrestrial DFWs (mostly field experiments from open grasslands and forest floors). Of these studies, 15 did not report all the response variables and 1 did not report all the necessary statistical data. Seventy eight of the studies from the original candidate pool, published between April 1984 and October 2014, met all of our criteria and were retained for the subsequent formal meta-analysis (Appendix); 32% ($n=25$) were based on single experiments, but most reported multiple independent observations (mean 2.73 per study, ranging between 2 and 8) for a total of 215 experimental observations included in the data set. The geographical distribution of the study areas within the data set was quite heterogeneous (Fig. 2): irrespective of the type of environment investigated, the most studies were carried out in North America (32 studies, corresponding to 41%), followed by Europe (12 studies, 15.4%), and Oceania (New Zealand and Micronesia: 9 studies, 11.5%, but not Australia). There was a general paucity of studies in the southern hemisphere and none from latitudes >64°N and >46°S.

In Fig. 3, the main features of the literature analysis are summarized. Investigations focusing on trophic cascades on litter decomposition steadily increased from 1976 onwards, in both aquatic and terrestrial ecosystems (Fig. 3A); field investigations were the most popular experimental approach,

Table 2 *A Priori*-Defined Categorical and Continuous Factors Hypothesized to Affect Detrital Trophic Cascades

Data Type	Rationale	Variable
(1) Habitat		
Latitude	In general, low latitudes are characterized by more intense primary consumer-basal resource interactions (Moles et al., 2011; Schemske et al., 2009), potentially reflecting in stronger trophic cascades. However, detritus-based trophic cascades may be weaker at low than at high latitudes, because (i) the tropics are generally characterized by a lower abundance of shredders and a higher activity of the heterotrophic microflora, limiting the contribution of invertebrate consumption to detritus processing (e.g. Boyero et al., 2011; Irons et al., 1994; Wantzen and Wagner, 2006; but see Cheshire et al., 2005) and (ii) leaf-litter from the tropics is a low-quality resource compared to temperate systems, resulting in a reduced palatability to shredders (Graça and Cressa, 2010).	• *Climate*, categorical, expressed as temperate or tropical
Type[a]	Detrital trophic cascades in lakes and other lentic environments are expected to be stronger than in lotic systems since in the latter water current-induced physical fragmentation might considerably contribute to leaf litter comminution (e.g. Webster and Benfield, 1986); in addition, in running waters passive advection of invertebrates can compensate for negative effects determined by predators on the local density of prey (Englund et al., 1999; Sih and Wooster, 1994).	
(2) Predator		
Phylogeny	Macroinvertebrate predators may induce stronger cascades compared to fish due to their higher local abundances and lower predation-related metabolic costs, ultimately reflecting on a higher predatory efficiency (Polis, 1999; Strong, 1992). In addition, freshwater invertebrate predators detect their prey primarily by mechano- or chemoreception, while fish adopt a less efficient visual perception to locate, beside macrobenthic invertebrates, a wide spectrum of planktonic and terrestrial prey (Wooster, 1994).	• *Phylogeny*, categorical: predators classified as invertebrate, vertebrate or both when multiple predatory species of both categories were included

Continued

Table 2 *A Priori*-Defined Categorical and Continuous Factors Hypothesized to Affect Detrital Trophic Cascades—cont'd

Data Type	Rationale	Variable
Trophic type	Omnivory is expected to produce less predictable, conflicting effects in food webs hindering trophic cascades, as they are predicated on discrete trophic levels (Polis and Strong, 1996; Strong, 1992).	• *Trophic type*, categorical: predators were classified as predator *sensu stricto* (C), omnivore (O) or both (CO) when consumers of both categories C and O were included[b]
(3) Litter quality		
Stoichiometry	The trophic performance of both the heterotrophic microflora and invertebrates generally increases from low- to high-quality leaf-litter, resulting in a faster processing of the latter (Ferreira et al., 2010). A low quality, slowly decomposing detrital resource may exert a double-fold negative effect on trophic cascades, by inducing generalist consumers to shift towards more edible trophic resources and by reducing the possibility of detecting treatment-induced changes in processing rates. The chemical composition of leaf detritus is acknowledged to reflect its trophic quality for consumers. Among others, the initial C:N ratio has been indicated as a good predictor of the detritus processing rates by freshwater invertebrates (Hladyz et al., 2009).	• *Initial C:N, C:P and N:P ratio*: continuous[c]

[a]Factor tested only for aquatic studies.
[b]Trophic strategy defined according to the information provided in the study and corroborated by a literature search.
[c]Species-specific C:N ratios were recorded from the original studies or provided by the authors; additional data were obtained by performing a supplementary search (meta-data in Mulder et al., 2013). Priority was given to studies performed by the same team at the same location or geographical area; when no information were available or only generic taxonomic information were provided, the data from Ostrofsky (1997) were used.
The number of categories and the information reported in each study used for categorization are indicated, together with the expected effect on the cascade strength index LLR_{DET}.

Figure 2 Spatial distribution of the 78 studies of our meta-analysis: terrestrial detrital food webs (DFWs from the brown world) in brown (grey in the print version) and aquatic detrital food webs (DFWs from the blue world) in blue (black in the print version); full list of studies in Appendix (Metareferences).

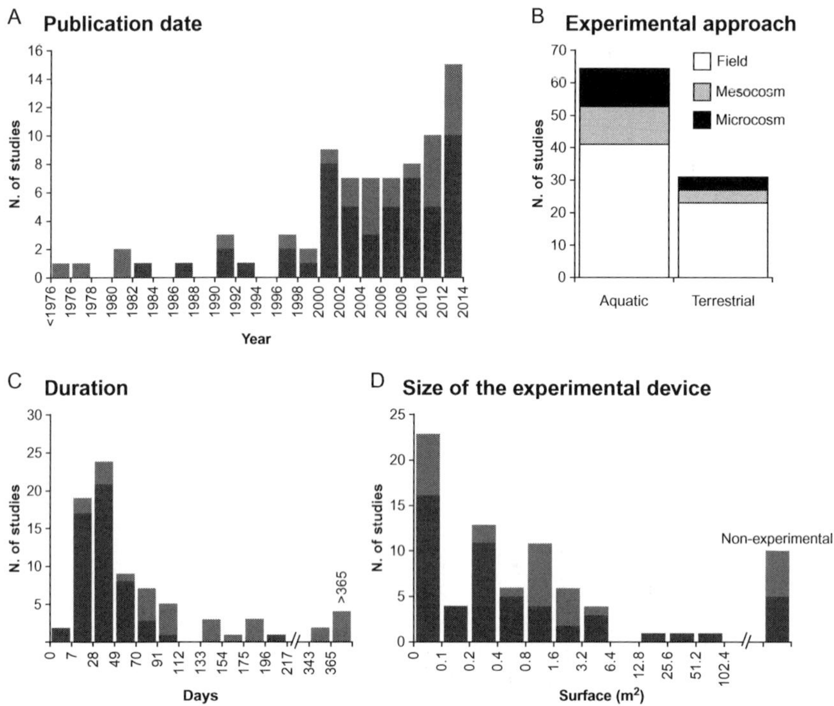

Figure 3 Main literature features. Clockwards: increasing focus on trophic cascades since 1976 (A), type of investigations (B), duration of investigations (C) and size of the experimental device (D).

while the relative contribution of mesocosm and microcosm experiments was comparable (Fig. 3B). The duration of aquatic studies was on average lower than in terrestrial investigations (Fig. 3C): terrestrial investigation was often exceeding 1 year in duration. Fifty one out of the 78 studies included in the data set were from freshwater environments (Fig. 4; Appendix: Table A1), with a considerable contribution from the United States (20 studies). Most of the freshwater studies were carried out in running waters and manipulated the occurrence of predators using cages, while only a few were performed in standing waters (Table A1). The majority of the 27 terrestrial studies were carried out in forests (18 studies). Overall, running waters provided 308 observations (61% of all observations), followed by 89 (18%) in forests, 62 (12%) in standing waters, and only 43 (9%) in grasslands (Fig. 4; Table A1). The aquatic studies focused mainly on the effects of predatory invertebrates. In contrast to aquatic studies, terrestrial investigations focusing on predatory invertebrates tested primarily entire predatory

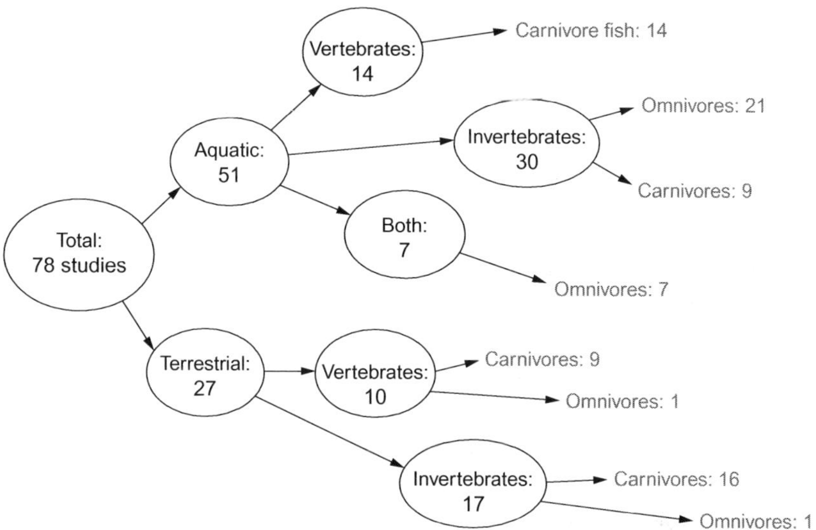

Figure 4 Partitioning of the studies included in our meta-analysis.

guilds, while 'omnivores' (obviously not the predators *sensu stricto*, *s.s.* hereafter, although to a certain extent predatory species as well) remained almost entirely neglected.

Predator impact on leaf-litter or dung-pat mass (hereafter 'detrital resource') was calculated as the unweighted natural log response ratio (Osenberg et al., 1997):

$$\mathrm{LRR_{DET}} = \mathrm{Ln}\left(M_{p+} / M_{p-}\right)$$

where M_{p+} and M_{p-} are the mean percent mass remaining in the presence and absence of predators, respectively. In general, the natural log response ratio measures the proportional change due to experimental manipulation (Luo et al., 2006; see also Sayer et al., 2012 for a more recent example). In our case, LRR will be greater than zero when the predator causes an increase in the remaining mass of leaf detritus, i.e., determines a decrease in the processing rate. Conversely, LRR will be less than zero when the predator determines an increase in the detritus processing rate. Similarly, predator's effect on prey was calculated as:

$$\mathrm{LRR_{PREY}} = \mathrm{Ln}\left(A_{p+} / A_{p-}\right)$$

where A_{p+} and A_{p-} are prey abundance in the presence and absence of predators, respectively (e.g., Woodward et al., 2008). Advantages and disadvantages of using log ratios in meta-analyses have been widely discussed (Borenstein et al., 2009; Hedges et al., 1999) but, in brief, a natural log ratio (i) is biologically meaningful, as it expresses the proportional change in the response variable; (ii) has good statistical properties—its sampling distribution is approximately normal (Curtis and Wang, 1998)—and (iii) is less biased than other metrics. Unweighted natural log ratios decrease the probability to detect differences only in inter-system comparisons (due to an increased Type II error rate) without influencing effect-size estimates (Gurevitch and Hedges, 1999; Hedges et al., 1999) and, like in several recent meta-analyses (Gruner et al., 2008; Poore et al., 2012), were used as conservative effect-size estimators.

Overall, the entire data set yielded 215, 188 and 103 measures of the response of detrital resource, the total abundance of invertebrates and the abundance of detritivorous prey, respectively. In addition, 181 and 102 estimations of LLR_{PREY}/LLR_{DET} absolute ratios were calculated. The grand mean effect of predators on leaf-litter over the whole set showed a negligible difference from zero, being LLR_{DET} equal to −0.038 with a 95% Confidence Interval (95% CI hereafter) of (−0.102; 0.022). In contrast, the mean effect of predators on prey was negative and significantly different from zero, either considering the total abundance of invertebrates [$LLR_{PREY}=-0.516$; 95%CI (−0.641; −0.401)] or only that of detritivorous invertebrates [$LLR_{PREY}=-0.611$; 95% CI (−0.699; −0.529)], indicating that the predators' effects exerted on the adjacent trophic level did not propagate downwards to the basal detrital resource regardless of the ecosystem type (Fig. 5, Table A2).

2.2 Effects of Categorical Variables

Trophic type was the only factor that influenced detrital resources: effect of the predators *s.s.* was positive and stronger than the negative effect exerted by omnivores (either alone or in coexistence with true predators) (Table 3). In general, effects on prey—either for whole invertebrate communities or detritivore assemblages—were negative in both aquatic and terrestrial habitats. For the latter, there was no difference between litterbag and dung-pat studies [total invertebrates: $LLR_{PREY}=-0.575$ versus −0.542, with respective 95% CIs (−0.801; −0.376) and (−0.747; −0.343)]. The total abundance of invertebrates was influenced by climate (contrast analysis $p=0.03$; Table 4). Omnivores, either alone or with predators, exerted a stronger negative effect than predators in both habitats, especially in

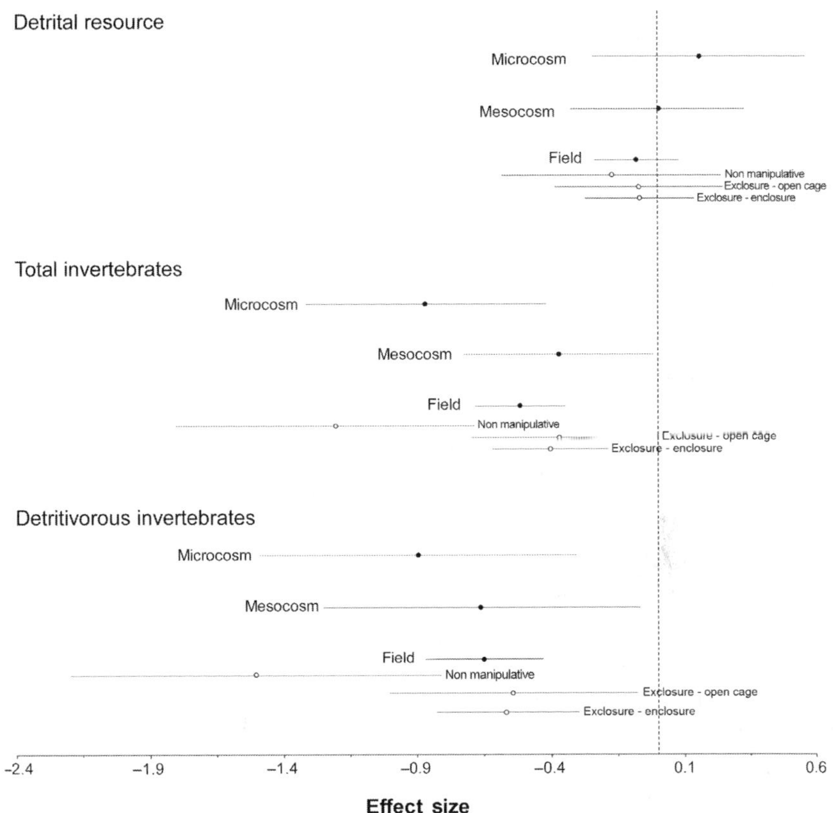

Figure 5 Effect size for the main categories of our meta-analysis. More details in the text and weighted averages and 95% Confidence Intervals per ecosystem type in our Table A1.

terrestrial systems. Similar effects were evident for detritivores (Table 5). The main variation was for the interaction 'trophic type' and 'habitat', where aquatic predators *s.s.* and omnivores showed stronger effects than terrestrial predators *s.s.*, again with dominant effects induced by terrestrial omnivores. Invertebrate predators *s.s.* exerted a stronger influence than vertebrate predators *s.s.* on the total abundance of invertebrates (contrast analysis $p=0.001$).

Of the three elemental ratios shown in Fig. 6, C:N and C:P were the only ones not correlated with one other ($r=-0.26$, $p=0.09$, 50 d.f.), so our analyses were restricted to these two. The influence of predators on the effect size of the detrital resource scaled positively with both ratios (Fig. 7A), as a saddle-shaped tri-dimensional surface was evident for the combined C:N:P ratios,

Table 3 Effects on the Detrital Resource

Parameters	*F*	*p*	Levels	Effect Size	95% CI	*n*
Intercept	2.87	0.09				
Habitat	0.12	0.73	(1) Aquatic	**−0.071**	−0.154; 0.005	151
			(2) Terrestrial	**0.039**	−0.049; 0.123	64
Climate	0.66	0.42	(1) Temperate	**−0.034**	−0.107; 0.041	172
			(2) Tropical	**−0.057**	−0.164; 0.043	43
Trophic type	4.04	**0.02**	(1) Predators *s.s.*	**0.087**	0.037; 0.135	126
			(2) Omnivores	**−0.179**	−0.326; −0.036	70
			(3) Predators *s.s.* and Omnivores	**−0.345**	−0.588; −0.135	19
Habitat × trophic type	0.47	0.49				
Habitat × climate	0.06	0.81				
Climate × trophic type	0.48	0.62				

Predators *sensu stricto* (*s.s.*) and omnivores aggregate all consumers; trophic type for trophic strategy as derived from literature.
Mean effect sizes and statistically significant differences ($p < 0.05$) are shown in bold.

Table 4 Effects on Total Invertebrate Prey

Parameters	*F*	*p*	Levels	Effect Size	95% CI	*n*
Intercept	43.05	**0.01**				
Habitat	4.01	0.07	(1) Aquatic	**−0.504**	−0.598; −0.415	137
			(2) Terrestrial	**−0.572**	−0.775; −0.399	51
Climate	6.21	**0.01**	(1) Temperate	**−0.576**	−0.677; −0.479	154
			(2) Tropical	**−0.281**	−0.463; −0.124	34
Trophic type	4.46	**0.02**	(1) Predators *s.s.*	**−0.365**	−0.463; −0.278	106
			(2) Omnivores	**−0.782**	−0.946; −0.626	65
			(3) Predators *s.s.* and Omnivores	**−0.514**	−0.861; −0.162	17
Habitat × trophic type	5.25	**0.03**	(1) Predators *s.s.*, Aquatic	**−0.312**	−0.428; −0.202	59
			(2) Omnivores, Aquatic	**−0.649**	−0.786; −0.521	78
			(3) Predators *s.s.*, Terrestrial	**−0.433**	−0.586; −0.302	47
			(4) Omnivores, Terrestrial	**−2.216**	−2.216; −2.216	4*
Habitat × climate	2.23	0.14				
Climate × trophic type	1.75	0.18				

Predators *sensu stricto* (*s.s.*) and omnivores aggregate all consumers. Studies on Omnivores, Terrestrial were not considered further in the analysis due to the low number of observations (*).
Mean effect sizes and statistically significant differences ($p < 0.05$) are shown in bold.

Table 5 Effects on Detritivorous Consumers

Parameters	*F*	*p*	Levels	Effect Size	95% CI	*n*
Intercept	41.11	**0.01**				
Habitat	2.53	0.12	(1) Aquatic	**−0.697**	−0.845; −0.539	77
			(2) Terrestrial	**−0.618**	−0.891; −0.319	26
Climate	0.28	0.61	(1) Temperate	**−0.685**	−0.823; −0.531	81
			(2) Tropical	**−0.648**	−0.987; −0.369	22
Trophic type	3.70	**0.04**	(1) Predators *s.s.*	**−0.534**	−0.681; −0.399	59
			(2) Omnivores	**−0.873**	−1.108; −0.648	37
			(3) Predators *s.s.* and Omnivores	**−0.855**	−1.656; −0.168	7*
Habitat × trophic type	7.45	**0.01**	(1) Predators *s.s.*, Aquatic	**−0.657**	−0.855; −0.448	37
			(2) Omnivores, Aquatic	**−0.735**	−0.958; −0.524	40
			(3) Predators *s.s.*, Terrestrial	**−0.327**	−0.468; −0.184	22
			(4) Omnivores, Terrestrial	**−2.216**	−2.216; −2.216	4*
Habitat × climate	0.01	0.92				
Climate × trophic type	0.75	0.48				

Predators *sensu stricto* (s.s.) and omnivores aggregate all consumers. Predators *s.s.* + omnivores and omnivores, terrestrial were not considered further in the analysis due to the low numbers of observations (*).
Mean effect sizes and statistically significant differences ($p < 0.05$) are shown in bold.

characterizing different cascade strengths with changing elemental ratios. Omnivore effects on the detrital resource also increased with nutrient ratios, resulting in a tri-dimensional plane, with maximum negative effects for omnivores in relation to detrital resources at higher C:N and lower N:P ratios (Fig. 7B). The overall effect of predators on the total invertebrate abundance increased with both C:N and N:P ratios, although the plane was now characterized by maximal negative effects for detrital resources at lower C:N ratios and higher N:P ratios. For omnivores, the efficiency in the effect transfer to basal resources was higher in comparison to true predators. Plotting the effects on the basal resources exerted by true predators and omnivores versus the C:N:P ratios revealed that C:N and N:P trends were often opposite (Fig. 7, bottom panels).

Figure 7 clearly demonstrates that the quality of the resource mediates predator-induced indirect effects in terrestrial and aquatic DFWs. The underlying trophic relationships can be visualized from a chemical perspective, seen that ecological stoichiometry shapes the occurrence of single

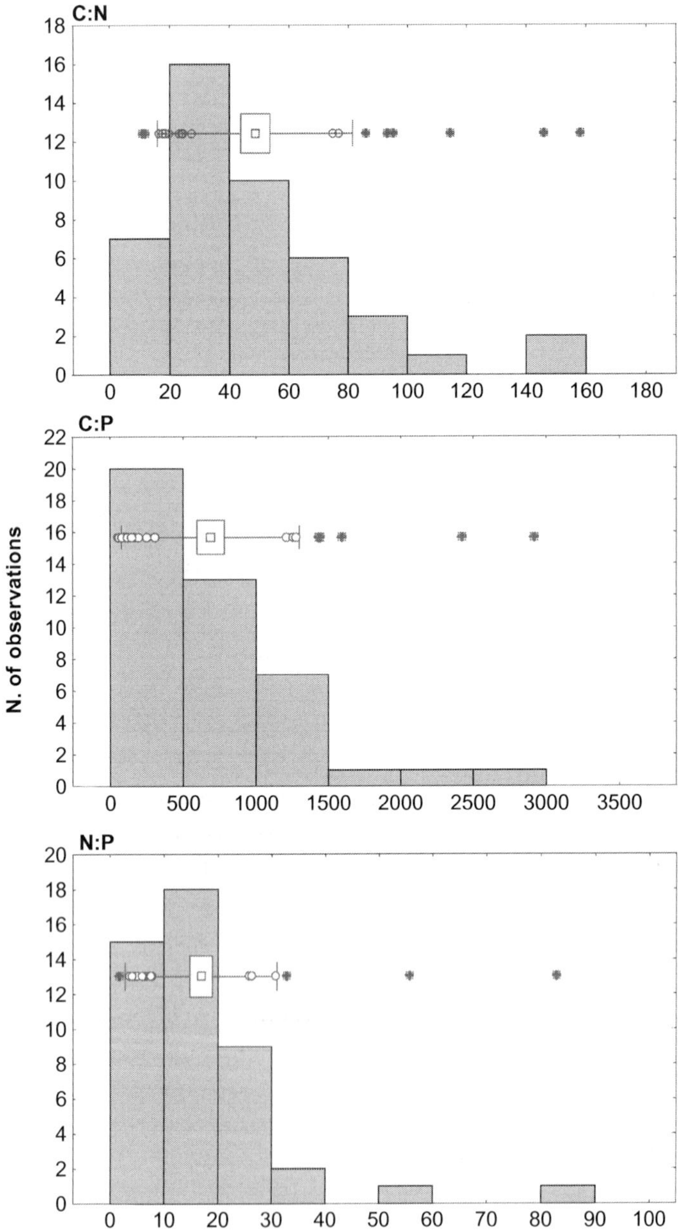

Figure 6 From top to bottom, number of observations on leaf-litter C:N, N:P and C:P initial ratios to assess the ecostoichiometric quality of the resources.

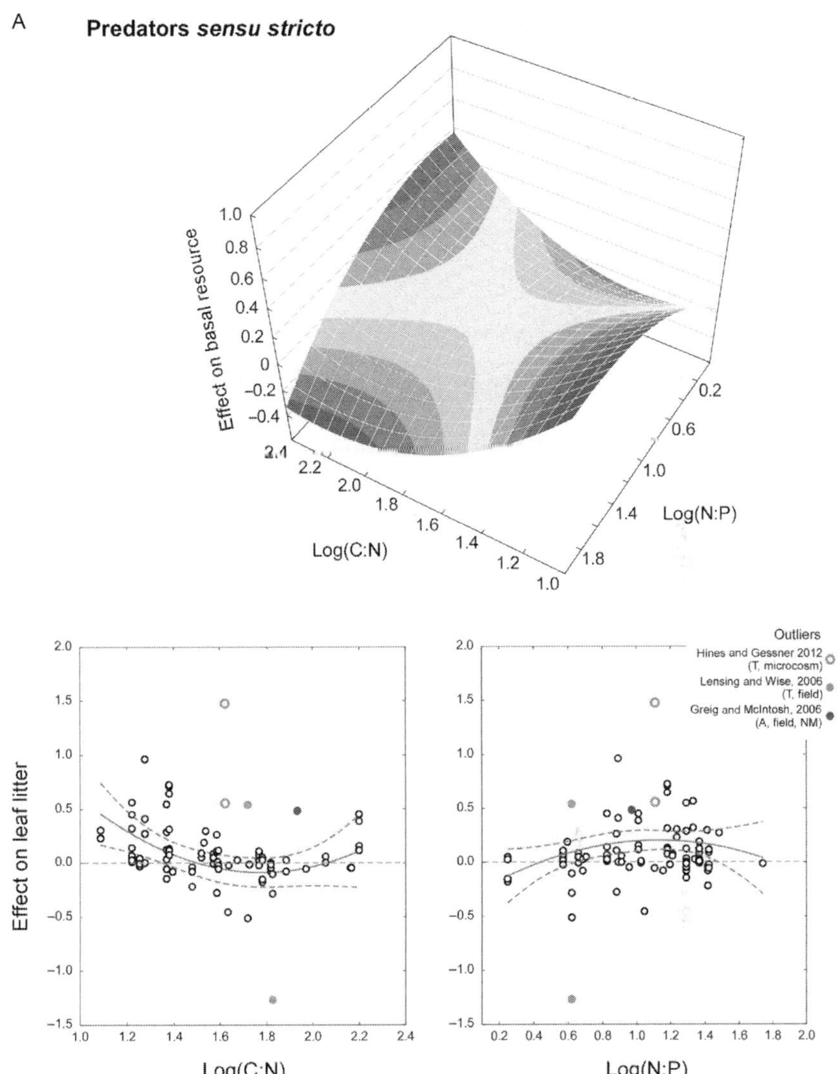

Figure 7 Fitted relationships between effect size and ecostoichiometric quality of the detrital resource for the predators *s.s.* (A). The upper panel shows tri-dimensional responses across the combined C:N:P gradient, and the lower panels show the response splitted along log(C:N), the left scatter plot, and log(N:P), the right scatter plot.

(Continued)

species and, hence, the entire community assemblages. An ecostoichiometrical framework would offer new ways to understand what drives cascading effects and shapes the organization of entire DFWs. Given the potential for consumers to influence basal resources in both

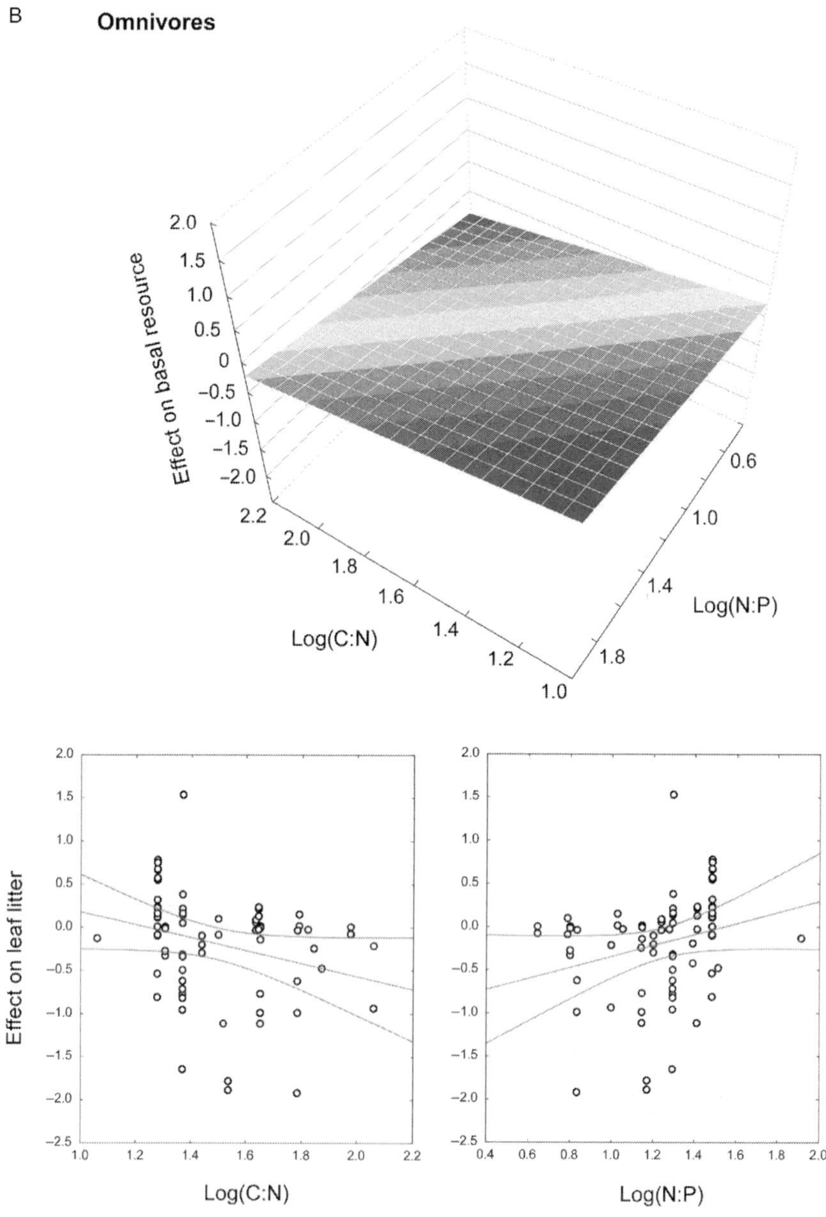

Figure 7—Cont'd (B): Fitted relationships between effect size and ecostoichiometric quality of the detrital resource for omnivores. As before the upper panel shows tri-dimensional responses across the combined C:N:P gradient, and the lower panels show the response splitted along log(C:N), the left scatter plot, and log(N:P), the right scatter plot.

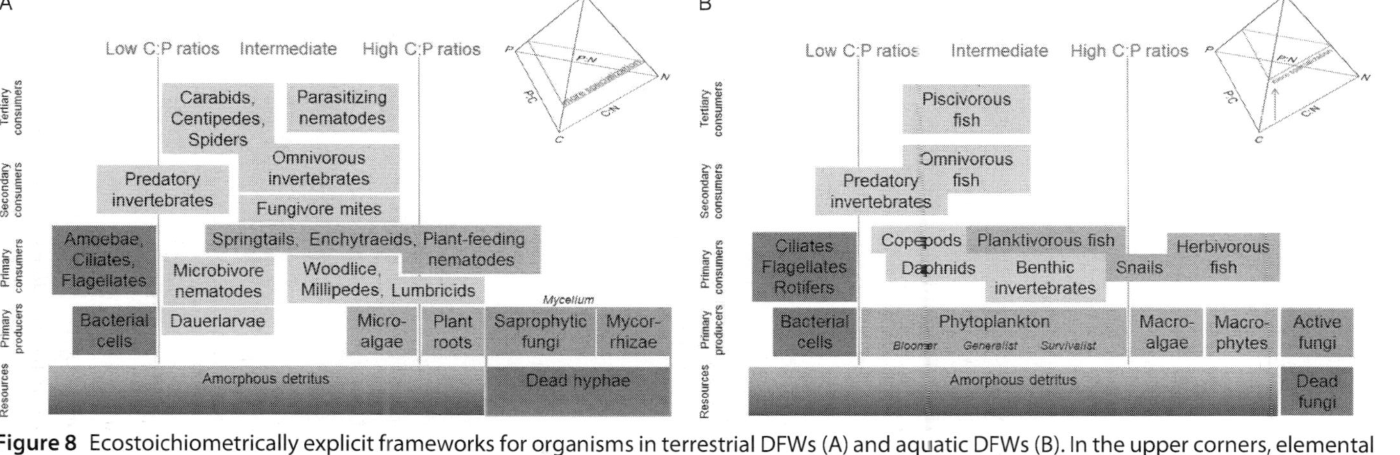

Figure 8 Ecostoichiometrically explicit frameworks for organisms in terrestrial DFWs (A) and aquatic DFWs (B). In the upper corners, elemental pyramids are shown. The base of these multiple ternary diagrams is a triangle plot which displays the proportion of the C, N and P variables, and their heights represent the efficiency of energy transfer between trophic levels (i.e. food-chain length). The pyramidal views and the skewed distributions of the organisms reflect the longer paths in terrestrial DFWs: omnivores occupying distinct trophic levels cannot switch to prey due to increased metabolic costs (Francisco Bozinovic as cited in Glazier, 2012) and the degree of omnivory for invertebrates of terrestrial DFWs is higher among the secondary consumers (in contrast to aquatic DFWs where omnivory is higher among primary consumers).

terrestrial and aquatic DFWs (Fig. 8A and B, respectively), C:N:P imbalances within a conceptual food web can be viewed by arranging guilds vertically according to trophic levels and horizontally according to feeding niches. The kind of represented niches can be very diverse, like fish consuming particulate detritus (Jepsen and Winemiller, 2002; Jones and Waldron, 2003; Wootton and Oemke, 1992) and plankton (Elser et al., 1998; Sommer et al., 2012). Preferential feeding on more basal resources is expected to be higher among primary consumers and might fluctuate, for instance, with shifted algal stoichiometry in relation to climate changes (cf. Yvon-Durocher et al., 2015). Romanuk et al. (2006) see consumption efficiency as inversely correlated with the aforementioned ability to access multiple resources and stability as directly correlated with the ability to access the same multiple resources. If so, then the shape of this triangular view (i.e. how broad the basal level is) reflects the availability of multiple resources (Fig. 8), hence the functional redundancy of an ecosystem. All together, these niches will result in site-specific networks, comparable to complex systems that exhibit an extraordinary homeostasis against failure (Solé et al., 2003; Sugihara, 1982). Such a kind of self-regulating networks can behave in contrast to the homeostasis regulation that acts as an environmental-driven constraint for organisms keeping chemical composition as constant as possible (Neufeld et al., 2007; Sterner and Elser, 2002).

When a DFW gets eroded from the bottom (e.g. due to changes in the environmental conditions that alter C:N:P ratios among basal resources), predators and omnivores might rapidly disappear (cf. Brennan et al., 2014), with consequences for both top-down and bottom-up processes. Food-web metrics reflect not only trait-mediated structures of ecological networks (Fig. A2) but may also be linked to stoichiometric imbalances between predators, omnivores and their basal resources (Table A3) due to either natural dynamics- or human-induced stress. For instance, food-web metrics should be used for a comparison between ecological networks only if these networks have a comparable taxonomical or functional resolution. It is often far easier to identify at higher resolution certain basal resources like algae and detritus in aquatic ecosystems (e.g. several diatom species, and dissolved and particulate organic matter like in Woodward et al., 2008) than may be the case in belowground soil systems (e.g. plant roots kept as a whole entity like in Sechi et al., 2015), making a robust comparison between DFWs with different levels of resolution difficult. Also, detritus can be resolved from a chemical or sedimentological perspective, for instance as progressive breakdown into steadily smaller particles linked to different sets of primary consumers (Cummins and Klug, 1979; Kitching, 2001), hence resolution matters even for non-living entities like detrital forms.

3. DISCUSSION

Invasive species, global warming, acidification, overfishing, pesticides and other anthropogenic drivers have especially strong impacts on the higher trophic levels in both aquatic and terrestrial food webs and can cause dramatic changes in the community structure of ecological networks, the functioning of the respective ecosystems and, by extension, the ecosystem services they deliver (e.g. Allan et al., 2005; Bohan et al., 2013; Carpenter et al., 2011; Rasmussen et al., 2012; Strong and Frank, 2010). The functional roles of predators within community assemblages, and the consequences of losing them, are both important but hard-to-quantify issues for the conservation, management and restoration of ecosystems and for the sustainability of the ecosystem services they provide—from biomass production in commercial or recreational fisheries up to pest control in agroecosystems (Dudgeon, 2010; Humphries and Winemiller, 2009; Vaughn, 2010).

Our results require us to re-examine both our initial assumptions and the wider conceptual framework in which the original literature review was cast. Among the diverse set of drivers we tested, only the predator's feeding mode explained a significant amount of variance and once partitioned further, only predators versus omnivores and the transfer ratio of effects across contiguous trophic levels exhibited significant relationships with detrital stoichiometry. This reveals the surprisingly important role played by indirect bottom-up forces, and is consistent with some other recent findings emerging from boreal standing waters (e.g. Morlon et al., 2014) and tropical rivers (Winemiller et al., 2014). We found that these underlying bottom-up constraints were not triggered by variations in the total initial mass of the detrital resource or the total fluxes of C, N or P, but by the stoichiometric ratios of these elements (as also reflected by several studies from Redfield, 1958, up to Yvon-Durocher et al., 2015). This is perhaps not surprising, given that P is the main limiting nutrient for a large fraction of organisms (Mulder and Elser, 2009; Sterner and Elser, 2002) and ecosystems (Vitousek, 1984; Vitousek et al., 1988). Inputs of anthropogenic N and P to natural and managed ecosystems worldwide are widespread, either from the atmosphere and/or run-off (Vet et al., 2014). Since they can alter—both directly and indirectly—the elemental quality of the detritus at the base of the food web, this could cause fundamental shifts in the structure and dynamics of entire ecosystems and the relative strength of cascades in the brown versus green pathways. As such, altered C:N:P ratios could affect human wellbeing through several supporting (nutrient cycling, primary production), regulating (climate regulation) and provisioning (fresh water, wood and fibre) ecosystem services, which could be very vulnerable to disruptions to the reciprocal and cascading interactions that drive them.

Detritus is an important basal resource that ultimately controls the cascading effects from higher trophic levels (Polis and Strong, 1996) and influences many other aspects of biodiversity and food-web structure (e.g. Elser et al., 1998; Hagen et al., 2012; Hillebrand et al., 2007; Moore et al., 2004; Shurin et al., 2002; Srivastava et al., 2009). Most studies have focused on its bottom-up effects, whereas very few have explored detrital stoichiometry effects at higher trophic levels, and most of those have only considered direct interactions with detritivores or microbes and their effects on decomposition rates (e.g. Cross et al., 2003, 2005; Frost et al., 2006; Hladyz et al., 2009; Krumins et al., 2006; Mooshammer et al., 2012). An even smaller subset have considered the detrital-based food-web responses to stoichiometric and environmental changes, either theoretically (DeAngelis, 1992; Elser et al., 2000, 2012; Kuijper et al., 2004) or empirically (Bradford et al., 2014; Mulder and Elser, 2009; Ott et al., 2014). The effect that the main predators in our study—fishes and invertebrates—exerted on their detritivorous prey was translated to detrital processing rates, revealing that trophic cascades were not inconsistent with the 'green world' hypothesis (Hairston et al., 1960). Further support for such cascading effects was provided by direct relationships between nutrients and LLR_{DET}: the generally positive correlation between nutrients and LLR_{PREY} could be suggestive of an indirect top-down effect otherwise hidden by prevailing bottom-up forces. Thus, freshwater DFWs dominated by predators might not differ so markedly from their terrestrial counterparts after all.

The microflora associated with particulate detritus may actually act as a 'hidden' trophic level, potentially acting as a buffer. Microorganisms can play such a dual role as both resources—enhancing detritus quality and palatability—and competitors of detritivorous invertebrates (Bärlocher, 1980; Chung and Suberkropp, 2009; Hättenschwiler et al., 2005). Thus, they represent an important route of energy transfer from the detrital resource that can be only partially intercepted by detritivores, determining a potential decoupling of the latter's control of detritus processing rates. Among the 78 investigations included in our data set, however, only a handful explicitly considered the changes in the biomass of the heterotrophic microflora (i.e. Majdi et al., 2014; Mancinelli et al., 2002; Schofield et al., 2001 for freshwater habitats; and Fukami et al., 2006; Miyashita and Niwa, 2006; Stanley and Ward, 2012 for terrestrial habitats), despite the increasing recognition of the importance of fungal biodiversity in regulating decomposition (Dang et al., 2005, 2009; Funck et al., 2013; Kominoski et al., 2009) and other services (Dighton, 2003).

The dynamic responses of microbial biomass to changing consumer abundance have long been known to dampen trophic cascades in terrestrial detrital systems (e.g. Mikola and Setälä, 1998; Taylor et al., 2010) and the recognition that microbial diversity affects decomposition rates (Dang et al., 2005) suggests that these heterotrophic systems might be controlled by a range of biotic drivers (e.g. Gessner et al., 2010; Kominoski et al., 2010; Srivastava et al., 2009). These also include basal resources (Kominoski et al., 2007, 2009; Swan and Palmer, 2004) and detritivores in the food web (Dangles and Malmqvist, 2004; McKie et al., 2008). The complexity of interactions may mitigate indirect predator-driven effects by providing a diverse array of alternative pathways in the food web (Hättenschwiler et al., 2005; Polis et al., 2000) and buffering effects could arise from consumer-resource C:N:P imbalances (Figs. 7 and 8).

Hence, instead of the traditional focus on food webs with scarce environmental information or low taxonomical resolution (as previously discussed in Goldwasser and Roughgarden, 1997), it may be more promising to use trait-driven ecostoichiometrical frameworks. Within such a functional perspective, our results suggest that it is not simply the ecosystem that shapes the food-web structure, but that both the ecosystem as the ecological network (like the multiple microbe–plant–animal interactions in Fig. A2) reflect, although in different ways, the same environmental filters (see the debate on 'everything is everywhere and the environment selects' in Martiny et al., 2006). The ecosystem also shapes the nutrient availability for both autotrophs and heterotrophs, and at the same time, the landscape influences the nutrient flows (Vannote et al., 1980). Fertilization is one way in which humans seek to improve ecosystem service delivery by overcoming stoichiometric imbalances at the base of the food web via technological solutions designed to enhance (short-term) nutrient availability in soil and litter (Hines et al., 2015; Jones et al., 2014). However, over longer terms, it can compromise ecosystem functioning and affect provisioning and/or supporting services (Mulder et al., 2015a; Wood et al., 2015). Fertilization and many other kind of human interventions alter the Redfield-like balance of detritus, especially in terrestrial ecosystems, which could start to impinge on crop yields (Jones et al., 2014; Mulder et al., 2011, 2015b). One typical example of such collateral damage can be found in Sayer et al. (2012), their Fig. 3, where several years of N-addition to a forest floor caused dramatic changes in the Ca content of litter. This obviously holds also for aquatic ecosystems. Overall, high trophic generalism of primary consumers in freshwater food webs, including herbivory and detritivory (MacNeil et al., 1997, 1999; Mancinelli, 2012), could impair the transfer across the food-web levels.

Paradoxically, our study both confirms the pivotal functional role played by omnivores in DFWs and suggests their inability to determine a trophic cascade. Indeed, these (facultative) predators exerted general, significant effect over both prey density and detritus processing. In contrast to predators *s.s.* (obligate carnivores); however, effects on the basal resource were consistent with direct exploitation of the detrital resource, consumed along with invertebrate prey, disrupting any cascading effect on the basal level. This view is supported by the similarity of the relationships between the C:N ratio and the two cascade indices LLR_{DET} and LLR_{PREY}. Furthermore, the LLR_{PREY}/LLR_{DET} ratio was smaller in comparison to predators *s.s.*, indicating that disruptive effects determined by omnivores are intense. It must be pointed out that the key role in processes played by taxa that are both competitive and functionally dominant (Creed et al., 2009; Romanuk et al., 2006) is confirmed here as well. Hence, the multiple aspects of ecological stoichiometry in investigated trait-mediated interactions have implications in ecosystem services that merit further research.

4. FUTURE ECOSYSTEM SERVICES RESEARCH

Ritchie et al. (2012) and Ripple et al. (2014) emphasized that insufficient attention was given to the trophic behaviours of large carnivores, but according to us even less attention is given to predatory invertebrates and small predatory vertebrates. This lack of knowledge is challenging for a correct assessment of ecosystem services. Primary production is an outstanding function providing key ecosystem services that can be significantly influenced by predators through indirect cascading mechanisms. Numerous reviews have suggested that trophic cascades are ubiquitous in grazing food webs (e.g., Borer et al., 2005; Shurin et al., 2002; Terborgh and Estes, 2010).

Hence, ecosystem services related to grazing systems may be regulated by indirect, predator-induced trophic cascades. This has strong implications for environmental effects related to atmospheric pollution and nutrient leaching from agroecosystems, as so many prey species are sensitive to acidification, P and N deposition (Blake and Downing, 2009; Brennan et al., 2014; Mulder et al., 2013). We indicated that leaf-litter decomposition, i.e., the particulate component of plant detritus occurring ubiquitously in both aquatic and terrestrial ecosystems, is regulated by the same type of indirect mechanisms. Decomposition is the 'brown' side of the 'green' primary production in the holistic perspective of ecosystem functioning, providing nutrients to autotrophic systems, indirect subsidies to both autotrophic and heterotrophic communities, and, finally, enabling primary production to provide goods and services to mankind. Decomposition is the missing link in the current attempts made to forecast ecosystem services within a 'true' ecosystem perspective.

Three important aspects relating to the quality of the detrital resource require further investigation in the light of our findings. First, we focused strictly on debris from autotrophs (Fig. 8), neglecting decomposition studies in animal carrion (*sensu* Wilson and Wolkovich, 2011). Second, although authors generally provided qualitative descriptions of multi-specific assemblages of the aquatic and riparian plants potentially contributing to in-water detritus, few studies used mixed-species litterbags, and with few exceptions, all were from terrestrial habitats. Recent investigations have focused on the implications of leaf-litter quality for benthic communities under mixed-species conditions (Ball et al., 2008; Leroy and Marks, 2006; Sanpera-Calbet et al., 2009), but no consensus has been reached as to how detritus composition and diversity affect processing rates (Gartner and Cardon, 2004; Schindler and Gessner, 2009) despite significant effects on shredder abundances (Kominoski and Pringle, 2009; Swan and Palmer, 2006). Third, detritus quality is greatly affected by the associated heterotrophic microflora soon after the conditioning process begins (Dighton, 2003; Gessner et al., 1999; Wurzbacher et al., 2010). In this study, the density data used to estimate LLR_{PREY} values were collected with reference to final sampling dates, when leaf detritus quality and palatability to invertebrate consumers is expected to be improved by microbial conditioning. Initial carbon to nutrient ratios may represent inadequate estimators of detritus quality, but if complemented by fungal biomass, ecological stoichiometry and microbiology they can be very effective predictors of detritus processing rates (e.g. Hladyz et al., 2009; Mulder et al., 2009). Unfortunately, quantitative information on detritus-associated microbial biomass was provided by only a small fraction of the studies in the database. In addition, these studies focused on microfungi, neglecting protozoans, which share with bacteria not only a key role in detritus conditioning at later stages of decomposition (Mille-Lindblom and Tranvik, 2003) but also compete with bacteria (Krumins et al., 2006; Yee et al., 2007) and mycorrhizal fungi (Bonkowski, 2004), and act even as pack-hunting intraguild predators as recently demonstrated by Geisen et al. (2016).

A quantitative synthesis of the available literature, like ours, can help to provide broad overviews on the generality of detrital trophic cascades, as a first step to a more rigorous mechanistic and quantitative understanding of how they translate to ecosystem services: it is clear that they do, but the exact form of the relationship is unknown (Raffaelli and Moller, 1999). In grazing systems, for instance, omnivores can induce trophic cascades, implying that their consumption of living basal resources is generally negligible compared to that of prey, but it will be the short-term response to resource pulses which will improve our understanding of the interrelationship between top-down and bottom-up processes (Holt, 2008; Leroux and Loreau, 2015). Leaf stoichiometry was suggested to regulate cascade strength by influencing the contribution of invertebrate detritivores to processing rate of detritus (Kitching, 2001), and lower palatability of living versus dead macrophytes as a function of fungal colonization for detritivores (Rong et al., 1995; Sterner and Elser, 2002). We demonstrated that at the same time: (i) predators *s.s.* decrease decomposition rates, slowing down the release of nutrients all systems, increasing the role of detritus as nutrient repository (acting as a kind of 'sponge' for primary producers), and playing a key regulating role in ecosystem functioning and (ii) omnivores can affect the decomposition rates as well, but being they mostly ecological engineers (e.g. crayfish, see Usio, 2000; Usio and Townsend, 2002, 2004), omnivores build the habitat in which they live, boost leaf-litter decomposition rates, highly increase comminution and enhance the trophic prey diversity and the resilience of the entire community.

Too often, supporting services like nutrient cycling are seen as top-down processes. A conventional perspective of a top-down nature and regulation of ecosystem services can be misleading for environmental regulation, and here we indicated that without considering bottom-up constraints related to the stoichiometry of the detrital resource the ecosystem picture is largely incomplete. Just as ecosystem services are interlinked and cannot be treated in isolation, organisms occupying adjacent trophic levels need to be analyzed in an ecosystem context: cascading effects can thus be seen as a new basis for understanding (and hence ultimately predicting and managing) provisioning, supporting and regulating services. Indeed, although rarely couched in the lexicon of ecosystem services, there are already plenty of examples of the use of knowledge of cascading interactions in non-detrital driven systems that are routinely used to manage ecosystem service delivery: size-selective marine fisheries and biomanipulation of lakes to enhance provisioning services as well as amenity and recreational value are some familiar examples. Our study represents a novel attempt to elucidate how similar frameworks could also be extended to the 'brown world' that underpins the production of the planet's agricultural crops as well as its major biogeochemical cycles.

ACKNOWLEDGEMENTS

We are indebted to all authors who kindly contributed additional information for this meta-analysis. David Dudgeon and Sukhmani Mantel were the first to provide unpublished data on the trophic ecology of *Macrobrachium hainanense*, followed by (listed without any particular order) Clement Lagrue, Daniel Hocking, Jeremy Jabiol, Eric Moody, Kristie Klose, Nabil Majdi, Jonathan Grey, Michelle Jackson and Tim Moulton. The picture of *Sarracenia purpurea* L. (Fig. A2D) was reproduced with kind permission of Michel Perron (http://floreduquebec.ca), while Nick Gotelli provided helpful advices on *Sarracenia* food webs. Guy Woodward, Isabelle Durance and an anonymous referee provided continuous feedbacks on our draft and suggestions by Alice Boit greatly improved our food-web visualization by NETWORK 3D. Research fundings by FIRB 2014–2015 to G.M. and by the EU Seventh Framework Programme to C.M. (SOLUTIONS grant 603437) are acknowledged. This chapter is dedicated to Sofia Mancinelli, thy eternal summer shall not fade.

APPENDIX

A.1 Data Sources and Selection Criteria

A literature search for freshwater studies testing the influence of predators on leaf-litter and dung mass loss was performed using multiple criteria (e.g. trophic cascade/detritus processing and top-down/leaf-litter decomposition) applied to the ISI Web of Science, BioAbstracts, PubMed, Agricola and JSTOR databases. The results were supplemented by studies included in review papers (e.g. Mancinelli et al., 2013) and by performing general searches on the World Wide Web (see Metareferences). We also collated stoichiometric observations on the plant species used in litterbags (Kattge et al., 2011; Mulder et al., 2013) and on the dung-pats (Wu and Sun, 2010). The literature search was completed by May 2015. Field and laboratory studies were included in the review only if they: (i) reported the effect on detritus processing as changes in remaining mass of litterbags (or leaf packs, *sensu* Bärlocher, 2005) and dung-pats, and reported the effect on prey abundance as changes in either mass (dry mass or ash-free dry mass) or numbers normalized per litterbag or per gram litterbag of invertebrates; and (ii) manipulated predators by excluding them (exclosure/open cages design), by varying their abundance (exclosure/enclosure design) or by comparing locations with and without predators (non-experimental design).

A.2 Quantifying Predator Effects

The literature search yielded 78 studies that met the criteria (Table A1 and Metareferences), published between April 1984 (Oberndorfer et al., 1984) and October 2014 (Liu et al., 2014). From these studies, data were extracted to estimate the response to predators of (i) leaf-litter—dung-pat's mass loss and (ii) prey abundance. Data extraction was performed on digital versions of the publications (PDF format); plots and diagrams were converted to numerical form after a fivefold enlargement using g3data ver. 1.5.1 (http://www.frantz.fi/software/g3data.php).

When leaf-litter processing was expressed as decomposition rate k, data were transformed according to a single exponential model (Olson, 1963): $M = M_0 \times e^{-kt}$ where M = mass remaining at the end of the experiment, M_0 = initial mass and t = duration of the experiment (in days). Prey response to predators was calculated on the total abundance of the invertebrate assemblage associated with the detrital resource, focusing, where data were available, on the abundance of macrophagous detritivore invertebrates, i.e., those directly involved in leaf-litter shredding and processing (e.g. Graça, 2001). When abundances were presented from multiple samplings during the experiment, data were collected for the last sampling time only, to avoid pseudoreplication.

If (i) multiple experiments were carried out in different locations or periods or (ii) the effect of more than one predator species was tested or (iii) predator manipulations were crossed with other treatments, data were collated for all the predator treatment versus predator-free control comparisons. In addition, when multiple predator densities were used, all treatments were contrasted with controls and included in the data set. Such an approach was chosen to minimize biases induced by the adoption of restrictive selection standards (Englund et al., 1999). Estimations were generally based on statistically independent observations; an exception was Reice (1991), who manipulated predators' abundance in a series of contiguous stream segments and produced 13 non-independent monthly estimations of both detritivores abundance and leaf-litter mass loss. An overall, unweighted mean response (Borenstein et al., 2009) was calculated for the whole experimental period. A second exception was represented by Moore et al. (2012), whose study consisted of two experimental phases: in each phase, leaf-litter decomposition was measured in several stream pools containing different densities of crayfish (15 and 8 pools, respectively). Analyses performed in disconnected pools have to be considered as independent; however, they require the estimation of effect sizes specific for correlational data (Borenstein et al., 2009). Given the general methodological approach adopted in the present study for the estimation of predator impacts (see later in this section), homogeneity effects were calculated by comparing for each experimental phase the leaf-litter mass loss and the invertebrate abundance in crayfish-free pools with the unweighted mean responses observed in all the pools containing crayfish.

A.3 Meta-Analytic Procedures

Data analyses were performed in five consecutive steps.

First, for each response variable (i.e. detrital resource, total abundance of invertebrates and abundance of detritivorous invertebrates), the occurrence of publication bias was checked graphically—by visual inspection of funnel plots and normal quantile plots—and statistically—using the Spearman rank correlation test (Møller and Jennions, 2001; Wang and Bushman, 1998).

Second, the occurrence of experimental biases was assessed. Whereas no specific information were available for aquatic habitats, in terrestrial systems Kampichler and Bruckner (2009) have suggested that leaf-litter mass loss estimated in litterbag studies may be affected by a number of confounding factors independent of soil animal effects. Here, we used general linear mixed models (GLMM) to test the influence on LLR_{DET} of two categorical variables related to the location of the experimental setup (three levels: field, mesocosm and microcosm) and the design of the experimental setup (three levels: exclosure-enclosure, exclosure-open and non-manipulative) and of two continuous variables related to the size of the caging devices and of the initial dry mass of litterbags or dung-pats used in the studies. The rationale for the choice of the four potential confounding variables is described in more detail in Table A2. Multiple observations extracted from the same study were controlled by adding the reference identity as a random factor to GLMM. Type III Sum of Squares and restricted maximum likelihood (REML) were used for model estimation. Additional GLMMs were run to test for the occurrence of experimental artefacts on responses of both total abundance of invertebrates associated with litter bags and abundance of detritivorous invertebrates. The analysis was repeated twice: the first time a GLMM was run on the whole data set without considering the factor 'experimental setup', as related specifically to studies carried out in the field; the factor was then considered in a second GLMM run on a reduced data set limited to field experiments. In both cases, continuous variables were log-transformed prior to analysis.

Third, after literature and methodological biases were checked and excluded, the grand mean LLR_{PREY} and LLR_{DET} values were calculated together with bootstrapped non-parametric 95% CI to provide an overall estimation of cascade strengths (Osenberg et al., 1999). Means were considered significantly different from zero if the CIs did not overlap with zero. CIs were used throughout the study to quantify variances around mean estimates.

Fourth, Linear Mixed Models were used to test the degree to which variation in LRRs was explained by categorical predictor variables. As in step 2, predictors were considered as fixed factors and the study as a random factor to account for non-independence of multiple measures. In addition, statistically and ecologically meaningful two-way interactions were included in the model as fixed factors.

Finally, for factors for which significant effects were detected, the influence of litter quality (expressed in terms of initial stoichiometric C:N:P ratios) was tested by fitting linear and non-linear regression models on log-transformed elemental ratios. Model comparison was carried out using a parsimonious procedure (Burnham and Anderson, 2002) based on the second-order Akaike Information Criterion AIC_c (e.g. Mancinelli, 2010; Longo and Mancinelli, 2014 and literature cited therein). In addition, to assess how effectively predator impacts on prey are transferred to leaf-litter processing, absolute LLR_{PREY}/LLR_{DET} ratios were estimated for each observation, and related to leaf-litter stoichiometry. The ratio is >1 when the impact is buffered, equal to 1 when directly transferred, or <1 when the impact is amplified between the two trophic levels. All analyses were performed using R version 3.2.0 (R Development Core Team, 2015).

Table A1 Data Summary

	Average Effect Size CI (0.05; 0.95)	*n*	References
Aquatic DFWs			
Stream water DFW			
Effect on detritus	−0.10 CI (−0.20; −0.01)	124	In field streams: −0.149 (−0.251 to −0.049) 95% CI Benstead et al. (2009), Bobeldyk and Lamberti (2008, 2010), Bondar and Richardson (2009, 2013), Creed and Reed (2004), Greig and McIntosh (2006), Holomuzki and Stevenson (1992), Jabiol et al. (2014), Klose and Cooper (2013), Konishi et al. (2001), Lagrue et al. (2014), Landeiro et al. (2008), Majdi et al. (2014), Malmqvist (1993), Mantel and Dudgeon (2004), March et al. (2001), Marshall et al. (2012), Moody and Sabo (2013), Moore et al. (2012), Moulton et al. (2010), Oberndorfer et al. (1984), Parkyn et al. (1997), Reice (1991), Rosemond et al. (1998, 2001), Ruetz et al. (2002, 2006), Schofield et al. (2001, 2004), Taylor and Hendricks (1987), Usio and Townsend (2002, 2004), Usio et al. (2006), Wach and Chambers (2007), Williams (2002), Woodward et al. (2008) and Zhang et al. (2004) In mesocosms: −0.104 (−0.465 to 0.225) 95% CI Atwood et al. (2014a), Bassar et al. (2010), Herrmann et al. (2012), Lecerf and Richardson (2011) and Usio (2000) In microcosms: 0.229 (0.084–0.377) 95% CI Dunoyer et al. (2014), Lagrue et al. (2014) and Malmqvist (1993)
Effect on detritivores	−0.70 CI (−0.86; −0.54)	64	In field streams between −0.809 and −0.455 for 95% CI Bondar and Richardson (2009), Greig and McIntosh (2006), Holomuzki and Stevenson (1992), Jabiol et al. (2014), Klose and Cooper (2013), Konishi et al. (2001), Lagrue et al. (2014), Landeiro et al. (2008), Majdi et al. (2014), Malmqvist (1993), March et al. (2001), Oberndorfer et al. (1984), Parkyn et al. (1997), Reice (1991), Rosemond et al. (1998), Ruetz et al. (2002, 2006), Schofield et al. (2001), (2004), Usio and Townsend (2002, 2004), Usio et al. (2006), Wach and Chambers (2007), Woodward et al. (2008) and Zhang et al. (2004) In all cosms between −1.496 and −0.729 for 95% CI Atwood et al. (2014a), Dunoyer et al. (2014), Malmqvist (1993) and Usio (2000)

Continued

Table A1 Data Summary—cont'd

	Average Effect Size CI (0.05; 0.95)	*n*	References
Effect on assemblage	−0.48 CI (−0.58 to −0.37)	115	In field streams between −0.61 and −0.36 for 95% CI Benstead et al. (2009), Bobeldyk and Lamberti (2008, 2010), Bondar and Richardson (2009, 2013), Creed and Reed (2004), Greig and McIntosh (2006), Holomuzki and Stevenson (1992), Jabiol et al. (2014), Klose and Cooper (2013), Konishi et al. (2001), Lagrue et al. (2014), Landeiro et al. (2008), Majdi et al. (2014), Malmqvist (1993), Mantel and Dudgeon (2004), March et al. (2001), Marshall et al. (2012), Moody and Sabo (2013), Moore et al. (2012), Oberndorfer et al. (1984), Parkyn et al. (1997), Reice (1991), Rosemond et al. (1998, 2001), Ruetz et al. (2002, 2006), Schofield et al. (2001, 2004), Usio and Townsend (2002, 2004), Usio et al. (2006), Wach and Chambers (2007), Woodward et al. (2008) and Zhang et al. (2004) In mesocosms between −0.55 and −0.21 for 95% CI Atwood et al. (2014a), Bassar et al. (2010), Herrmann et al. (2012), Lecerf and Richardson (2011) and Usio (2000) In microcosms between −1.08 and −0.65 for 95% CI Dunoyer et al. (2014), Lagrue et al. (2014) and Malmqvist (1993)
Calm water DFW			
Effect on detritus	0.05 CI (−0.04 to 0.14)	27	Overall, lakes, mesocosms and container plants taken together Atwood et al. (2014b), Greig et al. (2012), Jackson et al. (2014), Mancinelli et al. (2002, 2007), Phillips (2009) and Srivastava and Bell (2009)
Effect on detritivores	−0.66 CI (−0.99 to −0.35)	13	Overall, lakes, mesocosms and container plants taken together Atwood et al. (2014b), Jackson et al. (2014) and Mancinelli et al. (2002, 2007)
Effect on assemblage	−0.64 CI (−0.87 to −0.44)	22	Overall, lakes, mesocosms and container plants taken together Atwood et al. (2014b), Greig et al. (2012), Jackson et al. (2014), Mancinelli et al. (2002, 2007) and Phillips (2009)

Terrestrial DFWs			
Forest DFW			
Effect on detritus	−0.04 CI (−0.15 to 0.05)	39	Overall, soils and cosms taken together Beard (2001), Beard et al. (2003), Elkins et al. (1982), Fukami et al. (2006), Hocking and Babbitt (2014), Homyack et al. (2010), Huang et al. (2007), Lawrence and Wise (2000, 2004), Lensing and Wise (2006), Liu et al. (2014), McGlynn and Poirson (2012), Miyashita and Niwa (2006), Santos et al. (1981), Sin et al. (2008), Stanley and Ward (2012), Walton and Steckler (2005) and Wyman (1998)
Effect on detritivores	−0.63 CI (−0.64 to −0.41)	23	Overall, soils and cosms taken together Beard (2001), Beard et al. (2003), Fukami et al. (2006), Homyack et al. (2010), Huang et al. (2007), Lensing and Wise (2006), Liu et al. (2014), Miyashita and Niwa (2006), Sin et al. (2008) and Walton and Steckler (2005)
Effect on assemblage	−0.62 CI (−0.91 to −0.38)	36	Overall, soils and cosms taken together Beard (2001), Beard et al. (2003), Elkins et al. (1982), Fukami et al. (2006), Hocking and Babbitt (2014), Homyack et al. (2010), Huang et al. (2007), Lawrence and Wise (2000, 2004), Lensing and Wise (2006), Liu et al. (2014), Miyashita and Niwa (2006), Santos et al. (1981), Sin et al. (2008), Walton and Steckler (2005) and Wyman (1998)
Grassland DFW			
Effect on detritus	0.17 CI (0.05–0.33)	25	Overall, soils and mesocosms taken together Ewers et al. (2012), Hines and Gessner (2012), Kajak and Jakubczyk (1976, 1977), Kajak et al. (1991), Wu et al. (2011, 2014a,b) and Zhao et al. (2014)
Effect on detritivores	−0.53 CI (0.64 to −0.41)	3	Insufficient data (only Hines and Gessner, 2012; Zhao et al., 2014)
Effect on assemblage	−0.45 CI (−0.59 to −0.32)	15	Overall, soils and mesocosms taken together Hines and Gessner (2012), Kajak and Jakubczyk (1976, 1977), Kajak et al. (1991), Wu et al. (2014a,b) and Zhao et al. (2014)

A.4 Publication and Experimental Biases

The normal quantile plots of all the response variables analyzed showed no evidence of bias, as they were gap-free and fell within the 95% CI, indicating no significant bias (Fig. A1). Additionally, funnel plots of effect sizes versus sample size did not indicate any publication bias as expected if non-significant results with low replication would unlikely have been reported. Overall, plots were characterized by no significant asymmetry (Spearman R–O correlation as in the caption of Fig. A1).

In general, none of the potentially biasing factors was effective in explaining the variation in LLR_{DET} and LLR_{PREY} values. Overall, LLR_{DET} values increased inversely with the experimental scale, from field manipulations to mesocosms and to microcosms (−0.078, 0.001 and 0.159, respectively); within field studies, caging manipulations were characterized by less negative LLR_{DET} values in comparison to non-manipulative studies (enclosure-exclosure = −0.062, exclosure-open cage = −0.065, non-manipulative = −0.166). Effects of predators on the total abundance of invertebrate prey in microcosms were low in comparison to either mesocosm or field experiments (−0.864 versus −0.366 and −0.511, respectively), while non-manipulative experiments showed more negative responses in comparison to caging manipulations (non-manipulative = −1.256; enclosure-exclosure = −0.402; exclosure-open cage = −0.369). For detritivorous invertebrates, the influence of experimental location and design on predator effect size generally reproduced the patterns observed for the whole invertebrate assemblage. Overall, the interaction factor 'duration × litterbag mass' and the factors 'duration' and 'litterbag mass' *per se* on the abundance of detritivore invertebrates in field studies were characterized by the least non-significant effects (Table A2). Only very weak positive bivariate relationships with LLR_{PREY} values occurred ('duration': $r = 0.07$, $p = 0.48$; 'litterbag mass': $r = 0.03$, $p = 0.77$), suggesting that both factors might possibly affect detritivorous consumers by regulating their aggregation patterns (Kampichler and Bruckner, 2009).

Table A2 Potential Biasing Factors Tested

	Variables	Rationale	Type	Levels/Units
1	Location	Microcosm and mesocosm experiments are crucial in ecological research (Benton et al., 2007; Srivastava et al., 2004; Stewart et al., 2013). However, their restricted conditions affect both species' biology and their interaction with the environment, as they are usually conducted at temporal and spatial time scales that do not allow for the incorporation of important ecological processes. Specifically, they can constrain the movement of predators and prey, amplifying the strength of trophic interactions. Thus, cascades are expected to be stronger in mesocosm and microcosm experiments as compared to field experiments.	Categorical	3: Microsm, mesocosm, field
2	Design	TEST PERFORMED ONLY WITHIN THE LEVEL '*FIELD*' OF VARIABLE 1. Enclosure-exclosure constrains the movement of predators, while in exclosure-open end non-manipulative constrains do not occur.	Categorical	3: Enclosure-exclosure, exclosure-open cage, non-manipulative
3	Cage size	The database included studies using caging devices of different sizes. *Ceteris paribus*, predator effects were assessed in manipulated habitats characterized by different degrees of spatial	Continuous	Area of the device measured in m^2

Continued

Table A2 Potential Biasing Factors Tested—cont'd

	Variables	Rationale	Type	Levels/Units
		heterogeneity, as the latter scales with cage size (e.g. Englund and Cooper, 2003; Lähteenmäki et al., 2016), and is expected to peak in non-manipulative investigations. Cascade strength was expected to decrease with larger caging devices, because the higher spatial heterogeneity and availability of refuge for prey tends to decrease search efficiency of predators, thereby reducing their indirect effects on basal resources (Borer et al., 2005; Polis et al., 2000).		
4	Litterbag mass	Cascade strength may be negatively related with litterbag mass, as the latter scales with the interstitial space available to colonizing macroinvertebrates (Hassage and Stewart, 1991) and thus with the degree of refuge offered against predation (Ruetz et al., 2002, 2006).	Continuous	Initial dry mass of litterbags (or dung-pats) in grams
5	Time	Leaf-litter decomposition is characterized by high physical inertia and delayed response to variations in biotic and abiotic factors; the possibility of distinguishing significant inter-treatment effects is related with the duration of the study and other features of the experimental set up (Kampichler and Brucker, 2009).	Continuous	Duration of the experiment measured in days

Table A3 Food-Web Topology Metrics of the DFWs Shown in Fig. A2

		Above-Ground Food Web	Below-Ground DFW	Dung-Pat DFW	Pitcher DFW	Pond DFW	Stream DFW	
SpeciesCount	Species richness	154	135	32	91	77	142	Number of taxa (nodes) in a food web.
Total#Links	Total trophic links	370	1662	81	1834	958	1358	Total number of feeding interactions (links or edges between taxa) in a food web.
LinksPerSpecies	Link density	2.4	12.3	2.5	20.1	12.4	9.6	Mean number of feeding interactions (links or edges between taxa) per species.
Connectance	Food web connectance	0.02	0.09	0.08	0.22	0.16	0.07	Proportion of possible trophic links that are realized (trophic links are directional, such that 'A feeds on B' is a separate link from 'B feeds on A').
FracTop	Top taxa	0.01	0.41	0.06	0.07	0.39	0.49	Fraction of taxa that lack consumers.
FracIntermed	Intermediate taxa	0.99	0.37	0.22	0.91	0.60	0.43	Fraction of taxa that have both consumers and resources. These taxa are the most common non-target organisms affected by intensive application of insecticides or nematicides.
FracHerbiv	Herbivores	0.12	0.08	0.19	0.36	0.34	0.01	Fraction of taxa that feed only on basal taxa (including taxa that feed on detritus). This fraction incorporates in aquatic DFWs the taxa most sensitive to algicides and in terrestrial DFWs the taxa most sensitive to herbicides.

Continued

Table A3 Food-Web Topology Metrics of the DFWs Shown in Fig. A2—cont'd

		Above-Ground Food Web	Below-Ground DFW	Dung-Pat DFW	Pitcher DFW	Pond DFW	Stream DFW	
DegreeOmniv	Omnivores	0.21	0.75	0.09	0.24	0.58	0.69	Amount of taxa that feed on resource taxa that occur on more than one trophic level.
Prey:Predator	Ratio between resources and consumers	1.00	0.76	3.33	0.96	0.62	0.56	The number of basal and intermediate species divided by the number of predatory and intermediate species.
Total#Positions	Maximal trophic path length	9	8	4	5	5	5	Maximal length of the chain of feeding links connecting each pair of taxa, a measure of how many steps energy must take to get from an energy source to a focal taxon.

Besides for the aboveground network of Memmott et al. (2000) that cannot be regarded as a detritus-driven food web (far too many arthropod species pass their entire adult life on that single shrub), the five other DFWs show a strong direct correlation between the number of species and the number of links ($p=0.009$).

Sources: From left to right, the aboveground food web (Fig. A2A) as described by Memmot et al. (2000), the belowground DFW (Fig. A2B) by Sechi et al. (2015), the dung-pat DFW (Fig. A2C) by Valiela (1969), the *Sarracenia* phytotelmata DFW (in Fig. A2D in reduced form) by Baiser et al. (2012), the pond DFW (Fig. A2E) by Schneider (1997) and the stream DFW (Fig. A2F) by Woodward et al. (2008: their supplementary material).

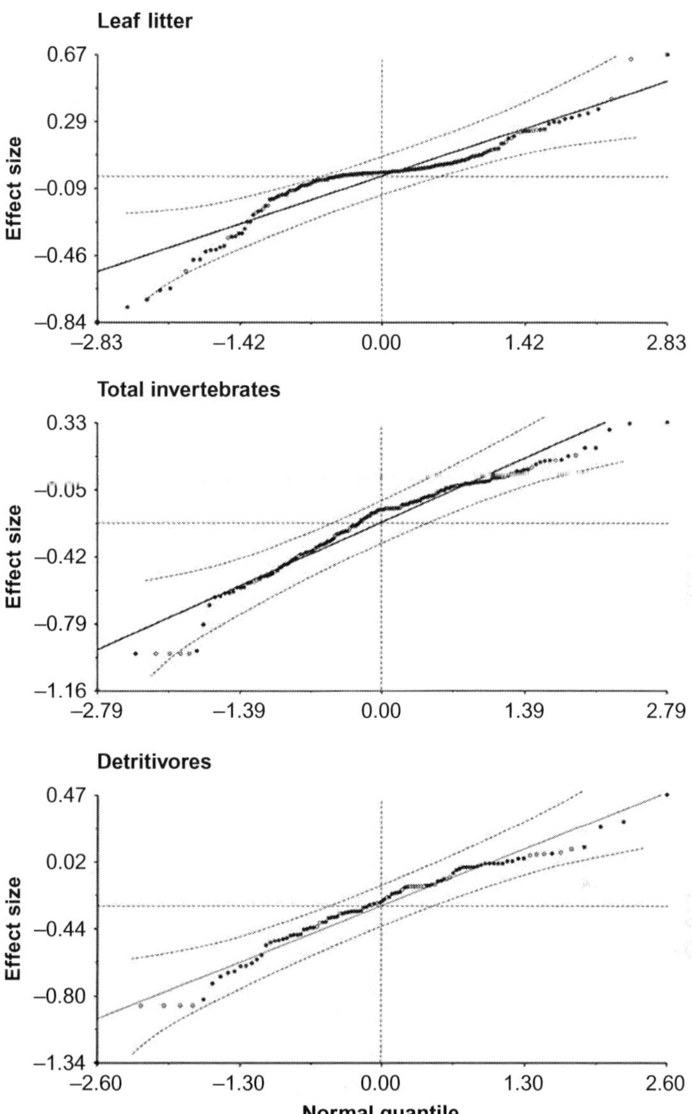

Figure A1 Normal quantile plots of all the response variables and their asymmetry tests; aquatic studies filled in black, terrestrial studies open circles. From top to bottom: leaf litter (Spearman R–O correlation: Rs = 0.114, p = 0.08, n = 216), total invertebrates (Rs = −0.118, p = 0.11, n = 188) and detritivores (Rs = 0.053, p = 0.58, n = 108). Noticeably, the pattern for detrital resources (upper panel) was characterized by a strongly hump-shaped pattern, suggesting that data belonged to different populations.

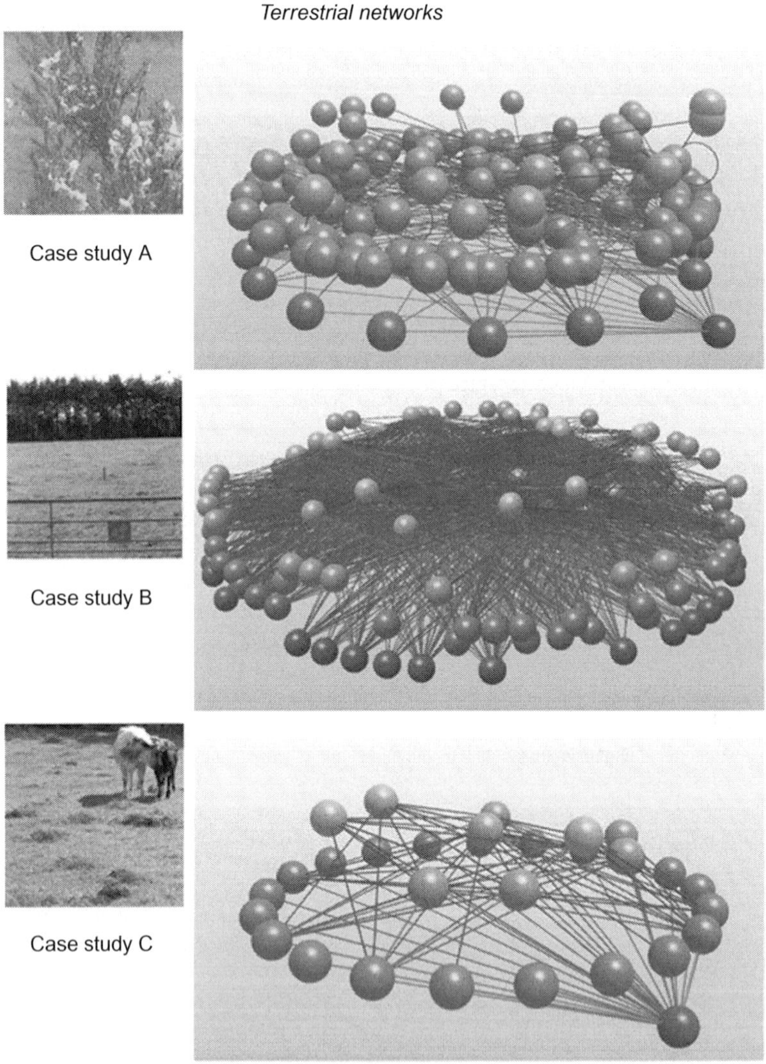

Terrestrial networks

Case study A

Case study B

Case study C

Figure A2 Visualization of case studies of DFWs described in Table A3; food webs are open and downloadable from web banks (Cohen, 2013) and repositories. From top to bottom and from this page to the next page: Topology of the multitrophic interactions in (A) one Scottish broom food web (Memmot et al., 2000; ECOWeB311), (B) a Dutch belowground DFW from the Dryad Digital Repository (http://dx.doi.org/10.5061/dryad.t5347; Sechi et al., 2015), (C) dung-pat DFW (Valiela, 1969; ECOWeB199), (D) phytotelmata DFW (Baiser et al., 2012; ECOWeB359), (E) temporary pond DFW (Schneider, 1997; GlobalFoodWeb263) and (F) Bere Stream DFW (Woodward et al., 2008). Given the relatively low taxonomic and functional resolution of the *Sarracenia* network reported in Baiser et al. (2012) the original network of 91 nodes was reduced to 52 by grouping undetermined bacterial taxa together. The change was made only for practical purposes, seen the well-known effects of taxonomic resolution on food-web metrics (Goldwasser and Roughgarden, 1997; Martinez 1991, 1993), at least in lentic and soil systems (Reuman et al., 2008 and Sechi et al., 2015, respectively).

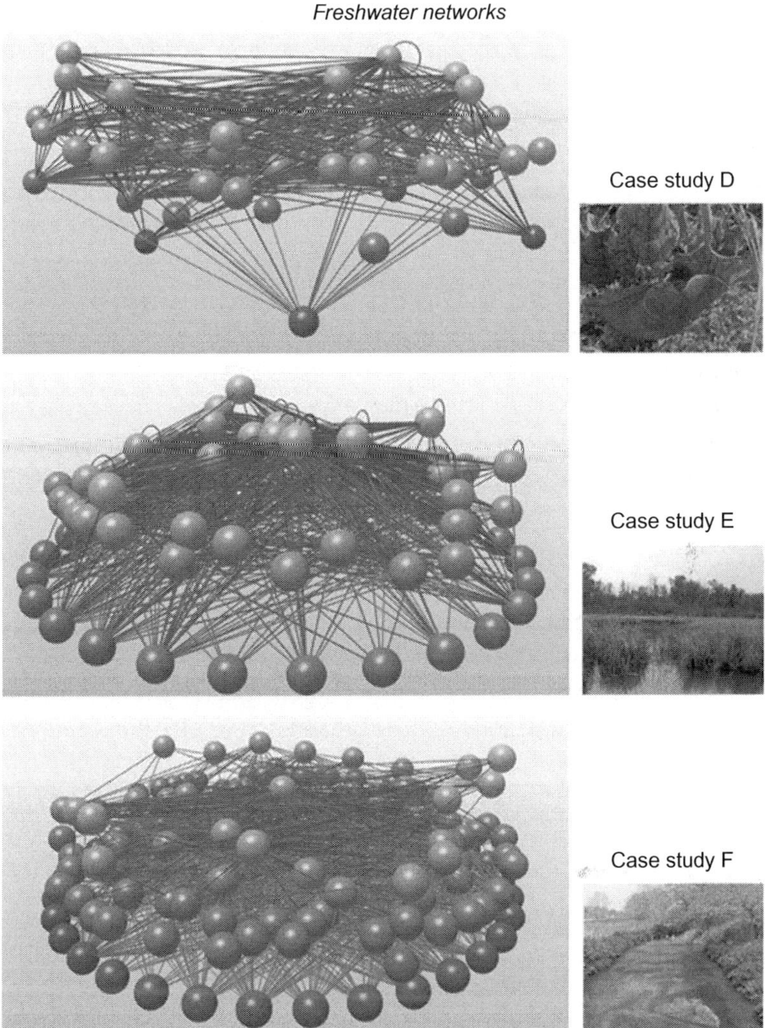

Figure A2—Cont'd The topology of each of these networks advocates that species occurrence (thus trophic links) reflects both the physiological response of a local population (here as node) as the extent to which local habitats meet the primary niche requirements of a population. Food-web analysis performed in NETWORK 3D: Red nodes (dark grey in the print version), basal resources (single, as the Scottish broom in (A) or bovine dung in (C), or multiple, as in (F) where freshwater autotrophs (algae) and heterotrophs (fungi) were highly resolved), orange nodes (grey in the print version), first-order consumers (detritivores and omnivores) and yellow nodes (light grey in the print version), predators *s.s.* (second-order consumers).

Metareferences

Atwood, T.B., Hammill, E., Richardson, J.S., 2014a. Trophic-level dependent effects on CO_2 emissions from experimental stream ecosystems. *Global Change Biol.* 20, 3386–3396.

Atwood, T.B., Hammill, E., Srivastava, D.S., Richardson, J.S., 2014b. Competitive displacement alters top-down effects on carbon dioxide concentrations in a freshwater ecosystem. *Oecologia* 175, 353–361.

Bassar, R.D., Marshall, M.C., López-Sepulcre, A., Zandonà, E., Auer, S.K., Travis, J., Pringle, C.M., Flecker, A.S., Thomas, S.A., Fraser, D.F., 2010. Local adaptation in Trinidadian guppies alters ecosystem processes. *Proc. Natl. Acad. Sci. USA* 107, 3616–3621.

Beard, K.H., 2001. The effect of a juvenile terrestrial frog, *Eleutherodactylus coqui* on the decomposer food web and leaf litter decomposition rate in the wet forests of Puerto Rico. *TRI News* 20, 18–21.

Beard, K.H., Eschtruth, A.K., Vogt, K.A., Vogt, D.J., Scatena, F.N., 2003. The effects of the frog *Eleutherodactylus coqui* on invertebrates and ecosystem processes at two scales in the Luquillo Experimental Forest, Puerto Rico. *J. Trop. Ecol.* 19, 607–617.

Benstead, J.P., March, J.G., Pringle, C.M., Ewel, K.C., Short, J.W., 2009. Biodiversity and ecosystem function in species-poor communities: Community structure and leaf litter breakdown in a Pacific island stream. *J. North Am. Benthol. Soc.* 28, 454–465.

Bobeldyk, A.M., Lamberti, G.A., 2008. A decade after invasion: Evaluating the continuing effects of rusty crayfish on a Michigan river. *J. Great Lakes Res.* 34, 265–275.

Bobeldyk, A.M., Lamberti, G.A., 2010. Stream food web responses to a large omnivorous invader, *Orconectes rusticus* (Decapoda, Cambaridae). *Crustaceana* 83, 641–657.

Bondar, C.A., Richardson, J.S., 2009. Effects of ontogenetic stage and density on the ecological role of the signal crayfish (*Pacifastacus leniusculus*) in a coastal Pacific stream. *J. North Am. Benthol. Soc.* 28, 294–304.

Bondar, C.A., Richardson, J.S., 2013. Stage-specific interactions between dominant consumers within a small stream ecosystem: Direct and indirect consequences. *Freshw. Sci.* 32, 183–192.

Creed, R.P., Reed, J.M., 2004. Ecosystem engineering by crayfish in a headwater stream community. *J. North Am. Benthol. Soc.* 23, 224–236.

Dunoyer, L., Dijoux, L., Bollache, L., Lagrue, C., 2014. Effects of crayfish on leaf litter breakdown and shredder prey: Are native and introduced species functionally redundant? *Biol. Invas.* 16, 1545–1555.

Elkins, N.Z., Whitford, W.G., 1982. The role of microarthropods and nematodes in decomposition in a semi-arid ecosystem. *Oecologia* 55, 303–310.

Ewers, C., Beiersdorf, A., Więski, K., Pennings, S.C., Zimmer, M., 2012. Predator/prey-interactions promote decomposition of low-quality detritus. *Wetlands* 32, 931–938.

Fukami, T., Wardle, D.A., Bellingham, P.J., Mulder, C.P.H., Towns, D.R., Yeates, G.W., Bonner, K.I., Durrett, M.S., Grant-Hoffman, M.N., Williamson, W.M., 2006. Above- and below-ground impacts of introduced predators in seabird-dominated island ecosystems. *Ecol. Lett.* 9, 1299–1307.

Greig, H.S., McIntosh, A.R., 2006. Indirect effects of predatory trout on organic matter processing in detritus-based stream food webs. *Oikos* 112, 31–40.

Greig, H.S., Kratina, P., Thompson, P.L., Palen, W.J., Richardson, J.S., Shurin, J.B., 2012. Warming, eutrophication, and predator loss amplify subsidies between aquatic and terrestrial ecosystems. *Global Change Biol.* 18, 504–514.

Herrmann, P.B., Townsend, C.R., Matthaei, C.D., 2012. Individual and combined effects of fish predation and bed disturbance on stream benthic communities: A streamside channel experiment. *Freshw. Biol.* 57, 2487–2503.

Hines, J., Gessner, M.O., 2012. Consumer trophic diversity as a fundamental mechanism linking predation and ecosystem functioning. *J. Anim. Ecol.* 81, 1146–1153.

Hocking, D.J., Babbitt, K.J., 2014. Effects of red-backed salamanders on ecosystem functions. *PLoS ONE* 9, e86854.

Holomuzki, J.R., Stevenson, R.J., 1992. Role of predatory fish in community dynamics of an ephemeral stream. *Can. J. Fish. Aqua. Sci.* 49, 2322–2330.

Homyack, J.A., Sucre, E.B., Haas, C.A., and Fox, T.R., 2010. Does *Plethodon cinereus* affect leaf litter decomposition and invertebrate abundances in mixed oak forest? *J. Herpetol.* 44, 447–456.

Huang, C.-Y., Wang, C.-P., Hou, P.-C.L., 2007. Toads (*Bufo bankorensis*) influence litter chemistry but not litter invertebrates and litter decomposition rates in a subtropical forest of Taiwan. *J. Trop. Ecol.* **23**, 161–168.

Jabiol, J., Cornut, J., Danger, M., Jouffroy, M., Elger, A., Chauvet, E., 2014. Litter identity mediates predator impacts on the functioning of an aquatic detritus-based food web. *Oecologia* **176**, 225–235.

Jackson, M.C., Jones, T., Milligan, M., Sheath, D., Taylor, J., Ellis, A., England, J., Grey, J., 2014. Niche differentiation among invasive crayfish and their impacts on ecosystem structure and functioning. *Freshw. Biol.* **59**, 1123–1135.

Kajak, A., Jakubczyk, H., 1976. The effect of intensive fertilization on the structure and productivity of meadow ecosystems. 18. Trophic relationships of epigeic predators. *Pol. Ecol. Studies* **2**, 219–229.

Kajak, A., Jakubczyk, H., 1977. Experimental studies of predation in the soil-litter interface. *Ecol. Bull.* **25**, 493–496.

Kajak, A., Chmielewski, K., Kaczmarek, M., Rembialkowska, E., 1991. Experimental studies on the effect of epigeic predators on matter decomposition processes in managed peat grasslands. *Pol. Ecol. Studies* **17**, 289–310.

Klose, K., Cooper, S.D., 2013. Complex impacts of an invasive omnivore and native consumers on stream communities in California and Hawaii. *Oecologia* **171**, 945–960.

Konishi, M., Nakano, S., Iwata, T., 2001. Trophic cascading effects of predatory fish on leaf litter processing in a Japanese stream. *Ecol. Res.* **16**, 415–422.

Lagrue, C., Podgorniak, T., Lecerf, A., Bollache, L., 2014. An invasive species may be better than none: Invasive signal and native noble crayfish have similar community effects. *Freshw. Biol.* **59**, 1982–1995.

Landeiro, V.L., Hamada, N., Melo, A.S., 2008. Responses of aquatic invertebrate assemblages and leaf breakdown to macroconsumer exclusion in Amazonian "terra firme" streams. *Fund. Appl. Limnol.* **172**, 49–58.

Lawrence, K.L., Wise, D.H., 2000. Spider predation on forest-floor collembola and evidence for indirect effects on decomposition. *Pedobiologia* **44**, 33–39.

Lawrence, K.L., Wise, D.H., 2004. Unexpected indirect effect of spiders on the rate of litter disappearance in a deciduous forest. *Pedobiologia* **48**, 149–157.

Lecerf, A., Richardson, J.S., 2011. Assessing the functional importance of large-bodied invertebrates in experimental headwater streams. *Oikos* **120**, 950–960.

Lensing, J.R., Wise, D.H., 2006. Predicted climate change alters the indirect effect of predators on an ecosystem process. *Proc. Natl. Acad. Sci. USA* **103**, 15502–15505.

Liu, S., Chen, J., He, X., Hu, J., Yang, X., 2014. Trophic cascade of a web-building spider decreases litter decomposition in a tropical forest floor. *Eur. J. Soil Biol.* **65**, 79–86.

Majdi, N., Boiché, A., Traunspurger, W., Lecerf, A., 2014. Predator effects on a detritus-based food web are primarily mediated by non-trophic interactions. *J. Anim. Ecol.* **83**, 953–962.
Malmqvist, B., 1993. Interactions in stream leaf packs - Effects of a stonefly predator on detritivores and organic matter processing. *Oikos* **66**, 454–462.
Mancinelli, G., Costantini, M.L., Rossi, L., 2002. Cascading effects of predatory fish exclusion on the detritus-based food web of a lake littoral zone (Lake Vico, central Italy). *Oecologia* **133**, 402–411.
Mancinelli, G., Costantini, M.L., Rossi, L., 2007. Top-down control of reed detritus processing in a lake littoral zone: Experimental evidence of a seasonal compensation between fish and invertebrate predation. *Internatl. Rev. Hydrobiol.* **92**, 117–134.
Mantel, S.K., Dudgeon, D., 2004. Effects of *Macrobrachium hainanense* predation on benthic community functioning in tropical Asian streams. *Freshw. Biol.* **49**, 1306–1319.
March, J.G., Benstead, J.P., Pringle, C.M., Ruebel, M.W., 2001. Linking shrimp assemblages with rates of detrital processing along an elevational gradient in a tropical stream. *Can. J. Fish. Aqua. Sci.* **58**, 470–478.
Marshall, M.C., Binderup, A.J., Zandonà, E., Goutte, S., Bassar, R.D., El-Sabaawi, R.W., Thomas, S.A., Flecker, A.S., Kilham, S.S., Reznick, D.N., Pringle, C.M., 2012. Effects of consumer interactions on benthic resources and ecosystem processes in a neotropical stream. *PLoS ONE* **7**, e45230.
McGlynn, T.P., Poirson, E.K., 2012. Ants accelerate litter decomposition in a Costa Rican lowland tropical rain forest. *J. Trop. Ecol.* **28**, 437–443.
Miyashita, T., Niwa, S., 2006. A test for top-down cascade in a detritus-based food web by litter-dwelling web spiders. *Ecol. Res.* **21**, 611–615.
Moody, E.K., Sabo, J.L., 2013. Crayfish impact desert river ecosystem function and litter-dwelling invertebrate communities through association with novel detrital resources. *PLoS ONE* **8**, e63274.
Moore, J.W., Carlson, S.M., Twardochleb, L.A., Hwan, J.L., Fox, J.M., Hayes, S.A., 2012. Trophic tangles through time? Opposing direct and indirect effects of an invasive omnivore on stream ecosystem processes. *PLoS ONE* **7**, e50687.
Moulton, T.P., Magalhaes-Fraga, S.A.P., Fuentes Brito, E., Barbosa, F.A., 2010. Macroconsumers are more important than specialist macroinvertebrate shredders in leaf processing in urban forest streams of Rio de Janeiro, Brazil. *Hydrobiologia* **638**, 55–66.

Oberndorfer, R.Y., McArthur, J.V., Barnes, J.R., Dixon, J., 1984. The effect of invertebrate predators on leaf litter processing in an alpine stream. *Ecology* **65**, 1325–1331.

Parkyn, S.M., Rabeni, C.F., Collier, K.J., 1997. Effects of crayfish (*Paranephrops planifrons*: Parastacidae) on in-stream processes and benthic faunas: A density manipulation experiment. *New Zealand J. Mar. Freshw. Res.* **31**, 685–692.

Phillips, I.D., Vinebrooke, R.D., Turner, M.A., 2009. Experimental reintroduction of the crayfish species *Orconectes virilis* into formerly acidified Lake 302S (Experimental Lakes Area, Canada). *Can. J. Fish. Aqua. Sci.* **66**, 1892–1902.

Reice, S.R., 1991. Effects of detritus loading and fish predation on leaf pack breakdown and benthic macroinvertebrates in a woodland stream. *J. North Am. Benthol. Soc.* **10**, 42–56.

Rosemond, A.D., Pringle, C.M., Ramirez, A., 1998. Macroconsumer effects on insect detritivores and detritus processing in a tropical stream. *Freshw. Biol.* **39**, 515–523.

Rosemond, A.D., Pringle, C.M., Ramirez, A., Paul, M.J., 2001. A test of top-down and bottom-up control in a detritus-based food web. *Ecology* **82**, 2279–2293.

Ruetz, C.R., Newman, R.M., Vondracek, B., 2002. Top-down control in a detritus-based food web: Fish, shredders, and leaf breakdown. *Oecologia* **132**, 307–315.

Ruetz, C.R., Breen, M.J., Vanhaitsma, D.L., 2006. Habitat structure and fish predation: Effects on invertebrate colonisation and breakdown of stream leaf packs. *Freshw. Biol.* **51**, 797–806.

Santos, P.F., Phillips, J., Whitford, W.G., 1981. The role of mites and nematodes in early stages of buried litter decomposition in a desert. *Ecology* **62**, 664–669.

Schofield, K.A., Pringle, C.M., Meyer, J.L., Sutherland, A.B., 2001. The importance of crayfish in the breakdown of rhododendron leaf litter. *Freshw. Biol.* **46**, 1191–1204.

Schofield, K.A., Pringle, C.M., Meyer, J.L., 2004. Effects of increased bedload on algal- and detrital-based stream food webs: Experimental manipulation of sediment and macroconsumers. *Limnol. Oceanogr.* **49**, 900–909.

Sin, H., Beard, K.H., Pitt, W.C., 2008. An invasive frog, *Eleutherodactylus coqui*, increases new leaf production and leaf litter decomposition rates through nutrient cycling in Hawaii. *Biol. Invas.* **10**, 335–345.

Srivastava, D.S., Bell, T., 2009. Reducing horizontal and vertical diversity in a foodweb triggers extinctions and impacts functions. *Ecol. Lett.* **12**, 1016–1028.

Stanley, M.C., Ward, D.F., 2012. Impacts of Argentine ants on invertebrate communities with below-ground consequences. *Biodiv. Cons.* **21**, 2653–2669.

Taylor, F., Hendricks, A.C., 1987. The influence of fish on leaf breakdown in a Virginia pond. *Freshw. Biol.* **18**, 45–51.

Usio, N., 2000. Effects of crayfish on leaf processing and invertebrate colonisation of leaves in a headwater stream: Decoupling of a trophic cascade. *Oecologia* **124**, 608–614.

Usio, N., Townsend, C.R., 2002. Functional significance of crayfish in stream food webs: roles of omnivory, substrate heterogeneity and sex. *Oikos* **98**, 512–522.

Usio, N., Townsend, C.R., 2004. Roles of crayfish: Consequences of predation and bioturbation for stream invertebrates. *Ecology* **85**, 807–822.

Usio, N., Suzuki, K., Konishi, M., Nakano, S., 2006. Alien vs. endemic crayfish: Roles of species identity in ecosystem functioning. *Archiv für Hydrobiologie* **166**, 1–21.

Wach, E., Chambers, R.M., 2007. Top-down effect of fish predation in Virginia headwater streams. *Northeastern Naturalist* **14**, 461–470.

Walton, B.M., Steckler, S., 2005. Contrasting effects of salamanders on forest-floor macro- and mesofauna in laboratory microcosms. *Pedobiologia* **49**, 51–60.

Williams, J.L., 2002. Effects of the tropical freshwater shrimp *Caridina weberi* (Atyidae) on leaf litter decomposition. *Biotropica* **34**, 616–619.

Woodward, G., Papantoniou, G., Edwards, F., Lauridsen, R.B., 2008. Trophic trickles and cascades in a complex food web: Impacts of a keystone predator on stream community structure and ecosystem processes. *Oikos* **117**, 683–692.

Wu, X., Duffy, J., Reich, P., Sun, S., 2011. A brown-world cascade in the dung decomposer food web of an alpine meadow: Effects of predator interactions and warming. *Ecol. Monogr.* **81**, 313–328.

Wu, X., Griffin, J.N., Sun, S., 2014a. Cascading effects of predator–detritivore interactions depend on environmental context in a Tibetan alpine meadow. *J. Anim. Ecol.* **83**, 546–556.

Wu, X., Zhang, C., Griffin, J.N., Sun, S., 2014b. The brown-world role of insectivores: Frogs reduce plant growth by suppressing detritivores in an alpine meadow. *Basic Appl. Ecol.* **15**, 66–74.

Wyman, R.L., 1998. Experimental assessment of salamanders as predators of detrital food webs: effects on invertebrates, decomposition and the carbon cycle. *Biodiv. Cons.* **7**, 641–650.

Zhang, Y., Richardson, J.S., Negishi, J.N., 2004. Detritus processing, ecosystem engineering and benthic diversity: A test of predator-omnivore interference. *J. Anim. Ecol.* **73**, 756–766.

Zhao, C., Wu, X., Griffin, J.N., Xi, X., Sun, S., 2014. Territorial ants depress plant growth through cascading non-trophic effects in an alpine meadow. *Oikos* **123**, 481–487.

REFERENCES

Allan, J.D., Castillo, M.M., 2007. Stream Ecology: Structure and Function of Running Waters. Springer, Dordrecht, The Netherlands.

Allan, J.D., Abell, R., Hogan, Z.E.B., Revenga, C., Taylor, B.W., Welcomme, R.L., Winemiller, K., 2005. Overfishing of inland waters. BioScience 55, 1041–1051.

Allison, S.D., 2005. Cheaters, diffusion and nutrients constrain decomposition by microbial enzymes in spatially structured environments. Ecol. Lett. 8, 626–635.

Baiser, B., Gotelli, N.J., Buckley, H.L., Miller, T.E., Ellison, A.M., 2012. Geographic variation in network structure of a nearctic aquatic food web. Glob. Ecol. Biogeogr. 21, 579–591.

Ball, B.A., Hunter, M.D., Kominoski, J.S., Swan, C.M., Bradford, M.A., 2008. Consequences of non-random species loss for decomposition dynamics: experimental evidence for additive and non-additive effects. J. Ecol. 96, 303–313.

Bärlocher, F., 1980. Leaf-eating invertebrates as competitors of aquatic hyphomycetes. Oecologia 47, 303–306.

Bärlocher, F., 1985. The role of fungi in the nutrition of stream invertebrates. Bot. J. Linn. Soc. 91, 83–94.

Bärlocher, F., 2005. Leaf mass loss estimated by litter bag technique. In: Graça, M.A.S., Bärlocher, F., Gessner, M.O. (Eds.), Methods to Study Litter Decomposition. Springer, Netherlands, pp. 37–42.

Bärlocher, F., Corkum, M., 2003. Nutrient enrichment overwhelms diversity effects in leaf decomposition by stream fungi. Oikos 101, 247–252.

Benstead, J.P., March, J.G., Pringle, C.M., Ewel, K.C., Short, J.W., 2009. Biodiversity and ecosystem function in species-poor communities: community structure and leaf litter breakdown in a Pacific island stream. J. N. Am. Benthol. Soc. 28, 454–465.

Benton, T.G., Solan, M., Travis, J.M.J., Sait, S.M., 2007. Microcosm experiments can inform global ecological problems. Trends Ecol. Evol. 22, 516–521.

Berlow, E.L., Navarrete, S.A., Briggs, C.J., Power, M.E., Menge, B.A., 1999. Quantifying variation in the strengths of species interactions. Ecology 80, 2206–2224.

Blake, T.W., Downing, J.A., 2009. Measuring atmospheric nutrient deposition to inland waters: evaluation of direct methods. Limnol. Oceanogr. Methods 7, 638–647.

Bohan, D.A., Raybould, A., Mulder, C., Woodward, G., Tamaddoni-Nezhad, A., Bluthgen, N., Pocock, M., Muggleton, S., Evans, D.M., Astegiano, J., Massol, F., Loeuille, N., Petit, S., Macfadyen, S., 2013. Networking agroecology: integrating the diversity of agroecosystem interactions. Adv. Ecol. Res. 49, 1–67.

Bondar, C.A., Richardson, J.S., 2009. Effects of ontogenetic stage and density on the ecological role of the signal crayfish (*Pacifastacus leniusculus*) in a coastal Pacific stream. J. N. Am. Benthol. Soc. 28, 294–304.

Bonkowski, M., 2004. Protozoa and plant growth: the microbial loop in soil revisited. New Phytol. 162, 617–631.
Borenstein, M., Hedges, L.V., Higgins, J.P.T., Rothstein, H.R., 2009. Introduction to Meta-Analysis. John Wiley & Sons, Chirchester.
Borer, E.T., Seabloom, E.W., Shurin, J.B., Anderson, K.E., Blanchette, C.A., Broitman, B., Cooper, S.D., Halpern, B.S., 2005. What determines the strength of a trophic cascade? Ecology 86, 528–537.
Borer, E.T., Halpern, B.S., Seabloom, E.W., 2006. Asymmetry in community regulation: effects of predators and productivity. Ecology 87, 2813–2820.
Boyero, L., Pearson, R.G., Gessner, M.O., Barmuta, L.A., Ferreira, V., Graça, M.A.S., Dudgeon, D., Boulton, A.J., Callisto, M., Chauvet, E., Helson, J.E., Bruder, A., et al., 2011. A global experiment suggests climate warming will not accelerate litter decomposition in streams but might reduce carbon sequestration. Ecol. Lett. 14, 289–294.
Bradford, M.A., Wood, S.A., Bardgett, R.D., Black, H.I.J., Bonkowski, M., Eggers, T., Grayston, S.J., Kandeler, E., Manning, P., Setälä, H., Jones, T.H., 2014. Discontinuity in the responses of ecosystem processes and multifunctionality to altered soil community composition. Proc. Natl. Acad. Sci. U.S.A. 111, 14478–14483.
Branch, T.A., Watson, R., Fulton, E.A., Jennings, S., McGilliard, C.R., Pablico, G.T., Ricard, D., Tracey, S.R., 2010. The trophic fingerprint of marine fisheries. Nature 468, 431–435.
Brennan, A., Woodward, G., Seehausen, O., Muñoz-Fuentes, V., Moritz, C., Guelmami, A., Abbott, R.J., Edelaar, P., 2014. Hybridization due to changing species distributions: adding problems or solutions to conservation of biodiversity during global change? Evol. Ecol. Res. 16, 475–491.
Burnham, K.P., Anderson, D.R., 2002. Model Selection and Multimodel Inference: A Practical Information—Theoretic Approach. Springer, Berlin.
Carpenter, S.R., Kitchell, J.F., 1988. Consumer control of lake productivity. BioScience 38, 764–769.
Carpenter, S.R., Kitchell, J.F., 1993. The Trophic Cascade in Lakes. Cambridge University Press, New York, NY.
Carpenter, S.R., Kitchell, J.F., Hodgson, J.R., 1985. Cascading trophic interactions and lake productivity. BioScience 35, 634–639.
Carpenter, S.R., Stanley, E.H., Vander Zanden, M.J., 2011. State of the world's freshwater ecosystems: physical, chemical, and biological changes. Annu. Rev. Environ. Resour. 36, 75–99.
Casini, M., Hjelm, J., Molinero, J.-C., Lövgren, J., Cardinale, M., Bartolino, V., Belgrano, A., Kornilovs, G., 2009. Trophic cascades promote threshold-like shifts in pelagic marine ecosystems. Proc. Natl. Acad. Sci. U.S.A. 106, 197–202.
Cebrian, J., 1999. Patterns in the fate of production in plant communities. Am. Nat. 154, 449–468.
Cebrian, J., Lartigue, J., 2004. Patterns of herbivory and decomposition in aquatic and terrestrial ecosystems. Ecol. Monogr. 74, 237–259.
Cheshire, K.I.M., Boyero, L.U.Z., Pearson, R.G., 2005. Food webs in tropical Australian streams: shredders are not scarce. Freshw. Biol. 50, 748–769.
Chung, N., Suberkropp, K., 2009. Contribution of fungal biomass to the growth of the shredder, *Pycnopsyche gentilis* (Trichoptera: Limnephilidae). Freshw. Biol. 54, 2212–2224.
Cohen, J.E., 1995. Population growth and Earth's human carrying capacity. Science 269, 341–346.
Cohen, J.E., 2013. Ecologists' Co-Operative Web Bank. Version 1.1—Machine-Readable Data Base of Food Webs. The Rockefeller University, New York.

Creed, R.P., Cherry, R.P., Pflaum, J.R., Wood, C.J., 2009. Dominant species can produce a negative relationship between species diversity and ecosystem function. Oikos 118, 723–732.
Cross, W.F., Benstead, J.P., Rosemond, A.D., Wallace, B.J., 2003. Consumer-resource stoichiometry in detritus-based streams. Ecol. Lett. 6, 721–732.
Cross, W.F., Benstead, J.P., Frost, P.C., Thomas, S.A., 2005. Ecological stoichiometry in freshwater benthic systems: recent progress and perspectives. Freshw. Biol. 50, 1895–1912.
Cummins, K.W., Klug, M.J., 1979. Feeding ecology of stream invertebrates. Annu. Rev. Ecol. Syst. 10, 147–172.
Curtis, P.S., Wang, X.Z., 1998. A meta-analysis of elevated CO_2 effects on woody plant mass, form and physiology. Oecologia 113, 299–313.
Dang, C.K., Chauvet, E., Gessner, M.O., 2005. Magnitude and variability of process rates in fungal diversity-litter decomposition relationships. Ecol. Lett. 8, 1129–1137.
Dang, C.K., Schindler, M., Chauvet, E., Gessner, M.O., 2009. Temperature oscillation coupled with fungal community shifts can modulate warming effects on litter decomposition. Ecology 90, 122–131.
Dangles, O., Malmqvist, B., 2004. Species richness-decomposition relationships depend on species dominance. Ecol. Lett. 7, 395–402.
DeAngelis, D., 1992. Dynamics of Nutrient Cycling and Food Webs. Chapman & Hall, London.
Dighton, J., 2003. Fungi in ecosystem processes. Mycology 17, 1–432.
Duarte, S., Pascoal, C., Cássio, F., Bärlocher, F., 2006. Aquatic hyphomycete diversity and identity affect leaf litter decomposition in microcosms. Oecologia 147, 658–666.
Dudgeon, D., 2010. Prospects for sustaining freshwater biodiversity in the 21st century: linking ecosystem structure and function. Curr. Opin. Environ. Sustain. 2, 422–430.
Elser, J.J., Chrzanowski, T.H., Sterner, R.W., Mills, K.H., 1998. Stoichiometric constraints on food-web dynamics: a whole-lake experiment on the Canadian Shield. Ecosystems 1, 120–136.
Elser, J.J., Fagan, W.F., Denno, R.F., Dobberfuhl, D.R., Folarin, A., Huberty, A., Interlandi, S., Kilham, S.S., McCauley, E., Schulz, K.L., Siemann, E.H., Sterner, R.W., 2000. Nutritional constraints in terrestrial and freshwater food webs. Nature 408, 578–580.
Elser, J.J., Loladze, I., Peace, A.L., Kuang, Y., 2012. Lotka re-loaded: modeling trophic interactions under stoichiometric constraints. Ecol. Model. 245, 3–11.
Englund, G., Cooper, S.D., 2003. Scale effects and extrapolation in ecological experiments. Adv. Ecol. Res. 33, 161–213.
Englund, G., Sarnelle, O., Cooper, S.D., 1999. The importance of data-selection criteria: meta-analyses of stream predation experiments. Ecology 80, 1132–1141.
Enquist, B.J., Norberg, J., Bonsor, S.P., Violle, C., Webb, C.T., Henderson, A., Sloat, L.L., Savage, V.M., 2015. Scaling from traits to ecosystems: developing a general Trait Driver Theory via integrating trait-based and metabolic scaling theories. Adv. Ecol. Res. 52, 249–318.
Estes, J.A., Terborgh, J., Brashares, J.S., Power, M.E., Berger, J., Bond, W.J., Carpenter, S.R., Essington, T.E., Holt, R.D., Jackson, J.B.C., Marquis, R.J., Oksanen, L., et al., 2011. Trophic downgrading of planet Earth. Science 333, 301–306.
Faithfull, C.L., Huss, M., Vrede, T., Bergström, A.K., 2011. Bottom–up carbon subsidies and top–down predation pressure interact to affect aquatic food web structure. Oikos 120, 311–320.
Ferreira, V., Gonçalves, A.N.A.L., Godbold, D.L., Canhoto, C., 2010. Effect of increased atmospheric CO_2 on the performance of an aquatic detritivore through changes in water temperature and litter quality. Glob. Cha. Biol. 16, 3284–3296.

Finn, J.A., 2001. Ephemeral resource patches as model systems for diversity-function experiments. Oikos 92, 363–366.

Folke, C., Carpenter, S., Walker, B., Scheffer, M., Elmqvist, T., Gunderson, L., Holling, C.S., 2004. Regime shifts, resilience, and biodiversity in ecosystem management. Annu. Rev. Ecol. Evol. Syst. 35, 557–581.

Fontaine, S., Barot, S., 2005. Size and functional diversity of microbe populations control plant persistence and long-term soil carbon accumulation. Ecol. Lett. 8, 1075–1087.

Fontaine, S., Bardoux, G., Abbadie, L., Mariotti, A., 2004. Carbon input to soil may decrease soil carbon content. Ecol. Lett. 7, 314–320.

Frank, K.T., Petrie, B., Shackell, N.L., 2007. The ups and downs of trophic control in continental shelf ecosystems. Trends Ecol. Evol. 22, 236–242.

Frost, P.C., Benstead, J.P., Cross, W.F., Hillebrand, H., Larson, J.H., Xenopoulos, M.A., Yoshida, T., 2006. Threshold elemental ratios of carbon and phosphorus in aquatic consumers. Ecol. Lett. 9, 774–779.

Fukami, T., Wardle, D.A., Bellingham, P.J., Mulder, C.P.H., Towns, D.R., Yeates, G.W., Bonner, K.I., Durrett, M.S., Grant-Hoffman, M.N., Williamson, W.M., 2006. Above- and below-ground impacts of introduced predators in seabird-dominated island ecosystems. Ecol. Lett. 9, 1299–1307.

Funck, J.A., Clivot, H., Felten, V., Rousselle, P., Guérold, F., Danger, M., 2013. Phosphorus availability modulates the toxic effect of silver on aquatic fungi and leaf litter decomposition. Aquat. Toxicol. 144–145, 199–207.

Gartner, T.B., Cardon, Z.G., 2004. Decomposition dynamics in mixed-species leaf litter. Oikos 104, 230–246.

Geisen, S., Rosengarten, J., Koller, R., Mulder, C., Urich, T., Bonkowski, M., 2016. Pack hunting by a common soil amoeba on nematodes. Environ. Microbiol. http://dx.doi.org/10.1111/1462-2920.12949. in press.

Gessner, M.O., Chauvet, E., Dobson, M., 1999. A perspective on leaf litter breakdown in streams. Oikos 85, 377–384.

Gessner, M.O., Swan, C.M., Dang, C.K., McKie, B.G., Bardgett, R.D., Wall, D.H., Hättenschwiler, S., 2010. Diversity meets decomposition. Trends Ecol. Evol. 25, 372–380.

Glazier, D.S., 2012. Temperature affects food-chain length and macroinvertebrate species richness in spring ecosystems. Freshw. Sci. 31, 575–585.

Goldwasser, L., Roughgarden, J., 1997. Sampling effects and the estimation of food web properties. Ecology 78, 41–54.

Graça, M.A.S., 2001. The role of invertebrates on leaf litter decomposition in streams—a review. Int. Rev. Hydrobiol. 86, 383–393.

Graça, M.A.S., Cressa, C., 2010. Leaf quality of some tropical and temperate tree species as food resource for stream shredders. Int. Rev. Hydrobiol. 95, 27–41.

Gruner, D.S., Smith, J.E., Seabloom, E.W., Sandin, S.A., Ngai, J.T., Hillebrand, H., Harpole, W.S., Elser, J.J., Cleland, E.E., Bracken, M.E.S., Borer, E.T., Bolker, B.M., 2008. A cross-system synthesis of consumer and nutrient resource control on producer biomass. Ecol. Lett. 11, 740–755.

Gurevitch, J., Hedges, L.V., 1999. Statistical issues in ecological meta-analyses. Ecology 80, 1142–1149.

Güsewell, S., Gessner, M.O., 2009. N:P ratios influence litter decomposition and colonization by fungi and bacteria in microcosms. Funct. Ecol. 23, 211–219.

Hagen, E.M., McCluney, K.E., Wyant, K.A., Soykan, C.U., Keller, A.C., Luttermoser, K.C., Holmes, E.J., Moore, J.C., Sabo, J.L., 2012. A meta-analysis of the effects of detritus on primary producers and consumers in marine, freshwater, and terrestrial ecosystems. Oikos 121, 1507–1515.

Hairston, N.G., Smith, F.E., Slobodkin, L.B., 1960. Community structure, population control, and competition. Am. Nat. 94, 421–425.

Hassage, R.L., Stewart, K.W., 1991. Use of substrate volume and void space to examine the presence of three stonefly species (Plecoptera) among stream habitats. Ann. Entomol. Soc. Am. 84, 309–315.

Hättenschwiler, S., Tiunov, A.V., Scheu, S., 2005. Biodiversity and litter decomposition in terrestrial ecosystems. Annu. Rev. Ecol. Syst. 36, 191–218.

Hawlena, D., Strickland, M.S., Bradford, M.A., Schmitz, O.J., 2012. Fear of predation slows plant-litter decomposition. Science 336, 1434–1438.

Hedges, L.V., Gurevitch, J., Curtis, P.S., 1999. The meta-analysis of response ratios in experimental ecology. Ecology 80, 1150–1156.

Hillebrand, H., Gruner, D.S., Bracken, M.E., Cleland, E.E., Elser, J.J., Harpole, W.S., Ngai, J.T., Seabloom, E.W., Shurin, J.B., Smith, J.E., 2007. Consumer versus resource control of producer diversity depends on ecosystem type and producer community structure. Proc. Natl. Acad. Sci. U.S.A. 104, 10904–10909.

Hines, J., Gessner, M.O., 2012. Consumer trophic diversity as a fundamental mechanism linking predation and ecosystem functioning. J. Anim. Ecol. 81, 1146–1153.

Hines, J., van der Putten, W.H., De Deyn, G.B., Wagg, C., Voigt, W., Mulder, C., Weisser, W.W., Engel, J., Melian, C., Scheu, S., Birkhofer, K., Ebeling, A., et al., 2015. Towards an integration of biodiversity-ecosystem functioning and food web theory to evaluate relationships between multiple ecosystem services. Adv. Ecol. Res. 53, 161–199.

Hladyz, S., Gessner, M.O., Giller, P.S., Pozo, J., Woodward, G., 2009. Resource quality and stoichiometric constraints on stream ecosystem functioning. Freshw. Biol. 54, 957–970.

Hodge, A., Robinson, D., Fitter, A., 2000. Are microorganisms more effective than plants at competing for nitrogen? Trends Plant Sci. 5, 304–308.

Högberg, M.N., Högberg, P., Myrold, D.D., 2007. Is microbial community composition in boreal forest soils determined by pH, C-to-N ratio, the trees, or all three? Oecologia 150, 590–601.

Holt, R.D., 2008. Theoretical perspectives on resource pulses. Ecology 89, 671–681.

Humphries, P., Winemiller, K.O., 2009. Historical impacts on river fauna, shifting baselines, and challenges for restoration. BioScience 59, 673–684.

Hunter, M.D., Price, P.W., 1992. Playing chutes and ladders: heterogeneity and the relative roles of bottom-up and top-down forces in natural communities. Ecology 73, 724–732.

Irons, J.G., Oswood, M.W., Stout, R.J., Pringle, C.M., 1994. Latitudinal patterns in leaf-litter breakdown—is temperature really important. Freshw. Biol. 32, 401–411.

Jennings, S., Smith, A.D.M., Fulton, E.A., Smith, D.C., 2014. The ecosystem approach to fisheries: management at the dynamic interface between biodiversity conservation and sustainable use. Ann. N. Y. Acad. Sci. 1322, 48–60.

Jepsen, D.B., Winemiller, K.O., 2002. Structure of tropical river food webs revealed by stable isotope ratios. Oikos 96, 46–55.

Jones, J.I., Waldron, S., 2003. Combined stable isotope and gut contents analysis of food webs in plant-dominated, shallow lakes. Freshw. Biol. 48, 1396–1407.

Jones, L., Provins, A., Holland, M., Mills, G., Hayes, F., Emmett, B., Hall, J., Sheppard, L., Smith, R., Sutton, M., Hicks, K., Ashmore, M., Haines-Young, R., Harper-Simmonds, L., 2014. A review and application of the evidence for nitrogen impacts on ecosystem services. Ecosyst. Serv. 7, 76–88.

Kampichler, C., Bruckner, A., 2009. The role of microarthropods in terrestrial decomposition: a meta-analysis of 40 years of litterbag studies. Biol. Rev. 84, 375–389.

Kaspari, M., Yanoviak, S.P., 2008. Biogeography of litter depth in tropical forests: evaluating the phosphorus growth rate hypothesis. Funct. Ecol. 22, 919–923.

Kattge, J., Díaz, S., Lavorel, S., Prentice, I., Leadley, P., Bonisch, G., Garnier, E., Westoby, M., Reich, P., Wright, I., Cornelissen, J., Violle, C., et al., 2011. TRY—a global database of plant traits. Glob. Cha. Biol. 17, 2905–2935.

Killham, K., 1994. Soil Ecology. Cambridge University Press, Cambridge, UK.

Kitching, R.L., 2001. Food webs in phytotelmata: "bottom-up" and "top-down" explanations for community structure. Annu. Rev. Entomol. 46, 729–760.

Kominoski, J.S., Pringle, C.M., 2009. Resource-consumer diversity: testing the effects of leaf litter species diversity on stream macroinvertebrate communities. Freshw. Biol. 54, 1461–1473.

Kominoski, J.S., Pringle, C.M., Ball, B.A., Bradford, M.A., Coleman, D.C., Hall, D.B., Hunter, M.D., 2007. Nonadditive effects of leaf litter species diversity on breakdown dynamics in a detritus-based stream. Ecology 88, 1167–1176.

Kominoski, J.S., Hoellein, T.J., Kelly, J.J., Pringle, C.M., 2009. Does mixing litter of different qualities alter stream microbial diversity and functioning on individual litter species? Oikos 118, 457–463.

Kominoski, J.S., Hoellein, T.J., Leroy, C.J., Pringle, C.M., Swan, C.M., 2010. Beyond species richness: expanding biodiversity–ecosystem functioning theory in detritus-based streams. River Res. Appl. 26, 67–75.

Krumins, J.A., Long, Z.T., Steiner, C.F., Morin, P.J., 2006. Indirect effects of food web diversity and productivity on bacterial community function and composition. Funct. Ecol. 20, 514–521.

Krumins, J.A., Van Oevelen, D., Bezemer, T.M., De Deyn, G.B., Hol, W.H.G., Van Donk, E., De Boer, W., De Ruiter, P.C., Middelburg, J.J., Monroy, F., Soetaert, K., Thébault, E., et al., 2013. Soil and freshwater and marine sediment food webs: their structure and function. BioScience 63, 35–42.

Kuijper, L.D.J., Kooi, B.W., Anderson, T.R., Kooijman, S.A.L.M., 2004. Stoichiometry and food-chain dynamics. Theor. Popul. Biol. 66, 323–339.

Laakso, J., Setälä, H., Palojärvi, A., 2000. Influence of decomposer food web structure and nitrogen availability on plant growth. Plant Soil 225, 153–165.

Lähteenmäki, S., Slade, E.M., Hardwick, B., Schiffler, G., Louzada, J., Barlow, J., Roslin, T., 2016. MESOCLOSURES—increasing realism in mesocosm studies of ecosystem functioning. Methods Ecol. Evol. 6, 916–924.

Lensing, J.R., Wise, D.H., 2006. Predicted climate change alters the indirect effect of predators on an ecosystem process. Proc. Natl. Acad. Sci. U.S.A. 103, 15502–15505.

Leroux, S.J., Loreau, M., 2008. Subsidy hypothesis and strength of trophic cascades across ecosystems. Ecol. Lett. 11, 1147–1156.

Leroux, S.J., Loreau, M., 2012. Dynamics of reciprocal pulsed subsidies in local and meta-ecosystems. Ecosystems 15, 48–59.

Leroux, S.J., Loreau, M., 2015. Theoretical perspectives on bottom–up and top–down interactions across ecosystems. In: Hanley, T.C., La Pierre, K.J. (Eds.), Ecology: Bottom-Up and Top-Down Interactions across Aquatic and Terrestrial Systems. Cambridge University Press, Cambridge, UK, pp. 3–27.

Leroy, C.J., Marks, J.C., 2006. Litter quality, stream characteristics and litter diversity influence decomposition rates and macroinvertebrates. Freshw. Biol. 51, 605–617.

Longo, E., Mancinelli, G., 2014. Size at the onset of maturity (SOM) revealed in length–weight relationships of brackish amphipods and isopods: an information theory approach. Estuar. Coast. Shelf Sci. 136, 119–128.

Lugo, A.E., Snedaker, S.C., 1974. The ecology of mangroves. Annu. Rev. Ecol. Syst. 5, 39–64.

Luo, Y.Q., Hui, D.F., Zhang, D.Q., 2006. Elevated CO_2 stimulates net accumulations of carbon and nitrogen in land ecosystems: a meta-analysis. Ecology 87, 53–63.

MacNeil, C., Dick, J.T.A., Elwood, R.E., 1997. The trophic ecology of freshwater *Gammarus* spp. (Crustacea: Amphipoda): problems and perspectives concerning the functional feeding group concept. Biol. Rev. 72, 349–364.

MacNeil, C., Dick, J.T.A., Elwood, R.W., 1999. The dynamics of predation on *Gammarus* spp. (Crustacea: Amphipoda). Biol. Rev. 74, 375–395.

Majdi, N., Boiché, A., Traunspurger, W., Lecerf, A., 2014. Predator effects on a detritus-based food web are primarily mediated by non-trophic interactions. J. Anim. Ecol. 83, 953–962.

Mancinelli, G., 2009. On the importance of body size in the colonisation of ephemeral resource patches by vagile consumers. Rendiconti Lincei 20, 139–151.

Mancinelli, G., 2010. Intraspecific, size-dependent variation in the movement behaviour of a brackish-water isopod: a resource-free laboratory experiment. Mar. Freshw. Behav. Physiol. 43, 321–337.

Mancinelli, G., 2012. On the trophic ecology of Gammaridea (Crustacea: Amphipoda) in coastal waters: a European-scale analysis of stable isotopes data. Estuar. Coast. Shelf Sci. 114, 130–139.

Mancinelli, G., Costantini, M.L., Rossi, L., 2002. Cascading effects of predatory fish exclusion on the detritus-based food web of a lake littoral zone (Lake Vico, central Italy). Oecologia 133, 402–411.

Mancinelli, G., Vignes, F., Sangiorgio, F., Mastrolia, A., Basset, A., 2009. On the potential contribution of microfungi to the decomposition of reed leaf detritus in a coastal lagoon: a laboratory and field experiment. Int. Rev. Hydrobiol. 94, 419–435.

Mancinelli, G., Sangiorgio, F., Scalzo, A., 2013. The effects of decapod crustacean macroconsumers on leaf detritus processing and colonization by invertebrates in stream habitats: a meta-analysis. Int. Rev. Hydrobiol. 98, 206–216.

Martinez, N.D., 1991. Artifacts or attributes? Effects of resolution on the Little Rock Lake food web. Ecol. Monogr. 61, 367–392.

Martinez, N.D., 1993. Effects of resolution on food web structure. Oikos 66, 403–412.

Martiny, J.B.H., Bohannan, B.J.M., Brown, J.H., Colwell, R.K., Fuhrman, J.A., Green, J.L., Horner-Devine, M.C., Kane, M., Krumins, J.A., Kuske, C.R., Morin, P.J., Naeem, S., et al., 2006. Microbial biogeography: putting microorganisms on the map. Nature Rev. Microbiol. 4, 102–112.

McKie, B.G., Woodward, G., Hladyz, S., Nistorescu, M., Preda, E., Popescu, C., Giller, P.S., Malmqvist, B., 2008. Ecosystem functioning in stream assemblages from different regions: contrasting responses to variation in detritivore richness, evenness and density. J. Anim. Ecol. 77, 495–504.

McQueen, D.J., Johannes, M.R.S., Post, J.R., Stewart, T.J., Lean, D.R.S., 1989. Bottom-up and top-down impacts on freshwater pelagic community structure. Ecol. Monogr. 59, 289–309.

Memmot, J., Martinez, N.D., Cohen, J.E., 2000. Predators, parasitoids and pathogens: species richness, trophic generality and body sizes in a natural food web. J. Anim. Ecol. 69, 1–15.

Menge, B.A., 2000. Top-down and bottom-up community regulation in marine rocky intertidal habitats. J. Exp. Mar. Biol. Ecol. 250, 257–289.

Mikola, J., Setälä, H., 1998. No evidence of trophic cascades in an experimental microbial-based soil food web. Ecology 79, 153–164.

Mille-Lindblom, C., Tranvik, L.J., 2003. Antagonism between bacteria and fungi on decomposing aquatic plant litter. Microb. Ecol. 45, 173–182.

Miyashita, T., Niwa, S., 2006. A test for top-down cascade in a detritus-based food web by litter-dwelling web spiders. Ecol. Res. 21, 611–615.

Moles, A.T., Bonser, S.P., Poore, A.G.B., Wallis, I.R., Foley, W.J., 2011. Assessing the evidence for latitudinal gradients in plant defence and herbivory. Funct. Ecol. 25, 380–388.

Møller, A.P., Jennions, M.D., 2001. Testing and adjusting for publication bias. Trends Ecol. Evol. 16, 580–586.
Moore, J.C., Berlow, E.L., Coleman, D.C., De Ruiter, P.C., Dong, Q., Hastings, A., Johnson, N.C., McCann, K.S., Melville, K., Morin, P.J., Nadelhoffer, K., Rosemond, A.D., et al., 2004. Detritus, trophic dynamics and biodiversity. Ecol. Lett. 7, 584–600.
Moore, J.W., Carlson, S.M., Twardochleb, L.A., Hwan, J.L., Fox, J.M., Hayes, S.A., 2012. Trophic tangles through time? Opposing direct and indirect effects of an invasive omnivore on stream ecosystem processes. PLoS One 7, e50687.
Mooshammer, M., Wanek, W., Schnecker, J., Wild, B., Leitner, S., Hofhansl, F., Blöchl, A., Hämmerle, L., Frank, A.H., Fuchslueger, L., 2012. Stoichiometric controls of nitrogen and phosphorus cycling in decomposing beech leaf litter. Ecology 93, 770–782.
Morlon, H., Kefi, S., Martinez, N.D., 2014. Effects of trophic similarity on community composition. Ecol. Lett. 17, 1495–1506.
Mulder, C., 2006. Driving forces from soil invertebrates to ecosystem functioning: the allometric perspective. Naturwissenschaften 93, 467–479.
Mulder, C., Elser, J.J., 2009. Soil acidity, ecological stoichiometry and allometric scaling in grassland food webs. Glob. Chang. Biol. 15, 2730–2738.
Mulder, C., Den Hollander, H.A., Vonk, J.A., Rossberg, A.G., Jagers op Akkerhuis, G.A.J.M., Yeates, G.W., 2009. Soil resource supply influences faunal size-specific distributions in natural food webs. Naturwissenschaften 96, 813–826.
Mulder, C., Boit, A., Bonkowski, M., De Ruiter, P.C., Mancinelli, G., Van der Heijden, M.G.A., Van Wijnen, H.J., Vonk, J.A., Rutgers, M., 2011. A belowground perspective on Dutch agroecosystems: how soil organisms interact to support ecosystem services. Adv. Ecol. Res. 44, 277–357.
Mulder, C., Ahrestani, F.S., Bahn, M., Bohan, D.A., Bonkowski, M., Griffiths, B.S., Guicharnaud, R.A., Kattge, J., Krogh, P.H., Lavorel, S., Lewis, O.T., Mancinelli, G., et al., 2013. Connecting the green and brown worlds: allometric and stoichiometric predictability of above- and belowground networks. Adv. Ecol. Res. 49, 69–175.
Mulder, C., Bennett, E.M., Bohan, D.A., Bonkowski, M., Carpenter, S.R., Chalmers, R., Cramer, W., Durance, I., Eisenhauer, N., Fontaine, C., Haughton, A.J., Hettelingh, J.-P., et al., 2015a. 10 Years later: revisiting priorities for science and society a decade after the Millennium Ecosystem Assessment. Adv. Ecol. Res. 53, 1–53.
Mulder, C., Hettelingh, J.-P., Montanarella, L., Pasimeni, M.R., Posch, M., Voigt, W., Zurlini, G., 2015b. Chemical footprints of anthropogenic nitrogen deposition on recent soil C:N ratios in Europe. Biogeosciences 12, 4113–4119.
Naeem, S., 2013. Biotic impoverishment. Elementa 1, 000015.
Neufeld, J.D., Wagner, M., Murrell, J.C., 2007. Who eats what, where and when? Isotope-labelling experiments are coming of age. ISME J. 1, 103–110.
Oberndorfer, R.Y., McArthur, J.V., Barnes, J.R., Dixon, J., 1984. The effect of invertebrate predators on leaf litter processing in an alpine stream. Ecology 65, 1325–1331.
Oksanen, L., Fretwell, S.D., Arruda, J., Niemelä, P., 1981. Exploitation ecosystems in gradients of primary productivity. Am. Nat. 118, 240–261.
Olson, J.S., 1963. Energy storage and the balance of producers and decomposers in ecological systems. Ecology 44, 322–331.
Osenberg, C.W., Sarnelle, O., Cooper, S.D., 1997. Effect size in ecological experiments: the application of biological models in meta-analysis. Am. Nat. 150, 798–812.
Osenberg, C.W., Sarnelle, O., Cooper, S.D., Holt, R.D., 1999. Resolving ecological questions through meta-analysis: goals, metrics, and models. Ecology 80, 1105–1117.
Ostrofsky, M.L., 1997. Relationship between chemical characteristics of autumn-shed leaves and aquatic processing rates. J. N. Am. Benthol. Soc. 16, 750–759.

Ott, D., Digel, C., Rall, B.C., Maraun, M., Scheu, S., Brose, U., 2014. Unifying elemental stoichiometry and metabolic theory in predicting species abundances. Ecol. Lett. 17, 1247–1256.

Pace, M.L., Cole, J.J., Carpenter, S.R., Kitchell, J.F., 1999. Trophic cascades revealed in diverse ecosystems. Trends Ecol. Evol. 14, 483–488.

Paine, R.T., 1980. Food webs: linkage, interaction strength and community infrastructure. J. Anim. Ecol. 49, 667–685.

Parkyn, S.M., Rabeni, C.F., Collier, K.J., 1997. Effects of crayfish (*Paranephrops planifrons*: Parastacidae) on in-stream processes and benthic faunas: a density manipulation experiment. N. Z. J. Mar. Freshw. Res. 31, 685–692.

Pauly, D., Christensen, V., Dalsgaard, J., Froese, R., Torres, F., 1998. Fishing down marine food webs. Science 279, 860–863.

Persson, L., 1999. Trophic cascades: abiding heterogeneity and the trophic level concept at the end of the road. Oikos 85, 385–397.

Pikitch, E.K., Rountos, K.J., Essington, T.E., Santora, C., Pauly, D., Watson, R., Sumaila, U., Boersma, P.D., Boyd, I.L., Conover, D.O., Cury, P., Heppell, S.S., et al., 2014. The global contribution of forage fish to marine fisheries and ecosystems. Fish Fish. 15, 43–64.

Pimm, S.L., 1982. Food Webs. Chapman and Hall, London.

Polis, G.A., 1999. Why are parts of the world green? Multiple factors control productivity and the distribution of biomass. Oikos 86, 3–15.

Polis, G.A., Strong, D.R., 1996. Food web complexity and community dynamics. Am. Nat. 147, 813–846.

Polis, G.A., Sears, A.L.W., Huxel, G.R., Strong, D.R., Maron, J., 2000. When is a trophic cascade a trophic cascade? Trends Ecol. Evol. 15, 473–475.

Poore, A.G.B., Campbell, A.H., Coleman, R.A., Edgar, G.J., Jormalainen, V., Reynolds, P.L., Sotka, E.E., Stachowicz, J.J., Taylor, R.B., Vanderklift, M.A., Duffy, E.J., 2012. Global patterns in the impact of marine herbivores on benthic primary producers. Ecol. Lett. 15, 912–922.

Power, M.E., 1992. Top-down and bottom-up forces in food webs: do plants have primacy? Ecology 73, 733–746.

Prather, C.M., Pelini, S., Laws, A., Rivest, E., Woltz, M., Bloch, C.P., Del Toro, I., Ho, C.-K., Kominoski, J., Newbold, T.A.S., Parsons, S., Joern, A., 2012. Invertebrates, ecosystem services and climate change. Biol. Rev. 88, 327–348.

Premke, K., Karlsson, J., Steger, K., Gudasz, C., Von Wachenfeldt, E., Tranvik, L.J., 2010. Stable isotope analysis of benthic fauna and their food sources in boreal lakes. J. N. Am. Benthol. Soc. 29, 1339–1348.

R Development Core Team, 2015. R: A Language and Environment for Statistical Computing. R Foundation for Statistical Computing, Vienna. Available at: http://www.r-project.org/.

Raffaelli, D., Moller, H., 1999. Manipulative field experiments in animal ecology: do they promise more than they can deliver? Adv. Ecol. Res. 30, 299–338.

Rasmussen, J.J., Wiberg-Larsen, P., Baattrup-Pedersen, A., Friberg, N., Kronvang, B., 2012. Stream habitat structure influences macroinvertebrate response to pesticides. Environ. Pollut. 164, 142–149.

Redfield, A.C., 1958. The biological control of chemical factors in the environment. Am. Sci. 46, 205–221.

Reice, S.R., 1991. Effects of detritus loading and fish predation on leaf pack breakdown and benthic macroinvertebrates in a woodland stream. J. N. Am. Benthol. Soc. 10, 42–56.

Reuman, D.C., Mulder, C., Raffaelli, D., Cohen, J.E., 2008. Three allometric relations of population density to body mass: theoretical integration and empirical tests in 149 food webs. Ecol. Lett. 11, 1216–1228.

Richardson, J.S., Hoover, T.M., Lecerf, A., 2009. Coarse particulate organic matter dynamics in small streams: towards linking function to physical structure. Freshw. Biol. 54, 2116–2126.

Ripple, I.J., Estes, J.A., Beschta, R.L., Wilmers, C.C., Ritchie, E.G., Hebblewhite, M., Berger, J., Elmhagen, B., Letnic, M., Nelson, M.P., Schmitz, O.J., Smith, D.W., et al., 2014. Status and ecological effects of the world's largest carnivores. Science 343, 1241484.

Ritchie, E.G., Elmhagen, B., Glen, A.S., Letnic, M., Ludwig, G., McDonald, R.A., 2012. Ecosystem restoration with teeth: what role for predators? Trends Ecol. Evol. 27, 265–271.

Romanuk, T.N., Beisner, B.E., Martinez, N.D., Kolasa, J., 2006. Non-omnivorous generality promotes population stability. Biol. Lett. 2, 374–377.

Rong, Q., Sridharl, K.R., Bärlocher, F., 1995. Food selection in three leaf-shredding stream invertebrates. Hydrobiologia 316, 173–181.

Rosemond, A.D., Pringle, C.M., Ramirez, A., Paul, M.J., 2001. A test of top-down and bottom-up control in a detritus-based food web. Ecology 82, 2279–2293.

Ruetz, C.R., Newman, R.M., Vondracek, B., 2002. Top-down control in a detritus-based food web: fish, shredders, and leaf breakdown. Oecologia 132, 307–315.

Ruetz, C.R., Breen, M.J., Vanhaitsma, D.L., 2006. Habitat structure and fish predation: effects on invertebrate colonisation and breakdown of stream leaf packs. Freshw. Biol. 51, 797–806.

Sanpera-Calbet, I., Lecerf, A., Chauvet, E., 2009. Leaf diversity influences in-stream litter decomposition through effects on shredders. Freshw. Biol. 54, 1671–1682.

Santos, P.F., Phillips, J., Whitford, W.G., 1981. The role of mites and nematodes in early stages of buried litter decomposition in a desert. Ecology 62, 664–669.

Sardans, J., Peñuelas, J., 2015. Trees increase their P:N ratio with size. Glob. Ecol. Biogeogr. 24, 147–156.

Sayer, E.J., Wright, J.S., Tanner, E.V.J., Yavitt, J.B., Harms, K.E., Powers, J.S., Kaspari, M., Garcia, M.N., Turner, B.L., 2012. Variable responses of lowland tropical forest nutrient status to fertilization and litter manipulation. Ecosystems 15, 387–400.

Schemske, D.W., Mittelbach, G.G., Cornell, H.W., Sobel, J.M., Roy, K., 2009. Is there a latitudinal gradient in the importance of biotic interactions? Annu. Rev. Ecol. Syst. 40, 245–269.

Schindler, M.H., Gessner, M.O., 2009. Functional leaf traits and biodiversity effects on litter decomposition in a stream. Ecology 90, 1641–1649.

Schmitz, O.J., Hawlena, D., Trussell, G.C., 2010. Predator control of ecosystem nutrient dynamics. Ecol. Lett. 13, 1199–1209.

Schneider, D.W., 1997. Predation and food web structure along a habitat duration gradient. Oecologia 110, 567–575.

Schofield, K.A., Pringle, C.M., Meyer, J.L., Sutherland, A.B., 2001. The importance of crayfish in the breakdown of rhododendron leaf litter. Freshw. Biol. 46, 1191–1204.

Sechi, V., Brussaard, L., De Goede, R.G.M., Rutgers, M., Mulder, C., 2015. Choice of resolution by functional trait or taxonomy affects allometric scaling in soil food webs. Am. Nat. 185, 142–149.

Setälä, H., 2002. Sensitivity of ecosystem functioning to changes in trophic structure, functional group composition and species diversity in belowground food webs. Ecol. Res. 17, 207–215.

Shurin, J.B., Borer, E.T., Seabloom, E.W., Anderson, K., Blanchette, C.A., Broitman, B., Cooper, S.D., Halpern, B.S., 2002. A cross-ecosystem comparison of the strength of trophic cascades. Ecol. Lett. 5, 785–791.

Shurin, J.B., Gruner, D.S., Hillebrand, H., 2006. All wet or dried up? Real differences between aquatic and terrestrial food webs. Proc. R. Soc. Lond. B 273, 1–9.

Sih, A., Wooster, D.E., 1994. Prey behavior, prey dispersal, and predator impacts on stream prey. Ecology 75, 1199–1207.
Solé, R.V., Ferrer-Cancho, R., Montoya, J.M., Valverde, S., 2003. Selection, tinkering, and emergence in complex networks: crossing the land of tinkering. Complexity 8, 20–33.
Sommer, U., 2008. Trophic cascades in marine and freshwater plankton. Int. Rev. Hydrobiol. 93, 506–516.
Sommer, U., Adrian, R., De Senerpont Domis, L., Elser, J.J., Gaedke, U., Ibelings, B., Jeppesen, E., Lürling, M., Molinero, J.C., Mooij, W.M., Van Donk, E., Winder, M., 2012. Beyond the Plankton Ecology Group (PEG) Model: mechanisms driving plankton succession. Annu. Rev. Ecol. Syst. 43, 429–448.
Srivastava, D.S., Kolasa, J., Bengtsson, J., Gonzalez, A., Lawler, S.P., Miller, T.E., Munguia, P., Romanuk, T., Schneider, D.C., Trzcinski, M.K., 2004. Are natural microcosms useful model systems for ecology? Trends Ecol. Evol. 19, 379–384.
Srivastava, D.S., Cardinale, B.J., Downing, A.L., Duffy, J.E., Jouseau, C., Sankaran, M., Wright, J.P., 2009. Diversity has stronger top-down than bottom-up effects on decomposition. Ecology 90, 1073–1083.
Stanley, M.C., Ward, D.F., 2012. Impacts of Argentine ants on invertebrate communities with below-ground consequences. Biodivers. Conserv. 21, 2653–2669.
Sterner, R.W., Elser, J.J., 2002. Ecological Stoichiometry—The Biology of Elements from Molecules to the Biosphere. Princeton University Press, Princeton and Oxford.
Stewart, R.I.A., Dossena, M., Bohan, D.A., Jeppesen, E., Kordas, R.L., Ledger, M.E., Meerhoff, M., Moss, B., Mulder, C., Shurin, J.B., Suttle, B., Thompson, R., et al., 2013. Mesocosm experiments as a tool for ecological climate-change research. Adv. Ecol. Res. 48, 69–179.
Strong, D.R., 1992. Are trophic cascades all wet? Differentiation and donor-control in speciose ecosystems. Ecology 73, 747–754.
Strong, D.R., Frank, K.T., 2010. Human involvement in food webs. Annu. Rev. Environ. Resour. 35, 1–23.
Suberkropp, K., Gulis, V., Rosemond, A.D., Benstead, J., 2010. Ecosystem and physiological scales of microbial responses to nutrients in a detritus-based stream: results of a 5-year continuous enrichment. Limnol. Oceanogr. 55, 149–160.
Sugihara, G., 1982. Niche Hierarchy: Structure, Organization and Assembly in Natural Communities. PhD. Dissertation, Princeton University, Princeton, NJ.
Swan, C.M., Palmer, M.A., 2004. Leaf diversity alters litter breakdown in a Piedmont stream. J. N. Am. Benthol. Soc. 23, 15–28.
Swan, C.M., Palmer, M.A., 2006. Preferential feeding by an aquatic consumer mediates non-additive decomposition of speciose leaf litter. Oecologia 149, 107–114.
Swift, M.J., Heal, O.W., Anderson, J.M., 1979. Decomposition in Terrestrial Ecosystems. University of California Press, Berkeley and Los Angeles, CA.
Tank, J.L., Rosi-Marshall, E.J., Griffiths, N.A., Entrekin, S.A., Stephen, M.L., 2010. A review of allochthonous organic matter dynamics and metabolism in streams. J. N. Am. Benthol. Soc. 29, 118–146.
Taylor, A.R., Pflug, A., Schröter, D., Wolters, V., 2010. Impact of microarthropod biomass on the composition of the soil fauna community and ecosystem processes. Eur. J. Soil Biol. 46, 80–86.
Terborgh, J., Estes, J.A., 2010. Trophic Cascades: Predators, Prey, and the Changing Dynamics of Nature. Island Press, Washington, DC.
Thompson, M.S.A., Bankier, C., Bell, T., Dumbrell, A.J., Gray, C., Ledger, M.E., Lehmann, K., McKew, B.A., Sayer, C.D., Shelley, F., Trimmer, M., Warren, S.L., et al., 2016. Gene-to-ecosystem impacts of a catastrophic pesticide spill: testing a multilevel bioassessment approach in a river ecosystem. Freshw. Biol. in press.

Tiegs, S.D., Peter, F.D., Robinson, C.T., Uehlinger, U., Gessner, M.O., 2008. Leaf decomposition and invertebrate colonization responses to manipulated litter quantity in streams. J. N. Am. Benthol. Soc. 27, 321–331.

Usio, N., 2000. Effects of crayfish on leaf processing and invertebrate colonisation of leaves in a headwater stream: decoupling of a trophic cascade. Oecologia 124, 608–614.

Usio, N., Townsend, C.R., 2002. Functional significance of crayfish in stream food webs: roles of omnivory, substrate heterogeneity and sex. Oikos 98, 512–522.

Usio, N., Townsend, C.R., 2004. Roles of crayfish: consequences of predation and bioturbation for stream invertebrates. Ecology 85, 807–822.

Valiela, I., 1969. An experimental study of the mortality factors of larval *Musca autumnalis* DeGeer. Ecol. Monogr. 39, 199–225.

Vannote, R.L., Minshall, G.W., Cummins, K.W., Sedell, J.R., Cushing, C.E., 1980. The river continuum concept. Can. J. Fish. Aquat. Sci. 37, 130–137.

Vaughn, C.C., 2010. Biodiversity losses and ecosystem function in freshwaters: emerging conclusions and research directions. BioScience 60, 25–35.

Vet, R., Artz, R.S., Carou, S., Shaw, M., Ro, C.-U., Aas, W., Baker, A., Bowersox, V.C., Dentener, F., Galy-Lacaux, C., Hou, A., Pienaar, J.J., et al., 2014. A global assessment of precipitation chemistry and deposition of sulfur, nitrogen, sea salt, base cations, organic acids, acidity and pH, and phosphorus. Atmos. Environ. 93, 3–100.

Vitousek, P., 1984. Litterfall, nutrient cycling and nutrient limitations in tropical forests. Ecology 65, 285–298.

Vitousek, P.M., Fahey, T., Johnson, D.W., Swift, M.J., 1988. Element interactions in forest ecosystems—succession, allometry and inputoutput budgets. Biogeochemistry 5, 7–34.

Wallace, J.B., Eggert, S.L., Meyer, J.L., Webster, J.R., 1999. Effects of resource limitation on a detrital-based ecosystem. Ecol. Monogr. 69, 409–442.

Wang, M.C., Bushman, B.J., 1998. Using normal quantile plots to explore meta-analytic data sets. Psychol. Methods 3, 46–54.

Wantzen, K.M., Wagner, R., 2006. Detritus processing by invertebrate shredders: a neotropical-temperate comparison. J. N. Am. Benthol. Soc. 25, 216–232.

Ward, C.L., McCann, K.S., Rooney, N., 2015. HSS revisited: multi-channel processes mediate trophic control across a productivity gradient. Ecol. Lett. 18, 1190–1197.

Wardle, D.A., 2010. Trophic cascades, aboveground-belowground linkages, and ecosystem functioning. In: Terborgh, J., Estes, J.A. (Eds.), Trophic Cascades: Predators, Prey, and the Changing Dynamics of Nature. Island Press, Washington, DC, pp. 203–218.

Webster, J.R., Benfield, E.F., 1986. Vascular plant breakdown In freshwater ecosystems. Annu. Rev. Ecol. Syst. 17, 567–594.

Wilson, E.E., Wolkovich, E.M., 2011. Scavenging: how carnivores and carrion structure communities. Trends Ecol. Evol. 26, 129–135.

Winemiller, K.O., Montaña, C.G., Roelke, D.L., Cotner, J.B., Montoya, J.V., Sanchez, L., Castillo, M.M., Layman, C.A., 2014. Pulsing hydrology determines top-down control of basal resources in a tropical river-floodplain ecosystem. Ecol. Monogr. 84, 621–635.

Wise, D.H., Snyder, W.E., Tuntibunpakul, P., Halaj, J., 1999. Spiders in decomposition food webs of agroecosystems: theory and evidence. J. Arachnol. 27, 363–370.

Wolkovich, E.M., Allesina, S., Cottingham, K.L., Moore, J.C., Sandin, S.A., De Mazancourt, C., 2014. Linking the green and brown worlds: the prevalence and effect of multichannel feeding in food webs. Ecology 95, 3376–3386.

Wood, S.A., Karp, D.S., DeClerck, F., Kremen, C., Naeem, S., Palm, C.A., 2015. Functional traits in agriculture: agrobiodiversity and ecosystem services. Trends Ecol. Evol. 30, 531–539.

Woodward, G., Hildrew, A.G., 2002. Food web structure in riverine landscapes. Freshw. Biol. 47, 777–798.

Woodward, G., Papantoniou, G., Edwards, F., Lauridsen, R.B., 2008. Trophic trickles and cascades in a complex food web: impacts of a keystone predator on stream community structure and ecosystem processes. Oikos 117, 683–692.

Wooster, D., 1994. Predator impacts on stream benthic prey. Oecologia 99, 7–15.

Wootton, J.T., Oemke, M.P., 1992. Latitudinal differences in fish community trophic structure, and the role of fish herbivory in a Costa Rican stream. Environ. Biol. Fishes 35, 311–319.

Wu, X., Sun, S., 2010. The roles of beetles and flies in yak dung removal in an alpine meadow of eastern Qinghai-Tibetan Plateau. Ecoscience 17, 146–155.

Wurzbacher, C.M., Bärlocher, F., Grossart, H.P., 2010. Fungi in lake ecosystems. Aquat. Microb. Ecol. 59, 125–149.

Yee, D.A., Yee, S.H., Kneitel, J.M., Juliano, S.A., 2007. Richness-productivity relationships between trophic levels in a detritus-based system: significance of abundance and trophic linkage. Oecologia 154, 377–385.

Yvon-Durocher, G., Dossena, M., Trimmer, M., Woodward, G., Allen, A.P., 2015. Temperature and the biogeography of algal stoichiometry. Glob. Ecol. Biogeogr. 24, 562–570.

CHAPTER FOUR

Towards an Integration of Biodiversity–Ecosystem Functioning and Food Web Theory to Evaluate Relationships between Multiple Ecosystem Services

Jes Hines[*,†,1], Wim H. van der Putten[‡,§], Gerlinde B. De Deyn[¶], Cameron Wagg[||], Winfried Voigt[#], Christian Mulder[**], Wolfgang W. Weisser[††], Jan Engel[#,††], Carlos Melian[‡‡], Stefan Scheu[§§], Klaus Birkhofer[¶¶], Anne Ebeling[#], Christoph Scherber[||||,##], Nico Eisenhauer[*,†]

[*]German Centre for Integrative Biodiversity Research (iDiv), Halle-Jena-Leipzig, Leipzig, Germany
[†]Institute of Biology, University Leipzig, Leipzig, Germany
[‡]Netherlands Institute of Ecology (NIOO), Wageningen, The Netherlands
[§]Laboratory of Nematology, Wageningen University, Wageningen, The Netherlands
[¶]Department of Soil Quality, Wageningen University, Wageningen, The Netherlands
[||]Institute of Evolutionary Biology and Environmental Studies, University of Zürich, Zürich, Switzerland
[#]Institute of Ecology, University of Jena, Jena, Germany
[**]National Institute for Public Health and Environment (RIVM), Bilthoven, The Netherlands
[††]Terrestrial Ecology Research Group, Department of Ecology and Ecosystem Management, School of Life Sciences, Technische Universität München, Freising, Germany
[‡‡]Swiss Federal Institute of Aquatic Science and Technology (Eawag), Kastanienbaum, Switzerland
[§§]J. F. Blumenbach Institute of Zoology and Anthropology, University of Göttingen, Göttingen, Germany
[¶¶]Department of Biology, Lund University, Lund, Sweden
[||||]Agroecology, Department of Crop Sciences, University of Göttingen, Göttingen, Germany
[##]Institute of Landscape Ecology, University of Münster, Münster, Germany
[1]Corresponding author: e-mail address: jes.hines@yahoo.com

Contents

1. Introduction 162
2. Contributions and Limitations of BEF and FWT 163
   2.1 BEF and Species Interactions Concepts 164
   2.2 FWT and Species Interactions Concepts 174
3. Principles for Integrating BEF and FWT 181
   3.1 Principle I: Interactions Occur between Taxonomic Units According to a Topology 181
   3.2 Principle II: Energy and Material Fluxes through Food Webs Provide a Common Currency for Assessing the Influence of Biodiversity on Ecosystem Functioning 182
   3.3 Principle III: Multiple Types of Species Interactions Affect Ecosystem Functioning 183

Advances in Ecological Research, Volume 53
ISSN 0065-2504
http://dx.doi.org/10.1016/bs.aecr.2015.09.001

4. Considering Trends in BEF–FWT Research for Better Management of Multiple ESs 184
5. Conclusions 187
Acknowledgements 187
References 187

Abstract

Ecosystem responses to changes in species diversity are often studied individually. However, changes in species diversity can simultaneously influence multiple interdependent ecosystem functions. Therefore, an important challenge is to determine when and how changes in species diversity that influence one function will also drive changes in other functions. By providing the underlying structure of species interactions, ecological networks can quantify connections between biodiversity and multiple ecosystem functions. Here, we review parallels in the conceptual development of biodiversity–ecosystem functioning (BEF) and food web theory (FWT) research. Subsequently, we evaluate three common principles that unite these two research areas by explaining the patterns, concentrations, and direction of the flux of nutrients and energy through the species in diverse interaction webs. We give examples of combined BEF–FWT approaches that can be used to identify vulnerable species and habitats and to evaluate links that drive trade-offs between multiple ecosystems functions. These combined approaches reflect promising trends towards better management of biodiversity in landscapes that provide essential ecosystem services supporting human well-being.

1. INTRODUCTION

Ecologists have long been fascinated by the diversity of species and the complexity of species interactions (Darwin, 1859; Elton, 1927). Today, we use the term biodiversity to describe and compare variation among taxa at multiple levels of ecological organization: between and within populations, species, phylogenies, functional groups, trophic levels, food web compartments, and even habitat patches that explain landscape diversity. Concern over the consequences of global changes in all levels of biodiversity has motivated examination of the relationship between biodiversity and ecosystem functioning (BEF; Naeem et al., 2012). BEF combines community and ecosystem ecology to examine how changes in diversity affect a broad suite of ecosystem functions (EFs) (Hooper et al., 2005) and the services ecosystems provide to humans (ESs) (Costanza et al., 1997, MEA, 2005). More than three decades of BEF experiments have demonstrated that changes in diversity within each level of organization can influence several focal EFs as well as ecosystem services (ESs) that influence human well-being (Balvanera et al., 2006; Cardinale et al., 2012; Hooper et al., 2005; Tilman et al., 2014).

Increasingly, we are realizing that changes in biodiversity can simultaneously influence multiple interdependent EFs and associated ESs, such as pollination, pest suppression, and carbon sequestration (Cardinale et al., 2012; Gamfeldt et al., 2013; Raudsepp-Hearne et al., 2010). Yet, we lack mechanistic understanding of how multiple EFs are connected to losses or gains of biodiversity that can simultaneously occur across several levels of ecological organization (Wardle et al., 2011). By explicitly providing the underlying network of species interactions, food web theory (FWT) can make these critical connections. At least five perspective papers published in the last decade have suggested that an explicit food web perspective is an important conceptual contribution to the understanding of BEF relationships (Duffy et al., 2007; Ives et al., 2005; Rooney and McCann, 2012; Thebault and Loreau, 2006; Thompson et al., 2012). These papers have emphasized that both horizontal diversity (within trophic level) and vertical diversity (between trophic levels) can influence focal EFs, such as production of biomass and resource depletion (Cardinale et al., 2006; Thebault and Loreau, 2006). Here, we extend this rationale and discuss how merging BEF and FWT approaches would also contribute to the understanding of the trade-offs and mechanisms driving relationships between biodiversity and multiple EFs, as well as the services ecosystems provide to humans. In the sections that follow, we describe our perspective on the development, convergence, and limitations of BEF and FWT (Section 2). Next we discuss three principles that unite the two research areas, generating testable hypotheses that can be used to evaluate relationships between biodiversity and multiple EFs (Section 3). While increasing biodiversity may increase ecosystem functioning (i.e. grassland community production), it may limit the contribution of focal species to some ESs (i.e. production of grain for food) that benefit humans. Therefore, we close by considering how development of combined BEF–FWT perspective has contributed to better management of multiple ESs (Section 4).

2. CONTRIBUTIONS AND LIMITATIONS OF BEF AND FWT

The development and convergence of BEF and FWT have proceeded through three conceptual phases (Fig. 1). While not intending to present a comprehensive review of all research in both sub-disciplines, these phases provide a road map outlining the parallel and convergent concepts developed in both research areas. The first phase describes research that, for the most part, has been completed. The second phase describes research that currently is being pursued, while the third phase describes a promising line of

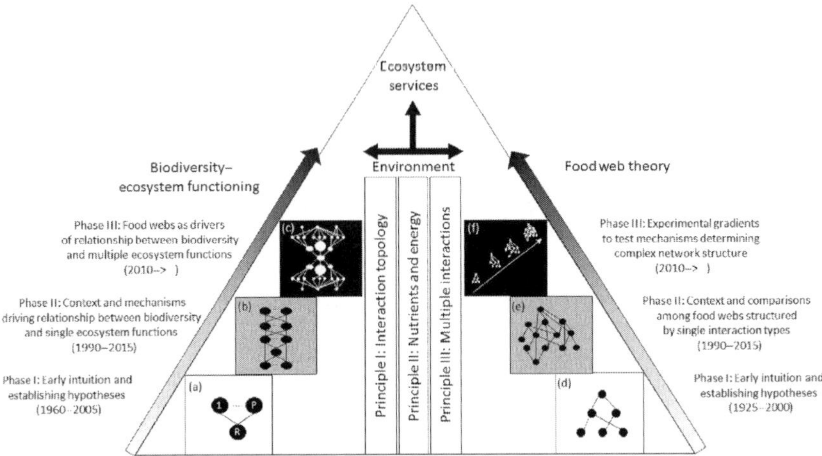

Figure 1 The models of biodiversity–ecosystem functioning (BEF) and food web theory (FWT) both utilize assumptions grounded in species interactions and flux of nutrients and energy. Three conceptual phases of research describe the development and convergence of these disciplines, which are united by three common principles allowing for the establishment of an integrative BEF–FWT framework (see text for description). Management of ecosystems providing multiple ecosystem services will benefit from an integrative approach that explicitly connects ecosystem functions and services to the network of species interactions that influence them.

inquiry that is still in its infancy. Assessment of these conceptual phases will allow us to consider how progress in the study of biodiversity, food webs, and ecosystem functioning may, or may not, be useful for management of species that provide essential ESs that benefit humans.

2.1 BEF and Species Interactions Concepts

2.1.1 First Phase BEF: Early Intuition and Establishing Hypotheses

As a research area, BEF is based on the intuition that ecosystems harbouring many species function differently than ecosystems with only few species. Experimental evidence for this intuition was lacking until initial experiments in agricultural (De Wit and Van Den Bergh, 1965) and natural grasslands (Berendse, 1983) demonstrated that plots with mixed plant species produced more biomass than monocultures of the same species (Hector et al., 1999; Roscher et al., 2004; Tilman, 1996). Disagreement surrounded the extent to which different experimental designs could test for mechanisms driving BEF relationships (Huston, 1997). Nonetheless, three hypotheses were proposed to explain increases in functioning resulting from higher diversity within a single trophic level: (1) complementarity effects, (2) sampling effects, and (3) selection effects. *Complementarity effects* evoke niche-based mechanisms to explain why unique attributes of each species reduce competition (Loreau and Hector, 2001) or facilitate associated species performances (Mulder et al., 2001) to enhance overall resource capture and use in mixed species communities. *Sampling effects* occur when a particularly effective species is more likely to be present in a more diverse community (Wardle, 1999). Sometimes considered in a similar category as sampling effects, *selection effects* transpire when the most effective species in monoculture also dominate ecosystem functioning in diverse communities, or reciprocally the most vulnerable species are diluted in diverse communities (Loreau and Hector, 2001). Because of the focus on plant diversity, each of these mechanisms focused on interactions within a trophic level, such as competition or facilitation, as the primary driver of ecosystem functioning. These hypotheses lead to a conceptual topology where species (i.e. producers (1-P in Fig. 1A)) are linked to a resource (R), and ecosystem functioning reflects the community's production of biomass or depletion of the resource (Fig. 1A). They also established an important initial paradigm shift in the understanding of biodiversity. That is, beyond being a response to environmental conditions, biodiversity now was also considered as a potential driver of ecosystem functioning (Schulze and Mooney, 1993).

Similar ideas regarding the influence of diversity on particular EFs were developed in multi-trophic systems, especially in the context of predator diversity effects on prey suppression in biocontrol (Pimentel, 1961). Emphasis was placed on discovery and naming of particular interaction pathways describing how consumers responded to, or caused, changes in plant diversity. For example, the terms 'associational resistance' and 'associational susceptibility' were coined and used to describe the indirect interaction by which the traits of neighbouring plants in more diverse communities do (Root, 1973) or do not (Atsatt and O'Dowd, 1976) influence the impact of herbivores on focal plants. Attention was focused on finding plant traits that served as underlying mechanisms behind associational resistance and associational susceptibility (reviewed in Barbosa et al., 2009). These included differences in plant chemistry (Karban and Maron, 2002), apparency (Perrin and Phillips, 1978), vegetation structure (Rauscher, 1981), and ability to attract predators (Dicke, 1994). Multi-trophic BEF research was not limited to plants and their interactions with aboveground consumers (Bardgett et al., 1999; Zak et al., 2003). Microbial-driven processes in soils were found to influence aboveground plant diversity and production by altering organic matter decomposition, developing mutualistic mycorrhizae–plant interactions, and modifying plant susceptibility to pathogens (Barbosa and Krischik, 1991; van der Heijden et al., 1998; Wolters et al., 2000). Despite the discovery of multiple potential interaction pathways, however, identifying general rules or predicting the effects of combinations of consumer species on EF proved to be difficult, especially in soil systems where soil fauna were highly omnivorous and played multiple ecological roles (Mikola and Setälä, 1998b). Therefore, although strong connections between aboveground and belowground consumers were established (De Deyn and Van der Putten, 2005), the context dependency behind the consumer–BEF relationships were not yet clear.

In the first phase of BEF research, the foundation was laid for examining the relationship between BEF in multi-trophic aboveground–belowground communities, and consumers were considered as both a response to, and a driver of, ecosystem functioning (Naeem et al., 1994). Species richness was indirectly (Tilman and Downing, 1994) and directly (Naeem and Li, 1997) manipulated, and debate focused on whether conclusions were biased by inferences drawn from particular experimental designs (Huston, 1997). This phase ended with a consensus statement that regardless of experimental design many, but not all, studies demonstrated an asymptotic relationship between biodiversity and ecosystem functioning such that functioning declined rapidly when species were lost from communities with low diversity (Hooper et al., 2005). The potential conservation applications and scale at which each mechanism operated, however, remained unresolved during this phase, particularly with respect to the influence of consumer diversity (Srivastava and Vellend, 2005).

2.1.2 Second Phase BEF: Context and Mechanisms Driving Relationship between Biodiversity and Single EFs

The second phase of BEF research moved beyond debates about experimental design and generated an explosion of studies used to evaluate the generality and context dependency of the relationship between biodiversity and ecosystem functioning (Cardinale et al., 2011). The type of diversity manipulated was considered as an important context for the influence of biodiversity on EF. For plants, not only species richness but also diversity at multiple levels of ecological organization, such as intra-specific genetic diversity, phylogenetic diversity, and functional trait diversity was also found to influence plant community production (Cadotte et al., 2008; Crutsinger et al., 2006; Flynn et al., 2011). In addition to grassland plants, the diversity of other groups of species, including consumers (Duffy, 2002), that range in body size from unicellular microbial systems (Bell et al., 2005) to trees (Rivest et al., 2015) was manipulated. Following up on the work of Mikola and Setälä (1998b), diversity of herbivores (Deraison et al., 2015; Duffy et al., 2003; Norberg, 2000), detritivores (Cardinale et al., 2002; Dangles et al., 2002), or predators (Cardinale et al., 2003; Finke and Denno, 2005; Straub and Snyder, 2006) were manipulated and ecosystem functioning was assessed by measuring depletion of resources in adjacent trophic levels. Ives et al. (2005) used basic Lotka–Volterra equations to identify a common vocabulary and conclusions between studies examining multitrophic and BEF interactions. Consumer diversity effects proved strong enough to cascade across multiple trophic levels in terrestrial (Wardle et al., 2005) and aquatic systems (Mancinelli and Mulder, 2015; Worm et al., 2003; but see O'Connor and Bruno, 2009), demonstrating that diversity effects on EF are not necessarily dependent upon study system or trophic level (Griffin et al., 2013).

The sensitivity of the response variables was considered as an additional factor that would influence BEF relationships, and the types of responses measured were expanded and compared (Allan et al., 2013; Balvanera et al., 2006; Hector and Bagchi, 2007). Ecosystem responses included soil nutrient cycling, decomposition, plant production, and soil water content, among others. Often not explicitly tied to ecosystem functioning, response of consumer community composition was assessed using several metrics including consumer species richness (Haddad et al., 2011), functional diversity (Best et al., 2014; Rzanny and Voigt, 2012), and consumer phylogenetic diversity (Lind et al., 2015). Consumers were found to be sensitive to manipulations of several types of plant diversity including plant species diversity (Haddad et al., 2009; Scherber et al., 2010), functional diversity (Symstad et al., 2000), and genetic diversity (Crutsinger et al., 2006). Their sensitivity to changes in plant diversity, however, was found to attenuate across trophic levels, with strongest effects of plant diversity on plant production and the abundance of their direct consumers, and diminished effects on higher trophic levels such as predators and omnivores (Haddad et al., 2009; Scherber et al., 2010). Considering these results together with studies manipulating consumer diversity, population dynamic models were used to demonstrate that bottom-up influences of plant diversity, and top-down influences of consumer diversity could interactively modify the relationship between biodiversity and focal EFs (Thebault and Loreau, 2003). This was confirmed by pioneering experimental tests, conducted predominantly in aquatic systems, which simultaneously manipulated diversity at multiple trophic levels (Bruno et al., 2008; Douglass et al., 2008; Fox, 2004; Gamfeldt et al., 2005; Jabiol et al., 2013).

Environmental conditions also were considered as a source of qualitative and quantitative variation BEF relationships. For example, plant diversity effects on ecosystem functioning were measured in experimental manipulations that simulated different global environmental change scenarios (Adair et al., 2009; Reich et al., 2001). The effects of biodiversity on ecosystem functioning were compared across different environmental contexts such as terrestrial, freshwater, and marine ecosystems (Cardinale et al., 2006; Covich et al., 2004), as well as primary producer or detritus-based ecosystems (Srivastava et al., 2009). In summary, this phase of research generated a wealth of case studies, which expanded the range of scenarios that could potentially influence BEF relationships.

To evaluate factors that influence magnitude and consistency of biodiversity effects on ecosystem functioning in this diverse array of experiments is a daunting task and this phase of research is currently in a period of synthesis (Cardinale et al., 2006). Meta-analyses generally support predicted biodiversity relationships for response variables such as plant productivity that are reported broadly across many experiments (Cardinale et al., 2012; Gamfeldt et al., 2015). Indeed, the influence of changes in diversity on plant production and decomposition can even surpass the magnitudes of impact imposed by other environmental change drivers such as climate warming, acidification, and nutrient pollution (Hooper et al., 2012; Tilman et al., 2012). However, when detailed responses are reported within single experiments, the influence of biodiversity on the magnitude and direction of effects have not proven as consistent (Allan et al., 2013). In a German grassland study, for example, plant diversity had positive effects on aboveground herbivore abundance, but no effect on belowground herbivore abundance (Allan et al., 2013). While this second phase of research answered questions about the strength and consistency of biodiversity effects across different environmental contexts, it also led to new inquiries as to how biodiversity affects connections between multiple EFs within a particular compartment (e.g. above- and belowground processes).

2.1.3 Third Phase BEF: Linking Multiple Functions and Scaling of Mechanisms

We are approaching a conceptual shift in BEF research. The variation in responses within experiments seen in the second phase demands an examination of how species influence connections between EFs. As with the conceptual shift in the first phase of BEF research, which established hypotheses explaining how biodiversity may not only be a response to environmental conditions but also a driver of EFs (Hillebrand and Matthiessen, 2009), we are generating new hypotheses about the influence of complexity in BEF (see Table 1 and Section 3). Now, rather than consumer species acting either as an additional response variable or as a driver of single functions, interactions between species may connect multiple EFs. This phase of research will seek a stronger understanding of connections between multiple response variables within a particular system (Bradford et al., 2014; Wagg et al., 2014).

Several quantitative approaches have been proposed to assess the simultaneous responses of multiple EFs (Byrnes et al., 2014; Lefcheck et al., 2015). These approaches examine correlations between functions, using data reduction approaches to generate multifunctionality metrics. Such metrics then can be compared across global data sets to assess whether results reflect generalizable insights about the relationship between biodiversity and ecosystem multifunctionality. For example, in a survey of 224 dryland ecosystems, 14 ecosystem responses were reduced into a single index of ecosystem multifunctionality, and increases in that multifunctionality index were associated with cooler temperatures and lower soil sand content (Maestre et al., 2012). Soil fauna and changes in net primary production by plants respond sensitively to desiccation in warmer drier soils, and they were implicated as possible drivers of ecosystem multifunctionality (Maestre et al., 2012). However, those responses were not reported directly in this study, illustrating that mechanisms behind multivariate responses sometimes can remain speculative in statistical analyses that involve dimensional reduction. Another approach is to embrace the complexity of consumer responses developed in phase two studies and consider connections between EFs as a component of complex food webs.

The groundwork for considering complexity-based approaches in BEF is built upon the observation that generalist predators and plants connect aboveground and belowground communities (Hooper et al., 2000; Scheu, 2001; Wardle et al., 2004). Beyond building a more complicated model, discovery of links connecting aboveground and belowground webs has emphasized that changes in species density and diversity in one food web compartment can alter the ecosystem functioning and services provided by species in another compartment (Bardgett and Van der Putten, 2014). Evaluations following this line of reasoning will benefit from quantitative methods typically used in FWT including, but not limited to, qualitative

Table 1 Three Common Principles Unite Biodiversity–Ecosystem Function (BEF) and Food Web Theory (FWT)

Principle	BEF	FWT	Hypotheses from Combined BEF–FWT	Methods	Application to Management of Multiple ES
I: Interactions occur between taxonomic units according to a topology	Unique aspect of each species allows coexistence and enhances resource capture of diverse communities using common resources (Complementarity effects)	Modular patterns of species interactions stabilize complex food webs (Compartmentalization effects)	CH1. Unique aspects of species within and between modules determine trade-offs and synergies in multiple ecosystem functions	Group detection (Gauzens et al., 2015)	Focus management on key modules within food webs to stabilize multiple ES. Prioritize conservation of key modules in space (Macfadyen et al., 2011; Montoya et al., 2015), or critical species connecting energy channels (Garay-Narváez et al., 2014; Terborgh et al., 2001)
II: Estimating fluxes of energy and materials through food web topology provides a common currency for assessing influence of biodiversity on ecosystem functioning	Diverse communities are more likely to include a species that enhances ecosystem functioning (Sampling effects)	Balances in transfer of biomass between trophic groups stabilizes food webs (Trophic effects)	CH2. Changes in diversity that limit uptake and transfer of biomass between trophic groups will influence multiple ecosystem functions	Ecosystem Network Analysis (Borrett and Lau, 2014)	Make management decisions based on the flux of energy through diverse food webs to stabilize multiple ES. Manage land-use intensity (Barnes et al., 2014), or harvesting of particular species (Fung et al., 2015) to enhance overall functionality

Continued

Table 1 Three Common Principles Unite Biodiversity–Ecosystem Function (BEF) and Food Web Theory (FWT)—cont'd

Principle	BEF	FWT	Hypotheses from Combined BEF–FWT	Methods	Application to Management of Multiple ES
III: Multiple types of species interactions influence ecosystem functioning	Dominance of species with traits that contribute positively to ecosystem functioning increases ecosystem functioning in diverse mixtures (Selection effects)	Balance of strong, weak, positive, and negative interactions stabilizes food webs (Interaction effects)	CH3. Trade-offs between multiple ecosystem functions are caused by dominance of species that have net positive species interactions with respect to one function but net negative interactions with respect to another function	Third-generation SEM (Grace et al., 2012)	Manage species interaction to enhance multiple ES. Prioritize timing of management actions based on its influence on direct and indirect interactions (Whalen et al., 2013) or prioritize conservation of multiple interactions themselves (Mougi and Kondoh, 2012)

Combined hypotheses (CH) result from development of a combined BEF–FWT perspective. These hypotheses are non-mutually exclusive, and here we highlight a few combinations that are well suited to test using quantitative methods developed using graph theoretic and systems theory. Results from these tests can be applied to ecological management strategies with the goal of enhancing and stabilizing multiple ecosystem services (ESs).

and quantitative descriptors of food web matrices (Bersier et al., 2002), group detection (Gauzens et al., 2015), ecosystem network analysis (Borrett and Lau, 2014; Ulanowicz, 2011), and third-generation structural equation modelling (SEM) (Grace et al., 2012). These tools can be used to characterize the structure and dynamics of whole ecosystems using an *interaction topology* to describe the *flux of nutrients and energy* through ecosystems. Although they place slightly different emphasis on the importance of structure and function, each tool establishes connections that mechanistically explain trade-offs and correlations between biodiversity and multiple EFs. Questions such as 'How often are species with positive effects on one function directly or indirectly connected to species that have negative influence on a second?' will be asked in this phase. Experiments here will reflect a convergence of BEF and FWT and test the relationship between the structure of complex food webs, biodiversity, and multiple EFs.

While we initially referred to the importance of above- and belowground compartments in the previous paragraph, similar relationships between food web structure and multiple EFs should exist in all ecosystems that are composed of discrete compartments. Traditionally, compartmentalized systems include aquatic ecosystems composed of benthic and pelagic compartments (Krause et al., 2003), coastal and riparian ecosystems composed of terrestrial and aquatic compartments (Polis and Hurd, 1996), agricultural fields composed of margins and croplands (Macfadyen et al., 2011), and any kind of ecosystem spanning environmental gradients that have thresholds in community interactions. This assortment of food web scaling allows us to think about how BEF relationships developed in small field plots may apply to changes in biodiversity at a landscape scale. This is an essential step to translate results from BEF experiments to broader scale management of ESs (Díaz et al., 2006; Kremen, 2005).

This third phase of BEF research, therefore, will move beyond evaluations of context-dependent effects to evaluate the influence of interactions among consumers in complex communities on multiple EFs (Fig. 1C). Consumers will be considered not only for their direct effects on resource uptake and production of biomass but also for their roles linking multiple EFs. The expectation is that the asymptotic relationship between BEF will be replaced by a non-saturating relationship when multiple EFs are considered, although trade-offs between some functions will limit the magnitude of this effect. To identify and evaluate such trade-offs, BEF will benefit from the conceptual advances being made in FWT, as described below. Network approaches will be used to test how relationships between changes in biodiversity and complex species interactions will influence ecosystem functioning and ES that influence human well-being.

2.1.4 Limitations of BEF with Respect to Understanding the Role of Consumers in Ecosystem Functioning

Much of the early debate about relationships between biodiversity and ecosystem functioning centred around the generality and strength of inferences that could be made regarding mechanisms revealed by particular experimental designs (Huston, 1997). Unfortunately, complex experimental designs have limited simultaneous manipulation of density (Griffin et al., 2008) and diversity at multiple trophic levels (but see experiments in aquatic systems emphasized above), which are necessary to evaluate the causal relationships between consumer community composition and ecosystem functioning. Complex experimental designs can also make it difficult to establish adequate replication needed to capture the shape of non-linear relationships between species–environment, species–species, and diversity–function relationships. Therefore, the classic BEF approach of manipulating diversity and measuring the response of ecosystem functioning has limitations that hinder the types of inferences made about the role of consumer as drivers of BEF relationships.

To overcome these limitations, experimental studies testing the influence of consumer diversity on EFs tend to take one of three approaches. First, some are conducted in simplified, but experimentally tractable meso- and micro-cosmos (O'Connor and Bruno, 2009; Setälä et al., 1998; Wardle et al., 2005). Second, others manipulate consumer diversity in more natural field settings without simultaneously manipulating plant diversity (Deraison et al., 2015; Schmitz, 2009). Such consumer diversity manipulations in field studies are often conducted in systems composed of monocultures of plants, such as agricultural fields (Snyder et al., 2006) or salt marshes (Finke and Denno, 2005), which makes it difficult to examine cause and effect relationships between drivers of plant and consumer diversity. The third alternative has been to manipulate consumer abundance using pesticides (Eisenhauer et al., 2011; Siemann and Weisser, 2004), which can be an effective way to control broad functional groups, but makes it difficult to determine the contribution of species diversity to responses, in part because biocides can have non-target effects (both direct and indirect) on other species. Despite the unique strengths and weaknesses of each of these three approaches, results frequently reveal unexpected indirect and non-trophic effects of consumers (Hawlena et al., 2012; Hines and

Gessner, 2012). For example, Losey and Denno (1998) found that predatory coccinellids hunting in plant canopies elicit a defence response in their aphid prey where the aphids drop to the ground causing them to be more susceptible to predation by ground-foraging carabids. Together, the combined impact of more diverse predator assemblages composed of coccinellids and carabids has a synergistic effect on pest suppression resulting from a change in prey behaviour that cannot be predicted by adding the direct consumption of the two predators alone (Losey and Denno, 1998). Consequently, theoretical approaches capable of modelling the outcome of multi-trophic interactions using complex effective competition matrices are difficult to parameterize from purely density-dependent approaches (Fowler, 2013).

Integrating more complex food web and network responses into studies that manipulate species diversity of a single trophic level may be a better direction because it allows quantification of multiple known interaction pathways and tests of when one can, or cannot, predict trade-offs or feedbacks among multiple EFs. However, in experimental studies, the limitations imposed by a lack of simultaneous manipulation of consumer communities remain. Furthermore, scaling of biodiversity effects inferred from small, short-term field plot experiments to assess the long-term stability of ecosystem functioning at landscape scales is a persistent challenge. Ultimately, pairing and comparing of multiple approaches including experimental manipulation of species abundance and diversity in the field, simulated extinctions, and dynamic food web models likely will provide the most robust understanding of biodiversity effects on multiple EFs.

2.2 FWT and Species Interactions Concepts

2.2.1 First Phase FWT: Early Intuition and Establishing Hypotheses

As a research area, the study of food webs is older than BEF, so we summarize a comparatively longer duration of inquiry in this first phase of research. Early FWT used graphic depictions of predator–prey interactions to illustrate the trophic pathways by which energy and biomass flow through ecosystems (Fig. 1D; Elton, 1927). These graphics generated hypotheses that there are emergent and generalizable properties of food web structure that allow populations and communities to be stable (Cohen, 1977; Pimm et al., 1991; Sugihara et al., 1989). Notably, there were strong conceptual divides between empiricists and theoreticians, as well as between those who preferred to study webs in terms of the natural history of species and those interested in physical and chemical attributes. Nonetheless, three main classes of hypotheses were suggested to influence the persistence and stability of food webs: (1) trophic structure effects, (2) interaction effects, and (3) compartmentalization effects. *Trophic structure effects* suggest that the amount of available biomass at a particular trophic level regulates food webs (Lindeman, 1942; Odum, 1969; Ulanowicz and Kemp, 1979). Debates focused on the importance of predation (top-down) as opposed to competition for resources (bottom-up) in stabilization of focal populations and communities at each trophic level (Hairston et al., 1960; Menge and Sutherland, 1987). The proportions of species in each trophic level also were thought to be scale invariant with respect to the number of species in the web (Cohen, 1977; but see Briand, 1983). *Interaction effects* occur when the number of interactions connecting species influence food web stability (MacArthur, 1955; Pimm, 1979). Addition of realistic patterns of interaction strengths revealed that highly connected species, such as generalists and omnivores, that have weak connections to many other species can stabilize food webs by relaxing predation pressure on populations at low densities, and allowing them to recover from disturbance (de Ruiter et al., 1995; Fagan, 1997; McCann et al., 1998). *Compartmentalization* effects transpire when sub-webs of species, often called modules, interact more with each other than with other species in the web. This clustering of interactions can limit the effects of disturbance to more localized modules within the food web (May, 1973; Yodzis, 1982). Compartmentalization effects can be limited by generalist species that link species in different modules, lending support to the idea that these effects may be weak in real systems (Pimm and Lawton, 1980). Some supporting evidence was found for each of these hypotheses, although the strength and consistency of effects with respect to the influence of changes in biodiversity on ecosystem functioning was not clear (Jones and Lawton, 1995; O'Neill, 2001).

As a field of expertise, FWT has placed less emphasis on consensus statements than BEF. The end of this phase was marked by a particularly insightful review by McCann (2000) that described the role of diversity–stability relationships in both BEF and FWT. With respect to FWT, McCann (2000) highlighted the influence of equilibrium dynamics on complexity–stability relationships as a key assumption that divided theoreticians and empiricists. Sadly, at around this time, the eminent ecologist Gary Polis died. Polis' work was providing the empirical evidence that was needed to challenge theoretical dogma suggesting that omnivory was rare and complex systems were unstable. He did so by quantifying the complexity and high degree of omnivory in desert food webs (Polis, 1991) and by documenting the strong influence of subsidies that cross traditionally subdivided landscape compartments (Polis et al., 1997). In time, reflection on his research reinforced the clear need for stronger integration of empirical and theoretical approaches in FWT.

2.2.2 Second Phase of FWT: Context and Comparisons among Food Webs Structured by Single Interaction Types

Inconsistencies between empirical and theoretical results led to an explosion of research testing the generality and context dependency of the relationship between food web structure and stability. The sensitivity food webs to disturbance was thought to depend upon how stability was defined and measured; there was an explosion of metrics used to describe network stability, including resilience, invasibility, persistence, permanence, coherence, and robustness (McCann, 2000; Pimm, 1984). Rather than being restricted to simple definitions of stable or unstable food webs a broader range of strategies leading to stability expanded our understanding of whole system dynamics in diverse food webs.

The generality and context dependency of each of the three main hypotheses also were tested. For example, debates focused on whether trophic effects were dependent upon study system, such as aquatic (Strong, 1992), aboveground (Shurin et al., 2006), and belowground (Mikola and Setälä, 1998a), or diversity of species within a trophic group (Hooper et al., 2005; Hunter and Price, 1992). The definition of interaction effects also was clarified to include and distinguish between trophic, indirect, and non-trophic interactions such as ecosystem engineering (Jones et al., 1994; Wootton, 1994). To determine the influence of different types of interactions on food web structure, traditional predator–prey interaction webs were compared with those structured by parasitism (Dunne et al., 2013; Lafferty et al., 2008) and mutualism (Bascompte and Jordano, 2007; Thebault and Fontaine, 2010). The existence and consequences of compartment effects also were debated among empiricist and theoreticians alike. In soil food webs, close interactions among species from different trophic levels were found to form compartments with divergent energetic pathways (Moore et al., 2005; Rooney et al., 2006). Some skepticism surrounded whether these effects reflected a general property of food webs because at least two lines of evidence suggested that sub-webs traditionally considered to be quite separate were found to be linked more closely than previously thought. Aquatic and terrestrial sub-webs were found to be linked by cross-habitat resource subsidies of plants (Nakano and Murakami, 2001; Polis et al., 1997) and animals (Dreyer et al., 2012).

Further, aboveground and belowground sub-webs were found to be linked by plants and generalist predators (Wardle, 2002). Previously, FWT focused on comparisons of aquatic, aboveground, or soil systems, and almost all BEF studies focus on either aquatic or terrestrial systems in isolation. Discovery of connections across compartments suggested that disturbances to any one part of the food web potentially could have much farther-reaching consequences than previously expected. Yet, development of suitable algorithms suggested that compartmentalization might be common in real food webs (Fortuna et al., 2010; Krause et al., 2003; Stouffer and Bascompte, 2011). Differences between studies demonstrating connections between compartments, and those demonstrating that compartmentalization was common reinforced interest in experimental studies examining causal drivers influencing the relationship between structure and function in food webs.

This second phase of FWT can be characterized by a strong emphasis on more finely and evenly resolved food webs, and comparisons between webs with different kinds of interactions (Fig. 1E; Ings et al., 2009). A growing number of food web databases facilitate sharing of food web data (Webs on the Web, ECOweB, Interaction Web Database-NCEAS) (Mulder, 2011). Outside of a limited set of examples, however, most assessments of food web structure are made from comparisons of detailed but unreplicated webs across ecosystems (Dunne et al., 2002). In contrast, considerably simplified food webs are the focus in replicated experiments (Denno et al., 2003; Menge et al., 2004). The increasing number of well-resolved food webs that are readily available in databases allows comparative tests of whether the consequences of disturbance can be generalized across all food webs, or if they differ for food webs in different environments.

2.2.3 Third Phase of FWT: Linking Multiple Interactions with Ecosystem Functioning

A key innovation in this phase of FWT research will be the use of experimental gradients to identify causal drivers of food web structure (Fig. 1F; Baiser et al., 2012; Thompson and Townsend, 2004; Tylianakis et al., 2007). FWT will benefit from BEF studies that use rigorous experimental designs to examine the relationship between biodiversity and ecosystem functioning. BEF experiments will also contribute detailed records of species diversity and nutrient fluxes to food web models that had previously focused on either species interactions or flux of nutrients through aggregated nodes. In this phase, therefore, consumer interactions will be considered not only for their direct effects on other consumers but also for their roles in providing and linking multiple EFs and services (i.e. pollination, pest suppression, and carbon sequestration).

The groundwork for this line of reasoning was built upon the realization that organismal growth and ecosystem dynamics are both constrained by the first principles of physics and chemistry (Brown et al., 2004). To show how species influence the flux of biomass through well-resolved food webs, network nodes in this phase will more commonly integrate traits such as mass and abundance (Cohen et al., 2009), metabolism (Barnes et al., 2014), or multivariate functional traits (Rzanny and Voigt, 2012; Rzanny et al., 2013). Stoichiometric traits (C:N:P) of plants and animals also could provide informative constraints of food web structure (Mulder et al., 2013; Ott et al., 2014). Predators and detritivores generally seem to have higher nutrient content than their prey (Martinson et al., 2008) and to maintain their body composition omnivores may supplement their low quality plant diets with higher nutrient prey (Denno and Fagan, 2003). Species with high nutrient content could be highly connected and central in the food web, effectively serving as keystone nodes that have a strong influence on both ecosystem functioning and food web stability. To our knowledge this expectation has not yet been tested in complex food webs. Notably, stoichiometry could turn out to be a key trait associated with complementarity effects in BEF research (Hillebrand et al., 2014). Therefore, metabolic theory and ecological stoichiometry theory, which describe the physiological and nutritional constraints of feeding interactions, provide important background for integrating BEF and complex interaction webs (Mulder and Elser, 2009; Mulder et al., 2013). These theories also permit explicit consideration of the scaling of interactions, from genes to individuals to ecosystems (Allen and Gillooly, 2009; Sterner and Elser, 2002). Consequently, it is likely that the traditional emphasis on aggregation of trophic groups will be relaxed in this phase. Instead, emphasis will be placed on the role of all levels of biodiversity in ecological networks that underlie the relationship between biodiversity and multiple EFs.

In summary, rather than relying entirely on comparative approaches to examine the consequences of different types of ecosystems (i.e. aquatic, aboveground, belowground) or interaction types (i.e. antagonistic vs. mutualistic or ecosystem engineering) on food web structure and stability, this phase of research will place a stronger emphasis on establishing causal drivers of changes in network structure and function. Relationships between complex ecological networks and ecosystem functioning will be evaluated by examining changes in the structure of species interaction webs across experimental gradients (Fig. 1F), by integrating species traits, and by including multiple interaction types into each web (Fontaine et al., 2011; Melian et al., 2009; Sanders et al., 2014; Suave et al., 2014). This phase of research, therefore, is likely to produce a better understanding of relationships between factors thought to influence food web structure (i.e. trophic effects, interaction effects, and compartmentalization effects) and factors associated with BEF relationships (complementarity effects, sampling effects, and selection effects). This understanding may help resolve long-standing debates about the relationship between interaction complexity, community stability and ecosystem functioning. As we look forward, we expect that this next phase of food web research will focus more strongly on scaling of multiple interaction types from local to global scales, and more directly link changes in network structure across all levels of ecological organization with the ability of ecosystems to maintain functions that provide services to human society.

2.2.4 Limitations of FWT with Respect to Understanding the Role of Food Webs in Ecosystem Functioning

The main limitation of FWT is that quantifying the influence of species interactions on ecosystem functioning remains deceptively difficult, due to challenges measuring the particular interaction (presence and strength), and subsequently establishing that the interaction is relevant for EF (Nowak, 2010). For example, some species-specific interactions, such as those among plants and pollinators (Burkle et al., 2013), or plants and some monophagous herbivores (Southwood and Leston, 1959), can be readily observed in the field and these interactions are, as a consequence, well established. However, documenting the presence of an interaction may not demonstrate its importance for ecosystem functioning. Feeding by an early-season herbivore may induce plant defences that increase resistance to herbivory later in the season (Faeth, 1986). In some cases, therefore, herbivory can enhance rather than limit plant productivity.

Feeding behaviour also may be cryptic, infrequent, and variable not only according to life stage (juveniles vs. adults) but also across seasons and years, making direct observations of many taxa challenging (Kaartinen and Roslin, 2012; Polis, 1991). Some chemical tracers, such as stable isotopes, lipid fatty acids, and molecular analysis of gut contents, can trace dominant energy channels and identify ingested prey to some degree of taxonomic resolution (Traugott et al., 2013). Even when sophisticated empirical methods are coupled with a quantification of prey availability, however, they may not reflect diet choice in other habitats where different prey species are available. Therefore, regardless of original method reporting the interaction, food webs based on potential interactions from the literature may not reflect realized feeding interactions in other habitats. For this reason, there is much interest in approaches that identify simple trait axes that can be used to distinguish the presence of a trophic interaction (Cohen and Newman, 1985; Eklöf et al., 2013; Williams and Martinez, 2000). Given the assumption that there are generalizable rules structuring food webs, machine learning systems can be used to detect patterns in webs and suggest where missing predator–prey links may be expected (Tamaddoni-Nezhad et al., 2013). Whether machine learning approaches can take the next step to accurately detect a full complement of positive and negative species interactions as well as their influence on ecosystem functioning remains an open question (Tamaddoni-Nezhad et al., 2015). Food webs developed using combined empirical and theoretical approaches constantly improve (and challenge) our understanding of food web structure.

Differences in sampling methods also may limit the application of the FWT to ecosystem functioning. Sampling methods, which may be important to assess the biology of each taxon, can make it difficult to assess density, biomass, and interaction strengths using common units of measurement for all taxa (Nowak, 2010). For example, pitfall traps (Birkhofer et al., 2008) and observations of flower visitation by pollinators (Ebeling et al., 2011) provide information on activity patterns rather than density or biomass *per se*. Conversely, the abundance of soil fauna sampled with soil cores (Kempson et al., 1963; MacFadyen, 1961) and aboveground fauna sampled with vacuum samplers (Brook et al., 2008) are more easily reported on a per unit area basis. Again, machine learning can be used to compare interactions gleaned from each approach (Tamaddoni-Nezhad et al., 2013). However, the per capita dry body mass of different soil faunal groups can range more than 10 orders of magnitude, from $<10^{-6}$ g for nematodes up to several grams for some earthworms (Sechi et al., 2015). Life history traits and foraging range can also vary by several orders of magnitude among species that coexist in the same habitat. Therefore, sampling that effectively captures the spatial distribution of each taxon at a scale that is comparable across taxa is challenging in many experimental plots where space is limited (De Deyn and Van der Putten, 2005; Kremen et al., 2007). Well-coordinated multi-investigator experiments that unite the efforts of scientists with a wide range of taxonomic and computational expertise (i.e. Roscher et al., 2004) have much to contribute to the further development of FWT.

3. PRINCIPLES FOR INTEGRATING BEF AND FWT

The models of BEF and FWT both describe species interactions and fluxes of nutrients and energy, and categorizing research as one approach or the other is not always a clear-cut distinction. For example, production of fish in aquatic habitats and pest suppression in terrestrial agriculture are two examples of particular ESs studied extensively using both approaches. Nevertheless, to understand when changes in biodiversity will directly, or indirectly, influence multiple ESs, it is useful to consider combined hypotheses (CH) that result from multiple possible groupings of hypotheses thought to explain diversity–functioning–stability relationships in BEF and FWT. Further, in this section we demonstrate that common quantitative frameworks can be established using three key principles that bridge assumptions behind both BEF and FWT. We acknowledge that none of these principles is new, but considered together in light of focal BEF and FWT hypotheses, they form a road map for hypotheses that guide the management of ESs essential for human well-being (Table 1).

3.1 Principle I: Interactions Occur between Taxonomic Units According to a Topology

Principle I may seem like a truism to ecologists in each area of expertise. Traditionally, however, the topologies of BEF and FWT interactions have been a bit different. The topologies of BEF studies often focus on species interactions within a single trophic level, or species that are connected by flux of nutrients and energy through simple interaction chains. The focus has been to describe taxa that coexist as competitors or facilitators feeding on the same resource (Fig. 1A and B). Complementarity effects, or unique aspects of each species, enhance an EF that reflects their resource consumption or collective accumulation of biomass (Table 1: Principle I-BEF). In contrast, because detailed diet information frequently is missing, food web studies often implicitly assume high levels of functional redundancy by aggregating species that share the same predators and the same prey into nodes that reflect 'trophic species' (Martinez, 1991; Williams and Martinez, 2000). The network topology is used to determine whether random or ordered loss of trophic species will trigger secondary extinctions. Therefore, trophic species generally are not directly associated with particular EFs (but see Gross and Cardinale, 2005). Instead, indexes describing the topology of species interactions (i.e. compartmentalization, connectivity, and omnivory) are associated with properties of food web stability, which are then indirectly associated with stability of ecosystem functioning as a whole, but without reference to particular functions (Table 1: Principle I-FWT).

Patterns of interactions among species will be an important predictor of when changes in species diversity that influence one focal EF will also affect other EFs (Table 1: CH1). Combined BEF–FWT approaches use group detection to consider the distribution of functionally redundant species within and between modules (Gauzens et al., 2015). Group detection in food webs can be applied to management of multiple ESs in several ways (Table 1). Detection of functionally unique compartments in space can be used to prioritize conservation of habitat patches that are particularly important for ESs or disservices. Such spatial compartments have been found in agricultural (Macfadyen et al., 2011) and salt marsh (Montoya et al., 2015) landscapes. Group detection also can identify particular species or resource inputs with high probability to influence multiple ESs. For example, a lot of attention has been placed on conservation of top predators that link energy channels due to their potential to drive 'ecological meltdowns' (Terborgh et al., 2001), and on regulation of pollutants and nutrients that causes dominance of one energy channel over another (Garay-Narváez et al., 2014; Scheffer et al., 2001).

3.2 Principle II: Energy and Material Fluxes through Food Webs Provide a Common Currency for Assessing the Influence of Biodiversity on Ecosystem Functioning

Principle II is important for understanding how biodiversity-mediated changes in the topology developed in Principle I contribute to ecosystem process rates. Combined BEF–FWT hypotheses propose that increases in biodiversity are not only more likely to include species that enhance ecosystem functioning within a trophic level (Principle II-BEF) but also more likely to include species that more efficiently transfer energy between trophic levels (Table 1: Principle II-FWT). This classic hypothesis of growth-defence trade-offs (Coley et al., 1985; Herms and Mattson, 1992) has rarely been considered in terms of flux of nutrients and energy in complex networks.

Combined BEF–FWT suggests that changes in diversity that limit uptake and transfer of biomass between trophic groups will influence multiple EFs (Table 1: CH2). Tests using dynamic equations (Carpenter et al., 1985; de Ruiter et al., 1995) or ecological network analysis (Borrett and Lau, 2014; Ulanowicz, 2011) can be used to examine how changes in biodiversity influence the stocks and flows of biomass, nutrients, and energy through food webs. Depending on the resource pool, these biodiversity induced changes in energy fluxes can be related to management of multiple ESs (Table 1). Robust management recommendations for land-use intensity rely on an understanding of ES trade-offs (Goldstein et al., 2012) that would benefit from an explicit BEF–FWT perspective. The conversion of land from rainforest to agricultural production of oil palm, for example, had strong effects on the efficiency of top predators, resulting in reductions in ecosystem functionality that were greater than effects of biodiversity loss alone (Barnes et al., 2014). Ecosystem network analysis that captures multi-trophic BEF relationships also can be used to prioritize conservation or harvesting of particular species. For example, production of fish in marine and freshwater systems is a focal ES that is well suited to combining BEF and trait-based food webs with quantitative links. A multi-trophic analysis from a large marine ecosystem demonstrated that selective harvesting of fish by body size, as opposed to unselective harvesting can change the biodiversity–ecosystem functioning relationship from unimodal to linear due to effects of releasing prey from larger predators (Fung et al., 2015). Such combined BEF–FWT evaluations give critical insights into when loss in functioning due to selective harvesting can overwhelm benefits gained from prey release (Fung et al., 2015). Consideration of how these trade-offs also influence multiple EFs will be important for developing an economic valuation of biodiversity.

3.3 Principle III: Multiple Types of Species Interactions Affect Ecosystem Functioning

Principle III states that multiple types of species interactions influence ecosystem functioning, including ecosystem engineering, parasitism, mutualism, and predation. This idea, which was presented in early models (May, 1972), and revisited recently (Mougi and Kondoh, 2012), serves as an important reminder about the diversity of interactions that influence food web structure. Principle II focuses on flux of nutrients and energy through the interaction topology. However, information is reported rarely about body size, and nutrient content for pathogens (Latz et al., 2012), parasites (Dunne et al., 2013), or pollinators (Woodward et al., 2005), despite the strong influence of these interactions on the maintenance of plant diversity (Klironomos, 2002), and BEF relationships (Maron et al., 2011; Schnitzer et al., 2011). To support principle III, therefore, we revisit the focal hypotheses of each research area. BEF research suggests that dominance of species with traits that contribute positively to ecosystem functioning selects for increased functioning in diverse mixtures (Table 1: Principle III BEF).

Combined BEF–FWT perspectives would then ask, 'Whose trait is it anyway?' by additionally considering the net effect of species interactions as an extended trait (Table 1: Principle III-FWT). A focal EF is then the net effect resulting from the sum of beneficial and antagonistic interactions described by network structure. Trade-offs between multiple EFs are caused by dominance of species that have net positive species interactions with respect to one function but net negative interactions with respect to another function (Table 1: CH3).

Third-generation SEM is well suited to evaluate CH3 (Grace et al., 2012). EFs can be conceptualized as latent variables that quantify the net effect of positive and negative interactions connected by an interaction topology (see Text Box 2 in Mulder et al., 2015). If data are collected in the same framework, SEM also stands to be a useful for comparing results from experimental manipulations with observational studies that examine relationships between consumer diversity and ecosystem functioning in natural ecosystems (Duffy et al., 2015; Mora et al., 2011). These results also could be used to manage multiple ESs (Table 1). Whalen et al. (2013) used SEM to examine associations between sea grass production and the direct and indirect effects of crustacean and gastropod mesograzers. They found temporal shifts in interactions among species that could be used to identify times where management actions such as nutrient regulation would be most effective. Here, additional information about multiple EFs and diverse species interactions would be particularly valuable. It has been suggested that interaction diversity itself should be a high conservation priority (Tylianakis et al., 2010; Memmott et al., 2007), as loss of multiple interaction types can have consequences for ecosystem functioning that precede loss of species diversity (Mougi and Kondoh, 2012; Valiente-Banuet et al., 2014). Establishing relationships between habitat conservation value and interaction networks would be needed to use this suggestion in practice (Heleno et al., 2012). Nonetheless, it is likely that more explicit consideration of species interactions as a source of complementarity effects that influences BEF (Eisenhauer, 2012; Poisot et al., 2013) and connects multiple EFs will improve decisions support for management of multiple ES.

4. CONSIDERING TRENDS IN BEF–FWT RESEARCH FOR BETTER MANAGEMENT OF MULTIPLE ESs

The concepts of ecosystem functioning and ecosystem services often use similar terminology and reasoning (Birkhofer et al., 2015; Mace et al., 2012; Mulder et al., 2015; Reyers et al., 2012). For example, results from soil food web studies that focus on mineralization, assimilation, and feeding rates to estimate energy fluxes (Moore and De Ruiter, 2012) can be used to assess ESs related to carbon sequestration (De Vries et al., 2013). In Section 3, we provided targeted examples demonstrating applications of combined BEF–FWT approaches. Here, we more broadly consider long-term trends in BEF–FWT research (see Section 2) as they apply to management of multiple ESs.

BEF and FWT research trends have led towards increased mechanistic understanding of detailed interaction webs (Fig. 1). However, outside of experimental settings detailed information about species interactions and ecosystem process rates often does not exist. It is tempting to say that the studies describing the complexity of species interactions reflect research mired in detail that cannot be applied to management of ESs. However, a potential difference between BEF and ES research suggests that detailed perspectives will prove to be useful. BEF research primarily focuses on how biodiversity influences functioning of communities (i.e. all dark green squares (black in the print version) species in Fig. 2), whereas ES allows for prioritization by stakeholders who may place differential value on particular services provided by separate species within the community (i.e. crop species indicated with a star) species in Fig. 2; Luck et al., 2009). Therefore, identifying and key trade-offs in BEF–FWT will provide important information for valuation of the ecological consequences

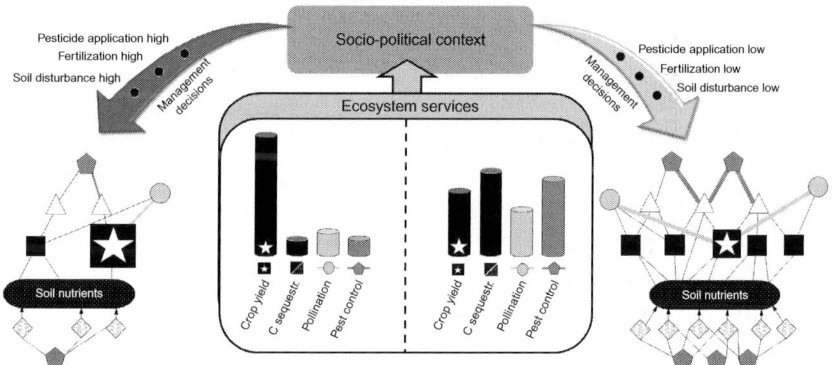

Figure 2 A unified BEF and FWT framework for management of multiple ecosystem services. This cartoon depicts connections between the diversity of species in a food webs and the management of multiple ecosystem services. Management decisions that focus purely on one ecosystem service such as crop yield can limit the balance of ecosystem services provided by other species in complex food webs (triangles-herbivores; pentagons-predators, circles-pollinators; diamonds-soil fauna). Socio-political context related to human population density, and stakeholder interests can influence feedbacks between ecosystem services and management of complex ecosystems.

ES trade-offs (i.e. Fig. 2 crop yield and C sequestration; De Groot et al., 2012; Gamfeldt et al., 2013; Lester et al., 2013).

While BEF focuses more on causal relationships between species in small-scale field plots, the focus of much ES research is on correlative patterns, often between land-use and ES, at larger spatial scales (Raudsepp-Hearne et al., 2010; Seppelt et al., 2011). Here, BEF–FWT research trends towards identifying causal drivers of thresholds in food web compartments could be useful for policy. Identifying factors that influence the connection between aboveground–belowground and terrestrial–aquatic networks is especially relevant because provenance of management agencies traditionally has been divided by food web compartment, trophic level, or ecosystem type. In the past policy for soil management was determined by different agencies than for air quality, and similarly policy for agricultural systems was made by separate governmental agencies than for ocean fisheries. Following trends in BEF–FWT research in recent decades, key agencies determining environmental policy, such as the US Environmental Protection Agency, DEFRA, the European Commission, and the Environmental Ministries of nations like Germany, have reorganized their policy-research programmes to reflect a more general consideration of ecosystem dynamics (TEEB, 2008; EPA, 2008). This is good news for those who propose to use biodiversity to manage the flux of nutrients across landscapes (Cardinale, 2011; Diaz and Rosenberg, 2008) and suggests that trends towards understanding BEF–FWT relationships across spatial scales should lead to more integrated ES policy.

Trends towards combined BEF–FWT research increasingly rely on quantitative network approaches. The challenges of modelling non-linear responses, feedbacks, and multiple interaction types in complex systems also apply to management of ES, which additionally considers the coupling of ecological dynamics to social systems (Levin et al., 2013). Management decisions that influence ES are often driven by expectations of multiple stakeholders and multiple management agencies that collectively define socio-political contexts (i.e. Fig. 2). These decisions can also engage scientists from several disciplines including sociologists, economists, geologists, and ecologists (Schröter et al., 2014). A key theme in the study of ESs, therefore, is the identification of holistic approaches that provide decision support for joint ways of thinking. Network approaches provide practical tools needed to examine factors that influence system stability (Levin et al., 2013) and can be used to make quantitative predictions that test a range of possible scenarios, reflecting socio-economic, political, and ecological interests (Marcot et al., 2006; McCann et al., 2006; Schmitt and Brugere, 2013). Ultimately, we expect that quantitative tools being used to combine BEF–FWT perspectives will support decision-making and assure broad-scale and long-term sustainability of resource use.

5. CONCLUSIONS

We conclude that network approaches are important tools that can be used to evaluate the contribution of diverse species assemblages to the maintenance of multiple EFs and ESs. The growth in network concepts over the last several decades, increasingly allows management decisions to be informed by more integrative approaches and evidence. In particular, our increased awareness of scaling, experimental replication of networks, and well-resolved webs that include multiple types of interactions are particularly valuable contributions to the understanding of the functions and services provided by diverse ecosystems. Although application still remains somewhat speculative, highly managed systems like agriculture (Macfadyen et al., 2011) and fisheries (Fung et al., 2015) currently provide the best examples of the potential for combined BEF and FWT approaches. Given the large-scale anthropogenic alteration of natural habitats (i.e. habitat destruction, biodiversity changes, nutrient pollution, and ocean acidification), we expect that understanding of species vulnerability and linkages developed in BEF experiments that adopt FWT approaches will provide valuable insights, which could be more broadly applied to the delivery, conservation, and restoration of ESs in the future.

ACKNOWLEDGEMENTS

This chapter was initiated at the 'Network Workshop' in the frame of the Jena Experiment, which is funded by the German Research Foundation (DFG, FOR 1451) with additional support from the Max Planck Society and the University of Jena. Further support to J.H. and N.E. came from the German Centre for Integrative Biodiversity Research (iDiv) Halle-Jena-Leipzig, funded by the German Research Foundation (FZT 118) and by the DFG (Ei 862/2-1, Ei 862/3-2, FOR 1451).

REFERENCES

Adair, E.C., Reich, P.B., Hobbie, S.E., Knops, J.M.H., 2009. Interactive effects of time, CO_2, N and diversity on total belowground carbon allocation and ecosystem carbon storage in a grassland community. Ecosystems 12, 1037–1052.

Allan, A., Weisser, W.W., Fischer, M., Schulze, E.-D., Weigelt, A., Roscher, C., Baade, J., Barnard, R.L., Beßler, H., Buchmann, N., Ebeling, A., Eisenhauer, N., et al., 2013.

A comparison of the strength of biodiversity effects across multiple functions. Oecologia 173, 223–237.

Allen, A.P., Gillooly, J.F., 2009. Towards an integration of ecological stoichiometry and the metabolic theory of ecology to better understand nutrient cycling. Ecol. Lett. 12, 369–384.

Atsatt, P.R., O'Dowd, D.J., 1976. Plant defense guilds. Science 193, 24–29.

Baiser, B., Gotelli, N.J., Buckley, H., Miller, T.E., Ellison, A., 2012. Geographic variation in network structure of a nearctic aquatic food web. Glob. Ecol. Biogeogr. 21, 579–591.

Balvanera, P., Pfisterer, A.B., Buchmann, N., He, J.-S., Nakashizuka, T., Raffaelli, D., Schmid, B., 2006. Quantifying the evidence for biodiversity effects on ecosystem functioning and services. Ecol. Lett. 9, 1146–1156.

Barbosa, P., Krischik, V., Jones, C.G., 1991. Microbial Mediation of Plant-Herbivore Interactions. John Wiley & Sons, Inc. New York, USA.

Barbosa, P., Hines, J.E., Kaplan, I., Martinson, H., Szczepaniec, A., Szendrei, Z., 2009. Associational resistance and susceptibility: having right or wrong neighbors. Annu. Rev. Ecol. Evol. Syst. 40, 1–20.

Bardgett, R.D., Van der Putten, W.H., 2014. Belowground biodiversity and ecosystem functioning. Nature 515, 505–511.

Bardgett, R.D., Mawdsley, J.L., Edwards, S., Hobbs, P.J., Rodwell, J.S., Davies, W.J., 1999. Plant species and nitrogen effects on soil biological properties of temperate upland grasslands. Funct. Ecol. 13, 650–660.

Barnes, A.D., Jochum, M., Mumme, S., Haneda, N.F., Farajallah, A., Widarto, T.H., Brose, U., 2014. Consequences of tropical land use for multitrophic biodiversity and ecosystem functioning. Nat. Commun. 5, 5351.

Bascompte, J., Jordano, P., 2007. Plant-animal mutualistic networks: the architecture of biodiversity. Annu. Rev. Ecol. Evol. Syst. 38, 567–593.

Bell, T., Newman, J.A., Silverman, B.W., Turner, S.L., Lilley, A.K., 2005. The contribution of species richness and composition to bacterial services. Nature 436, 1157–1160.

Berendse, F., 1983. Interspecific competition and niche differentiation between *Plantago lanceolata* and *Anthoxanthum odoratum* in a natural hayfield. J. Ecol. 71, 379–390.

Bersier, L.-F., Banašek-Richter, C., Cattin, M.-F., 2002. Quantitative descriptors of food-web matrices. Ecology 83, 2394–2407.

Best, R.J., Chaudoin, A.L., Bracken, M.E.S., Graham, M.H., Stachowicz, J.J., 2014. Plant-animal diversity relationships in a rocky intertidal system depend on invertebrate body size and algal cover. Ecology 95, 1308–1322.

Birkhofer, K., Wise, D.H., Scheu, S., 2008. Subsidy from the detrital food web, but not microhabitat complexity, affects the role of generalist predators in an aboveground herbivore food web. Oikos 117, 494–500.

Birkhofer, K., Diehl, E., Andersson, J., Ekroos, J., Früh-Müller, A., Machnikowski, F., Mader, V.L., Nilsson, L., Sasaki, K., Rundlöf, M., Wolters, V., Smith, H.G., 2015. Ecosystem services—current challenges and opportunities for ecological research. Front. Ecol. Evol. 2, 1–12.

Borrett, S.R., Lau, M.K., 2014. enaR: an r package for ecosystem network analysis. Methods Ecol. Evol. 5, 1206–1213.

Bradford, M.A., Wood, S.A., Bardgett, R.D., Black, H., Bonkowski, M., Eggers, T., Grayson, S.J., Kandeler, E., Manning, P., Setälä, H., Jones, T.H., 2014. Discontinuity in the responses of ecosystem processes and multifunctionality to altered soil community composition. Proc. Natl. Acad. Sci. U.S.A. 111, 14478–144483.

Briand, F., 1983. Environmental control of food web structure. Ecology 64, 253–263.

Brook, A.J., Woodcock, B.A., Sinka, M., Vanbergen, A.J., 2008. Experimental verification of suction sampler capture efficiency in grasslands of differing vegetation height and structure. J. Appl. Ecol. 45, 1357–1363.

Brown, J.H., Gillooly, J.F., Allen, A.P., Savage, V.M., West, G.B., 2004. Toward a metabolic theory of ecology. Ecology 85, 1171–1789.

Bruno, J.F., Boyer, K.E., Duffy, J.E., Lee, S.C., 2008. Relative and interactive effects of plant and grazer richness in a benthic marine community. Ecology 89, 2518–2528.

Burkle, L.A., Marlin, J.C., Knight, T.M., 2013. Plant-pollinator interactions over 120 years: loss of species, co-occurrence, and function. Science 339, 1611–1615.

Byrnes, J.E.K., Gamfeldt, L., Isbell, F., Lefcheck, J.S., Griffin, J.N., Hector, A., Cardinale, B.J., Hooper, D.U., Dee, L.E., Duffy, J.E., 2014. Investigating the relationship between biodiversity and ecosystem multifunctionality: challenges and solutions. Methods Ecol. Evol. 5, 111–124.

Cadotte, M.W., Cardinale, B.J., Oakley, T.H., 2008. Evolutionary history and the effect of biodiversity on plant productivity. Proc. Natl. Acad. Sci. U.S.A. 105, 17012–17017.

Cardinale, B.J., 2011. Biodiversity improves water quality through niche partitioning. Nature 472, 86–89.

Cardinale, B.J., Palmer, M.A., Collins, S.L., 2002. Species diversity enhances ecosystem functioning through interspecific facilitation. Nature 415, 426–429.

Cardinale, B.J., Harvey, C.T., Gross, K., Ives, A.R., 2003. Biodiversity and biocontrol: emergent impacts of a multi-enemy assemblage on pest suppression and crop yield in a agroecosystem. Ecol. Lett. 6, 857–865.

Cardinale, B.J., Srivastava, D., Duffy, J.E., Wright, J.P., Downing, A.L., Sankaran, M., Jouseau, C., 2006. Effects of biodiversity on the functioning of trophic groups and ecosystems. Nature 443, 989–992.

Cardinale, B.J., Matulich, K.L., Hooper, D.U., Byrnes, J.E.K., Duffy, J.E., Gamfeldt, L., Balvanera, P., O'Connor, M.I., Gonzalez, A., 2011. The functional role of producer diversity in ecosystems. Am. J. Bot. 98, 572–592.

Cardinale, B.J., Duffy, J.E., Gonzalez, A., Hooper, D.U., Perrings, C., Venail, P., Narwani, A., Mace, G.M., Tilman, D., Wardle, D., Kinzig, A.P., Daily, G.C., et al., 2012. Biodiversity loss and its impact on humanity. Nature 486, 59–67.

Carpenter, S.R., Kitchell, J.F., Hodgson, J.R., 1985. Cascading trophic interactions and lake productivity. Bioscience 35, 634–639.

Cohen, J.E., 1977. Ratio of prey to predators in community food webs. Nature 270, 165–167.

Cohen, J.E., Newman, C.M., 1985. A stochastic theory of community food webs. I. Models and aggregated data. Proc. R. Soc. Lond. B Biol. Sci. 224, 421–448.

Cohen, J.E., Schittler, D.N., Raffaelli, D.G., Reuman, D.C., 2009. Food webs are more than the sum of their tri-trophic parts. Proc. Natl. Acad. Sci. U.S.A. 106, 22335–22340.

Coley, P.D., Bryant, J.P., Chapin III, F.S., 1985. Resource availability and plant antiherbivore defense. Science 230, 895–899.

Costanza, R., d'Arge, R., De Groot, R., Farber, S., Grasso, M., Hannon, B., Limburg, K., Naeem, S., O'Neill, R.V., Paruelo, J., Raskin, R.G., Sutton, P., Van den Belt, M., 1997. The value of the world's ecosystem services and natural capital. Nature 387, 253–260.

Covich, A.P., Austen, M.C., Barlocher, F., Chauvet, E., Cardinale, B.J., Biles, C.L., Inchausti, P., Dangles, O., Solan, M., Gessner, M.O., Statzner, B., Moss, B., 2004. The role of biodiversity in the functioning of freshwater and marine benthic ecosystems. Bioscience 54, 767–775.

Crutsinger, G.M., Collins, M.D., Fordyce, J.A., Gompert, Z., Nice, C.C., Sanders, N.J., 2006. Plant genotypic diversity predicts community structure and governs an ecosystem process. Science 313, 966–968.

Dangles, O., Jonsson, M., Malmqvist, B., 2002. The importance of detritivore species diversity for maintaining stream ecosystem functioning following the invasion of a riparian plant. Biol. Invasions 4, 441–446.

Darwin, C., 1859. On the Origin of Species by Means of Natural Selection. John Murray, London.

De Deyn, G.B., Van der Putten, W.H., 2005. Linking aboveground and belowground diversity. Trends Ecol. Evol. 20, 625–633.

De Groot, R., Brander, R.L., Van der Ploeg, S., Costanza, R., Bernard, F., Braat, L., Christie, M., Crossman, N., Ghermandi, A., Hein, L., Hussain, S., Kumar, P., et al., 2012. Global estimates of the value of ecosystems and their services in monetary units. Ecosyst. Serv. 1, 50–61.

Denno, R.F., Fagan, W.F., 2003. Might nitrogen limitation promote omnivory among carnivorous arthropods? Ecology 84, 2522–2531.

Denno, R.F., Gratton, C., Döbel, H.G., Finke, D.L., 2003. Predation risk affects relative strength of top-down and bottom-up impacts on insect herbivores. Ecology 84, 1032–1044.

Deraison, H., Badenhausser, I., Börger, L., Gross, N., 2015. Herbivore effect traits and their impact on plant community biomass: an experimental test using grasshoppers. Funct. Ecol. 29, 650–661.

De Ruiter, P.C., Neutel, A.M., Moore, J.C., 1995. Energetics, patterns of interaction strengths, and stability in real ecosystems. Science 269, 1257–1260.

De Vries, F.T., Thébault, E., Liiri, M., Birkhofer, K., Tsiafouli, M.A., Bjørnlund, L., Jørgensen, H.B., Brady, M.V., Christensen, S., De Ruiter, P.C., Hertefeldt, T., Frouz, J., et al., 2013. Soil food web properties explain ecosystem services across European land use systems. Proc. Natl. Acad. Sci. U.S.A. 110, 14296–14301.

De Wit, C.T., Van den Bergh, J.P., 1965. Competition between herbage plants. Neth. J. Agric. Sci. 13, 212–221.

Diaz, R.J., Rosenberg, R., 2008. Spreading dead zones and consequences for marine ecosystems. Science 321, 926–929.

Díaz, S., Fargione, J., Chapin III, F.S., Tilman, D., 2006. Biodiversity loss threatens human well-being. PLoS Biol. 4, e277.

Dicke, M., 1994. Local and systemic production of volatile herbivore-induced terpenoids: their role in plant carnivore mutualism. J. Plant Physiol. 143, 465–472.

Douglass, J.G., Duffy, J.E., Bruno, J.F., 2008. Herbivore and predator diversity interactively affect ecosystem properties in an experimental marine community. Ecol. Lett. 11, 598–608.

Dreyer, J., Hoekman, D., Gratton, C., 2012. Lake-derived midges increase abundance of shoreline terrestrial arthropods via multiple trophic pathways. Oikos 121, 252–258.

Duffy, J.E., 2002. Biodiversity and ecosystem function: the consumer connection. Oikos 99, 201–219.

Duffy, J.E., Richardson, J.P., Canuel, E.A., 2003. Grazer diversity effects on ecosystem functioning in sea grass beds. Ecol. Lett. 6, 637–645.

Duffy, J.E., Cardinale, B.J., France, K.E., McIntyre, P.B., Thébault, E., Loreau, M., 2007. The functional role of biodiversity in ecosystems: incorporating trophic complexity. Ecol. Lett. 10, 522–538.

Duffy, J.E., Reynolds, P.L., Bostrom, C., Coyer, J.A., Cusson, M., Donadi, S., Douglass, J.G., Eklöf, J.S., Engelen, A.H., Eriksson, B.K., Fredriksen, S., Gamfeldt, L., et al., 2015. Biodiversity mediates top-down control in eelgrass ecosystems: a global comparative-experimental approach. Ecol. Lett. 18, 696–705.

Dunne, J.A., Williams, R.J., Martinez, N.D., 2002. Network structure and biodiversity loss in food webs: robustness increases with connectance. Ecol. Lett. 5, 558–567.

Dunne, J.A., Lafferty, K.D., Dobson, A.P., Hechinger, R.F., Kuris, A.M., Martinez, N.D., McLaughlin, J.P., Mouritsen, K.N., Koulin, R., Reise, K., Stouffer, D.B., Thieltges, D.W., et al., 2013. Parasites affect food web structure primarily through increased diversity and complexity. PLoS Biol. 11, e1001579.

Ebeling, A., Klein, A.M., Tscharntke, T., 2011. Plant-flower visitor interaction webs: temporal stability and pollinator specialization increases along an experimental plant diversity gradient. Basic Appl. Ecol. 12, 300–309.

Eisenhauer, N., 2012. Aboveground-belowground interactions as a source of complementarity effects in biodiversity experiments. Plant and Soil 351, 1–22.

Eisenhauer, N., Milcu, A., Allan, E., Nitschke, N., Scherber, C., Temperton, V., Weigelt, A., Weisser, W.W., Scheu, S., 2011. Impact of above- and below-ground invertebrates on temporal and spatial stability of grassland of different diversity. J. Ecol. 99, 572–582.

Eklöf, A., Jacob, U., Kopp, J., Bosch, J., Castro-Urgal, R., Chacoff, N.P., Dalsgaard, B., De Sassi, C., Galetti, M., Guimarães, P.R., Lomáscolo, S.B., González, A.M.M., et al., 2013. The dimensionality of ecological networks. Ecol. Lett. 16, 577–583.

Elton, C., 1927. Animal Ecology. MacMillan, New York, USA.

EPA: United States Environmental Protection Agency Office of Research and Development, 2008. Ecological Research Program Multi-Year Plan FY2008-2014. U.S. Environmental Protection Agency, Office of Research and Development, Washington D. C., USA.

Faeth, S.H., 1986. Indirect interactions between temporally separated herbivores mediated by the host plant. Ecology 67, 479–494.

Fagan, W.F., 1997. Omnivory as a stabilizing feature of natural communities. Am. Nat. 150, 554–567.

Finke, D.L., Denno, R.F., 2005. Predator diversity and the functioning of ecosystems: the role of intraguild predation in dampening trophic cascades. Ecol. Lett. 8, 1299–1306.

Flynn, D.F.B., Mirotchnick, N., Jain, M., Palmer, M.I., Naeem, S., 2011. Functional and phylogenetic diversity as predictors of biodiversity-ecosystem function relationships. Ecology 92, 1573–1581.

Fontaine, C., Guimarães, P.R., Kéfi, S., Loeuille, N., Memmott, J., Van der Putten, W.H., Van Veen, F.J., Thébault, E., 2011. The ecological and evolutionary implications of merging different types of networks. Ecol. Lett. 14, 1170–1181.

Fortuna, M.A., Stouffer, D.B., Olesen, J.M., Jordano, P., Mouillot, D., Krasnov, B.R., Poulin, R., Bascompte, J., 2010. Nestedness versus modularity in ecological networks: two sides of the same coin? J. Anim. Ecol. 79, 811–817.

Fowler, M.S., 2013. The form of direct interspecific competition modifies secondary extinction patterns in multi-trophic food webs. Oikos 122, 1730–1738.

Fox, J.W., 2004. Modelling the joint effects of predator and prey diversity on total prey biomass. J. Anim. Ecol. 73, 88–96.

Fung, T., Farnsworth, K.D., Reid, D.G., Rossberg, A.G., 2015. Impact of biodiversity loss on production in complex marine food webs mitigated by prey-release. Nat. Commun. 6, 6657.

Gamfeldt, L., Hillebrand, H., Jonsson, P.R., 2005. Species richness changes across two trophic levels simultaneously affect prey and consumer biomass. Ecol. Lett. 8, 696–703.

Gamfeldt, L., Snäll, T., Bagchi, R., Jonsson, M., Gustafsson, L., Kjellander, P., Ruiz-Jaen, M.C., Fröberg, M., Stendahl, J., Philipson, C.D., Mikusiński, G., Andersson, E., et al., 2013. Higher levels of multiple ecosystem services are found in forests with more tree species. Nat. Commun. 4, 1340.

Gamfeldt, L., Lefcheck, J.S., Byrnes, J.E.K., Cardinale, B.J., Duffy, J.E., Griffin, J.H., 2015. Marine biodiversity and ecosystem functioning: what's known and what's next? Oikos 124, 252–265.

Garay-Narváez, L., Flores, J.D., Arim, M., Ramos-Jiliberto, R., 2014. Food web modularity and biodiversity promote species persistence in polluted environments. Oikos 123, 583–588.

Gauzens, B., Thébault, E., Lacroix, G., Legendre, S., 2015. Trophic groups and modules: two levels of group detection in food webs. J. R. Soc. Interface 12, 20141176. http://dx.doi.org/10.1098/rsif.2014.1176.

Goldstein, J.H., Caldarone, G., Duarte, T.K., Ennaanay, D., Hannahs, N., Mendoza, G., Polasky, S., Wolny, S., Daily, G.C., 2012. Integrating ecosystem-service trade-offs into land-use decisions. Proc. Natl. Acad. Sci. U.S.A. 109, 7565–7570.

Grace, J.B., Schoolmaster Jr., D.R., Guntenspergen, G.R., Little, A.M., Mitchell, B.R., Miller, K.M., Schweiger, E.W., 2012. Guidelines for a graph-theoretic implementation of structural equation modeling. Ecosphere 3, 73.

Griffin, J.H., de la Haye, K.L., Hawkins, S.J., Thompson, R.C., Jenkins, S.R., 2008. Predator diversity and ecosystem functioning: density modifies the effect of resource partitioning. Ecology 89, 298–305.

Griffin, J.H., Byrnes, J.E.K., Cardinale, B.J., 2013. Effects of predator richness on prey suppression: a meta-analysis. Ecology 94, 2180–2187.

Gross, K., Cardinale, B.J., 2005. The functional consequences of random vs. ordered species extinctions. Ecol. Lett. 8, 409–418.

Haddad, N.M., Crutsinger, G.M., Gross, K., Haarstad, J., Knops, J.M.H., Tilman, D., 2009. Plant species loss decreases arthropod diversity and shifts trophic structure. Ecol. Lett. 12, 1029–1039.

Haddad, N.M., Crutsinger, G.M., Gross, K., Haarstad, J., Tilman, D., 2011. Plant diversity and the stability of foodwebs. Ecol. Lett. 14, 42–46.

Hairston, N.G., Smith, F.E., Slobodkin, L.B., 1960. Community structure, population control, and competition. Am. Nat. 94, 421–424.

Hawlena, D., Strickland, M.S., Bradford, M.A., Schmitz, O.J., 2012. Fear of predation slows plant-litter decomposition. Science 336, 1434–1438.

Hector, A., Bagchi, R., 2007. Biodiversity and ecosystem multifunctionality. Nature 448, 188–190.

Hector, A., Schmid, B., Beierkuhnlein, C., Caldeira, M.C., Diemer, M., Dimitrakopoulos, P.G., Finn, J.A., Freitas, H., Giller, P.S., Good, J., Harris, R., Högberg, P., et al., 1999. Plant diversity and productivity experiments in European grasslands. Science 286, 1123–1127.

Heleno, R., Devoto, M., Pocock, M., 2012. Connectance of species interaction networks and conservation value: is it any good to be well connected? Ecol. Indic. 14, 7–10.

Herms, D.A., Mattson, W.J., 1992. The dilemma of plants: to grow or defend. Q. Rev. Biol. 67, 283–335.

Hillebrand, H., Matthiessen, B., 2009. Biodiversity in a complex world: consolidation and progress in functional biodiversity research. Ecol. Lett. 12, 1405–1419.

Hillebrand, H., Cowles, J.M., Lewandowska, A., Van de Waal, D.B., Plum, C., 2014. Think ratio! A stoichiometric view on biodiversity ecosystem functioning research. Basic Appl. Ecol. 15, 465–474.

Hines, J., Gessner, M.O., 2012. Consumer trophic diversity as a fundamental mechanism linking predation and ecosystem functioning. J. Anim. Ecol. 81, 1146–1153.

Hooper, D.U., Bignell, D.E., Brown, V.K., Brussaard, L., Dangerfield, J.M., Wall, D.H., Wardle, D.A., Coleman, D.C., Giller, K.E., Lavelle, P., Van der Putten, W.H., De Ruiter, P.C., et al., 2000. Interactions between aboveground and belowground biodiversity in terrestrial ecosystems: patterns, mechanisms, and feedbacks. Bioscience 50, 1049–1061.

Hooper, D.U., Chapin III, F.S., Ewel, J.J., Hector, A., Inchausti, P., Lavorel, S., Lawton, J.-H., Lodge, D.M., Loreau, M., Naeem, S., Schmid, B., Setälä, H., et al., 2005. Effects of biodiversity on ecosystem functioning: a consensus of current knowledge. Ecol. Monogr. 75, 3–35.

Hooper, D.U., Adair, E.C., Cardinale, B.J., Byrnes, J.E.K., Hungate, B.A., Matulich, K.L., Gónzalez, A., Duffy, J.E., Gamfeldt, L., O'Connor, M.I., 2012. A global synthesis reveals biodiversity loss as a major driver of ecosystem change. Nature 486, 105–108.

Hunter, M.D., Price, P.W., 1992. Playing chutes and ladders—heterogeneity and the relative roles of bottom-up and top-down forces in natural communities. Ecology 73, 724–732.
Huston, M.A., 1997. Hidden treatments in ecological experiments: re-evaluating the ecosystem function of biodiversity. Oecologia 110, 449–460.
Ings, T.C., Montoya, J.M., Bascompte, J., Blüthgen, N., Brown, L., Dormann, C.F., Edwards, F., Figueroa, D., Jacob, U., Jones, J.I., Lauridsen, R.B., Ledger, M.E., et al., 2009. Ecological networks—beyond food webs. J. Anim. Ecol. 78, 253–269.
Ives, A.R., Cardinale, B.J., Snyder, W.E., 2005. A synthesis of subdisciplines: predator-prey interactions, and biodiversity and ecosystem functioning. Ecol. Lett. 8, 102–116.
Jabiol, J., McKie, B.G., Bruder, A., Bernadet, C., Gessner, M.O., Chauvet, E., 2013. Trophic complexity enhances ecosystem functioning in an aquatic detritus-based model system. J. Anim. Ecol. 82, 1042–1051.
Jones, C., Lawton, J. (Eds.), 1995. Linking Species and Ecosystems. Chapman & Hall, Inc., London, p. 387.
Jones, C., Lawton, J., Shachak, M., 1994. Organisms as ecosystem engineers. Oikos 69, 373–386.
Kaartinen, R., Roslin, T., 2012. High temporal consistency in quantitative food web structure in the face of extreme species turnover. Oikos 121, 1771–1782.
Karban, R., Maron, J., 2002. The fitness consequences of interspecific eavesdropping between plants. Ecology 83, 1209–1213.
Kempson, D., Lloyd, M., Ghelardi, R., 1963. A new extractor for woodland litter. Pedobiologia 3, 1–21.
Klironomos, J.N., 2002. Feedback with soil biota contributes to plant rarity and invasiveness in communities. Nature 417, 67–70.
Krause, A.E., Frank, K.A., Mason, D.M., Ulanowicz, R.E., Taylor, W.W., 2003. Compartments revealed in food-web structure. Nature 426, 282–285.
Kremen, C., 2005. Managing ecosystem services: what do we need to know about their ecology. Ecol. Lett. 8, 468–479.
Kremen, C., Williams, N.M., Aizen, M.A., Gemmill-Herren, B., LeBuhn, G., Minckley, R., Packer, L., Potts, S.G., Roulston, T., Steffan-Dewenter, I., Vazquez, D.P., Winfree, R., et al., 2007. Pollination and other ecosystem services produced by mobile organisms: a conceptual framework for the effects of land-use change. Ecol. Lett. 10, 299–314.
Lafferty, K.D., Allesina, S., Arim, M., Briggs, C.J., De Leo, G., Dobson, A.P., Dunne, J.A., Johnson, P.T., Kuris, A.M., Marcogliese, D.J., Martinez, N.D., Memmott, J., et al., 2008. Parasites in food webs: the ultimate missing links. Ecol. Lett. 11, 533–546.
Latz, E., Eisenhauer, N., Rall, B.C., Allan, E., Roscher, C., Scheu, S., Jousset, A., 2012. Plant diversity improves protection against soil-borne pathogens by fostering antagonistic bacteria communities. J. Ecol. 100, 597–604.
Lefcheck, J.S., Byrnes, J.E.K., Isbell, F., Gamfeldt, L., Griffin, J.H., Eisenhauer, N., Hensel, M.J.H., Hector, A., Cardinale, B.J., Duffy, J.E., 2015. Biodiversity enhances ecosystem multifunctionality across trophic levels and habitats. Nat. Commun. 6, 6936.
Lester, S.E., Costello, C., Halpern, B.S., Gaines, S.D., White, C., Barth, J., 2013. Evaluating tradeoffs among ecosystem services to inform marine spatial planning. Mar. Policy 38, 80–89.
Levin, S., Xepapadeas, T., Crépin, A.-S., Norberg, J., de Zeeuw, A., Folke, C., Hughes, T., Arrow, K., Barrett, S., Daily, G., Ehrlich, P., Kautsky, N., et al., 2013. Social-ecological systems as complex adaptive systems: modeling and policy implications. Environ. Dev. Econ. 18, 111–132.
Lind, E., Vincent, J.B., Weiblen, G.D., Cavender-Bares, J., Borer, E.T., 2015. Trophic phylogenetics: evolutionary influences on body size, feeding, and species associations in grassland arthropods. Ecology 96, 998–1009.

Lindeman, R.L., 1942. The trophic-dynamic aspect of ecology. Ecology 23, 399–418.
Loreau, M., Hector, A., 2001. Partitioning selection and complementarity in biodiversity experiments. Nature 412, 72–76.
Losey, J.E., Denno, R.F., 1998. The escape response of pea aphids to foliar-foraging predators: factors affecting dropping behaviour. Ecol. Entomol. 23, 53–61.
Luck, G.W., Harrington, R., Harrison, P.A., Kremen, C., Berry, P.M., Bugter, R., Dawson, T.R., De Bello, F., Dìaz, S., Feld, C.K., Haslett, J.R., Hering, D., et al., 2009. Quantifying the contribution of organisms to the provision of ecosystem services. Bioscience 59, 223–235.
MacArthur, R.H., 1955. Fluctuations of animal populations and a measure of community stability. Ecology 36, 533–536.
Mace, G.M., Norris, K., Fitter, A.H., 2012. Biodiversity and ecosystem services: a multilayered relationship. Trends Ecol. Evol. 27, 19–26.
MacFadyen, A., 1961. Improved funnel type extractor for soil arthropods. J. Anim. Ecol. 30, 171–184.
Macfadyen, S., Gibson, R.H., Symondson, W.O.C., Memmott, J., 2011. Landscape structure influences modularity patterns in farm food webs: consequences for pest control. Ecol. Appl. 21, 516–524.
Maestre, F.T., Quero, J.L., Gotelli, N.J., Escudero, A., Ochoa, V., Delgado-Baquerizo, M., García-Gómez, M., Bowker, M.A., Soliveres, S., Escolar, C., García-Palacios, P., Berdugo, M., et al., 2012. Plant species richness and ecosystem multifunctionality in global drylands. Science 335, 214–218.
Mancinelli, G., Mulder, C., 2015. Detrital dynamics and cascading effects on supporting ecosystem services. Adv. Ecol. Res. 53, 97–160.
Marcot, B.G., Steventon, J.D., Sutherland, G.D., McCann, R.K., 2006. Guidelines for developing and updating Bayesian belief networks applied to ecological modeling and conservation. Can. J. Forest Res. 36, 3063–3074.
Maron, J.L., Marler, M., Klironomos, J.N., Cleveland, C.C., 2011. Soil fungal pathogens and the relationship between plant diversity and productivity. Ecol. Lett. 14, 36–41.
Martinez, N.D., 1991. Artifacts or attributes—effects of resolution on the Little-Rock Lake food web. Ecol. Monogr. 61, 367–392.
Martinson, H.M., Schneider, K., Gilbert, J., Hines, J., Hambäck, P., Fagan, W.F., 2008. Detritivory: stoichiometry of a neglected trophic level. Ecol. Res. 23, 487–491.
May, R.M., 1972. Will a large complex system be stable? Nature 238, 413–414.
May, R.M., 1973. Stability and Complexity in Model Ecosystems. Princeton University Press, Princeton, USA.
McCann, K.S., 2000. The diversity-stability debate. Nature 405, 228–233.
McCann, K.S., Hastings, A., Huxel, G.R., 1998. Weak trophic interactions and the balance of nature. Nature 395, 794–798.
McCann, R.K., Marcot, B.G., Ellis, R., 2006. Bayesian belief networks: applications in ecology and natural resource management. Can. J. Forest Res. 36, 3053–3062.
MEA, 2005. Ecosystems and Human Well-Being: Synthesis. Island Press, Washington, DC.
Melian, C.J., Bascompte, J., Jordano, P., Křivan, V., 2009. Diversity in a complex ecological network with two interaction types. Oikos 118, 122–130.
Memmott, J., Gibson, R., Gigante Carvalheiro, L., Henson, K., Huttel Heleno, R., Lopezaraiza Mikel, M., Pearce, S., 2007. The conservation of ecological interactions. In: Stewart, A.J.A., New, T.R., Lewis, O.T. (Eds.), Proceedings of the Royal Entomological Society's 23rd Symposium on Insect Conservation Biology. CABI Publishing, Wallingford, UK, pp. 226–244. (Chapter 10).
Menge, B.A., Sutherland, J.P., 1987. Community regulation: variation in disturbance, competition, and predation in relation to environmental stress and recruitment. Am. Nat. 130, 730–757.

Menge, B.A., Blanchette, C.A., Raimondi, P., Freidenburg, T., Gaines, S., Lubchenco, J., Lohse, D., Hudson, G., Foley, M., Pamplin, J., 2004. Species interaction strength: testing model prediction s along an upwelling gradient. Ecol. Monogr. 74, 663–684.
Mikola, J., Setälä, H., 1998a. No evidence of trophic cascades in an experimental microbial-based soil food web. Ecology 79, 153–164.
Mikola, J., Setälä, H., 1998b. Relating species diversity to ecosystem functioning: mechanistic backgrounds and experimental approach with a decomposer food web. Oikos 83, 180–194.
Montoya, D., Yallop, M.L., Memmott, J., 2015. Functional group diversity increases with modularity. Nat. Commun. 6, 7379.
Moore, J.C., De Ruiter, P.C., 2012. Energetic Food Webs: An Analysis of Real and Model Ecosystems. Oxford University Press, Oxford.
Moore, J.C., McCann, K., de Ruiter, P.C., 2005. Modeling trophic pathways, nutrient cycling, and dynamic stability in soils. Pedobiologica 49, 499–510.
Mora, C., Aburto-Oropeza, O., Bocos, A.A., Ayotte, P.M., Banks, S., Bauman, A.G., Beger, M., Bessudo, S., Booth, D.J., Brokovich, E., Brooks, A., Chabanet, P., et al., 2011. Global human footprint on the linkages between biodiversity and ecosystem functioning in reef fishes. PLoS Biol. 9, e1000606.
Mougi, A., Kondoh, M., 2012. Diversity of interaction types and ecological community stability. Science 337, 349–351.
Mulder, C., 2011. World wide food webs: power to feed ecologists. Ambio 40, 335–337.
Mulder, C., Elser, J.J., 2009. Soil acidity, ecological stoichiometry and allometric scaling in grassland food webs. Glob. Cha. Biol. 15, 2730–2738.
Mulder, C.P.H., Uliassi, D.D., Doak, D.F., 2001. Physical stress and diversity-productivity relationships: the role of positive interactions. Proc. Natl. Acad. Sci. U.S.A. 98, 6704–6708.
Mulder, C., Ahrestani, F.S., Bahn, M., Bohan, D.A., Bonkowski, M., Griffiths, B.S., Guicharnaud, R.A., Kattge, J., Krogh, P.H., Lavorel, S., Lewis, O.T., Mancinelli, G., et al., 2013. Connecting the green and brown worlds: allometric stoichiometric predictability of above- and below-ground networks. Adv. Ecol. Res. 49, 69–175.
Mulder, C., Bennett, E.M., Bohan, D.A., Bonkowski, M., Carpenter, S.R., Chalmers, R., Cramer, W., Durance, I., Eisenhauer, N., Fontaine, C., Haughton, A.J., Hettelingh, J.-P., et al., 2015. 10 Years later: revisiting priorities for science and society a decade after the Millennium Ecosystem Assessment. Adv. Ecol. Res. 53, 1–53.
Naeem, S., Li, S., 1997. Biodiversity enhances ecosystem reliability. Nature 380, 507–509.
Naeem, S., Thompson, L.J., Lawler, S.P., Lawton, J.H., Woodfin, R.M., 1994. Declining biodiversity can alter the performance of ecosystems. Nature 368, 734–737.
Naeem, S., Duffy, J.E., Zavaleta, E., 2012. The functions of biological diversity in an age of extinction. Science 336, 1401–1406.
Nakano, S., Murakami, M., 2001. Reciprocal subsidies: dynamic interdependence between terrestrial and aquatic food webs. Proc. Natl. Acad. Sci. U.S.A. 98, 166–170.
Norberg, J., 2000. Resource-niche complementarity and autotrophic compensation determines ecosystem-level responses to increased cladoceran species richness. Oecologia 122, 264–272.
Nowak, M., 2010. Estimating interaction strengths in nature: experimental support for an observational approach. Ecology 91, 2394–2405.
O'Connor, M.I., Bruno, J.F., 2009. Predator richness has no effect in a diverse marine food web. J. Anim. Ecol. 78, 732–740.
Odum, E.P., 1969. The strategy of ecosystem development. Science 164, 262–270.
O'Neill, R.V., 2001. Is it time to bury the ecosystem concept? (with full military honors, of course!). Ecology 82, 3275–3284.

Ott, D., Digel, C., Rall, B.C., Maraun, M., Scheu, S., Brose, U., 2014. Unifying elemental stoichiometry and metabolic theory in predicting species abundance. Ecol. Lett. 17, 1247–1256.

Perrin, R.M., Phillips, M.L., 1978. Some effects of mixed cropping on the population dynamics of insect pests. Entomol. Exp. Appl. 24, 385–393.

Pimentel, D., 1961. Species diversity and insect population outbreaks. Ann. Entomol. Soc. Am. 54, 76–86.

Pimm, S.L., 1979. The structure of food webs. Theor. Popul. Biol. 16, 144–158.

Pimm, S.L., 1984. The complexity and stability of ecosystems. Nature 307, 321–326.

Pimm, S.L., Lawton, J.H., 1980. Are food webs divided into compartments? J. Anim. Ecol. 49, 879–898.

Pimm, S.L., Lawton, J.H., Cohen, J.E., 1991. Food web patterns and their consequences. Nature 350, 669–674.

Poisot, T., Mouquet, N., Gravel, D., 2013. Trophic complementarity drives the biodiversity-ecosystem functioning relationship in food webs. Ecol. Lett. 16, 853–861.

Polis, G.A., 1991. Complex trophic interactions in deserts an empirical critique of food-web theory. Am. Nat. 138, 123–155.

Polis, G.A., Hurd, S.D., 1996. Allochthonous input across habitats, subsidized consumers, and apparent trophic cascades: examples from the ocean-land interface. In: Polis, G.A., Winemiller, K.O. (Eds.), Food Webs: Integration of Patterns and Dynamics. Chapman and Hall, New York, USA, pp. 275–285.

Polis, G.A., Anderson, W.B., Holt, R.D., 1997. Toward an integration of landscape and food web ecology: the dynamics of spatially subsidized food webs. Annu. Rev. Ecol. Syst. 28, 289–316.

Raudsepp-Hearne, C., Peterson, G.D., Bennett, E.M., 2010. Ecosystem service bundles for analyzing tradeoffs in diverse landscapes. Proc. Natl. Acad. Sci. U.S.A. 107, 5242–5247.

Rauscher, M.D., 1981. The effect of native vegetation on the susceptibility of *Aristolochia reticulata* (Aristolochiaceae) to herbivore attack. Ecology 62, 1187–1195.

Reich, P.B., Knops, J., Tilman, D., Craine, J., Ellsworth, D., Tjoelker, M., Lee, T., Wedin, D., Naeem, S., Bahauddin, D., Hendrey, G., Jose, S., Wrage, K., Goth, J., Bengston, W., 2001. Plant diversity enhances ecosystem responses to elevated CO_2 and nitrogen deposition. Nature 410, 809–812.

Reyers, B., Polasky, S., Tallis, H., Mooney, H.A., Larigauderie, A., 2012. Finding common ground for biodiversity and ecosystem services. Bioscience 62, 503–507.

Rivest, D., Pacquette, A., Shipley, B., Reich, P.B., Messier, C., 2015. Tree communities rapidly alter soil microbial resistance and resilience to drought. Funct. Ecol. 29, 570–578.

Rooney, N., McCann, K., 2012. Integrating food web diversity, structure and stability. Trends Ecol. Evol. 27, 40–46.

Rooney, N., McCann, K., Gellner, G., Moore, J.C., 2006. Structural asymmetry and the stability of diverse food webs. Nature 442, 265–269.

Root, R.B., 1973. Organization of a plant-arthropod association in simple and diverse habitats: the fauna of collards (*Brassica oleracea*). Ecol. Monogr. 43, 95–124.

Roscher, C., Schumacher, J., Baade, J., Wilcke, W., Gleixner, G., Weisser, W.W., Schmid, B., Schulze, E.-D., 2004. The role of biodiversity for element cycling and trophic interactions: an experimental approach in a grassland community. Basic Appl. Ecol. 5, 107–121.

Rzanny, M., Voigt, W., 2012. Complexity of multitrophic interactions in a grassland ecosystem depends on plant species diversity. J. Anim. Ecol. 81, 614–627.

Rzanny, M., Kuu, A., Voigt, W., 2013. Bottom-up and top-down forces structuring consumer communities in an experimental grassland. Oikos 122, 967–976.

Sanders, D., Jones, C.G., Thébault, E., Bouma, T.J., Van der Heide, T., Van Belzen, J., Barot, S., 2014. Integrating ecosystem engineering and food webs. Oikos 123, 513–524.

Scheffer, M., Carpenter, S.R., Foley, J.A., Folke, C., Walker, B., 2001. Catastrophic shifts in ecosystems. Nature 413, 591–596.

Scherber, C., Eisenhauer, N., Weisser, W.W., Schmid, B., Voigt, W., Fischer, M., Schulze, E.-D., Roscher, C., Weigelt, A., Allan, E., Bessler, H., Bonkowski, M., et al., 2010. Bottom-up effects of plant diversity on multitrophic interactions in a biodiversity experiment. Nature 468, 553–556.

Scheu, S., 2001. Plants and generalist predators as links between the below-ground and above-ground system. Basic Appl. Ecol. 2, 3–13.

Schmitt, L.H.M., Brugere, C., 2013. Capturing ecosystem services, stakeholders' preferences and trade-offs in coastal aquaculture decisions: a Bayesian belief network application. PLoS One 8, e75956.

Schmitz, O.J., 2009. Effects of predator functional diversity on grassland ecosystem function. Ecology 90, 2339–2345.

Schnitzer, S.A., Klironomos, J.N., Hille Ris Lambers, J., Kinkel, L.L., Reich, P.B., Xiao, K., Rillig, M.C., Sikes, B.A., Callaway, R.M., Mangan, S.A., Van Nes, E.H., Scheffer, M., 2011. Soil microbes drive the classic plant divereristy-productivity pattern. Ecology 92, 296–303.

Schröter, M., Vander Zanden, E.H., Van Oudenhoven, A.P.E., Remme, R.P., Serna-Chavez, H.M., de Groot, R.S., Opdam, P., 2014. Ecosystem services as a contested concept: a synthesis of critique and counter-arguments. Conserv. Lett. 7, 514–523.

Schulze, E.-D., Mooney, H.A. (Eds.), 1993. Biodiversity and Ecosystem Function. Springer-Verlag, New York.

Sechi, V., Brussaard, L., De Goede, R.G.M., Rutgers, M., Mulder, C., 2015. Choice of resolution by functional trait or taxonomy affects allometric scaling in soil food webs. Am. Nat. 185, 142–149.

Seppelt, R., Dormann, C.F., Eppink, F.V., Lautenbach, S., Schmidt, S., 2011. A quantitative review of ecosystem service studies: approaches, shortcomings and the road ahead. J. Appl. Ecol. 48, 630–636.

Setälä, H., Laakso, J., Mikola, J., Huhta, V., 1998. Functional diversity of decomposer organisms in relation to primary production. Appl. Soil Ecol. 9, 25–31.

Shurin, J.B., Gruner, D.S., Hillebrand, H., 2006. All wet or dried up? Real differences between aquatic and terrestrial food webs. Proc. R. Soc. Lond. B Biol. Sci. 273, 1–9.

Siemann, E., Weisser, W.W., 2004. Testing the role of insects for ecosystem functioning. In: Weisser, W.W., Siemann, E. (Eds.), Insects and Ecosystem Function. Springer Ecological Studies, Berlin, pp. 383–401.

Snyder, W.E., Snyder, G.B., Finke, D.L., Straub, C.S., 2006. Predator biodiversity strengthens herbivore suppression. Ecol. Lett. 9, 789–796.

Southwood, T.R.E., Leston, D., 1959. Land and Water Bugs of the British Isles. Frederick Warne & Co. Ltd, London & New York.

Srivastava, D., Vellend, M., 2005. Biodiversity-ecosystem function research: is it relevant to conservation. Annu. Rev. Ecol. Evol. Syst. 36, 267–294.

Srivastava, D.S., Cardinale, B.J., Downing, A.L., Duffy, J.E., Jouseau, C., Sankaran, M., Wright, M.G., 2009. Diversity has stronger top-down than bottom-up effects on decomposition. Ecology 90, 1073–1083.

Sterner, R.W., Elser, J.J., 2002. Ecological Stoichiometry: The Biology of Elements from Molecules to the Biosphere. Princeton University Press, Princeton, MA.

Stouffer, D.B., Bascompte, J., 2011. Compartmentalization increases food-web persistence. Proc. Natl. Acad. Sci. U.S.A. 108, 3648–3652.

Straub, C.S., Snyder, W.E., 2006. Species identity dominates the relationship between predator biodiversity and herbivore suppression. Ecology 87, 277–282.

Strong, D.R., 1992. Are trophic cascades all wet? Differentiation and donor-control in speciose ecosystems. Ecology 73, 747–754.

Suave, A., Fontaine, C., Thebault, E., 2014. Structure-stability relationships in networks combining mutualistic and antagonistic interactions. Oikos 123, 378–384.

Sugihara, G., Schoenly, K., Trombla, A., 1989. Scale invariance in food web properties. Science 245, 48–52.

Symstad, A., Siemann, E., Haarstad, J., 2000. An experimental test of the effect of plant functional group diversity on arthropod diversity. Oikos 89, 243–253.

Tamaddoni-Nezhad, A., Milani, G.A., Raybould, A., Muggleton, S., Bohan, D.A., 2013. Construction and validation of food webs using logic-based machine learning and text mining. Adv. Ecol. Res. 49, 225–289.

Tamaddoni-Nezhad, A., Bohan, D., Raybould, A., Muggleton, S.H., 2015. Towards machine learning of predictive models from ecological data. In: Proceedings of the 24th International Conference on Inductive Logic Programming. Springer-Verlag, pp. 159–173. LNAI 9046.

TEEB, 2008. The Economics of Ecosystems & Biodiversity (TEEB): An Interim Report. European Communities. A Banson Production, Cambridge, UK.

Terborgh, J., Lopez, L., Nuñez, P., Rao, M., Shahabuddin, G., Orihuela, G., Riveros, M., Ascanio, R., Adler, G.H., Lambert, T.D., Balbas, L., 2001. Ecological meltdown in predator-free forest fragments. Science 294, 1923–1925.

Thebault, E., Fontaine, C., 2010. Stability of ecological communities and the architecture of mutualistic and trophic networks. Science 329, 853–856.

Thebault, E., Loreau, M., 2003. Food-web constraints on biodiversity-ecosystem functioning relationships. Proc. Natl. Acad. Sci. U.S.A. 100, 14949–14954.

Thebault, E., Loreau, M., 2006. The relationship between biodiversity and ecosystem functioning in food webs. Ecol. Res. 21, 17–25.

Thompson, R.M., Townsend, C.R., 2004. Land-use influences on New Zealand stream communities: effects on species composition, functional organization, and food-web structure. N. Z. J. Mar. Freshw. Res. 38, 595–608.

Thompson, R.M., Brose, U., Dunne, J.A., Hall, R.O.J., Hladyz, S., Kitching, R.L., Martinez, N.D., Rantala, H., Romanuk, T.N., Stouffer, D.B., Tylianakis, J.M., 2012. Food webs: reconciling the structure and function of biodiversity. Trends Ecol. Evol. 27, 689–697.

Tilman, D., 1996. Biodiversity: population versus ecosystem stability. Ecology 77, 350–363.

Tilman, D., Downing, J.A., 1994. Biodiversity and stability in grasslands. Nature 367, 363–365.

Tilman, D., Reich, P.B., Isbell, F., 2012. Biodiversity impacts ecosystem productivity as much as resources, disturbance, or history. Proc. Natl. Acad. Sci. U.S.A. 109, 10394–10397.

Tilman, D., Isbell, F., Cowles, J.M., 2014. Biodiversity and ecosystem functioning. Annu. Rev. Ecol. Evol. Syst. 45, 471–493.

Traugott, M., Kamenova, S., Ruess, L., Seeber, J., Plantegenest, M., 2013. Empirically characterising trophic networks: what emerging DNA based methods, stable isotope and fatty acid analyses can offer. Adv. Ecol. Res. 49, 177–224.

Tylianakis, J.M., Tscharntke, T., Lewis, O.T., 2007. Habitat modification alters the structure of trophical host-parasitoid food webs. Nature 445, 202–205.

Tylianakis, J.M., Laliberté, E., Nielsen, A., Bascompte, J., 2010. Conservation of species interaction networks. Biol. Conser. 143, 2270–2279.

Ulanowicz, R.E., 2011. Quantitative methods for ecosystem network analysis and its application to coastal ecosystems. In: Wolanski, E., McLusky, D.S. (Eds.), Treatise on Estuarine and Coastal Science. Academic Press, Waltham, pp. 35–57.

Ulanowicz, R.E., Kemp, W.M., 1979. Toward canonical trophic aggregations. Am. Nat. 114, 871–883.

Valiente-Banuet, A., Aizen, M.A., Alcántara, J.M., Arroyo, J., Cocucci, A., Galetti, M., García, M.B., García, D., Gómez, J.M., Jordano, P., Medel, R., Navarro, L., et al.,

2014. Beyond species loss: the extinction of ecological interactions in a changing world. Funct. Ecol. 29, 299–307.

Van der Heijden, M., Klironomos, J.N., Ursic, M., Moutoglis, P., Streitwolf-Engel, R., Boller, T., Wiemken, A., Sanders, I.R., 1998. Mycorrhizal fungal diversity determines plant biodiversity, ecosystem variability and productivity. Nature 396, 69–72.

Wagg, C., Bender, S.F., Widmer, F., Van der Heijden, M.G.A., 2014. Soil biodiversity and soil community composition determine ecosystem multifunctionality. Proc. Natl. Acad. Sci. U.S.A. 111, 5266–5270.

Wardle, D., 1999. Is 'sampling effect' a problem for experiments investigating biodiversity-ecosystem function relationships? Oikos 87, 403–407.

Wardle, D.A., 2002. Communities and Ecosystems: Linking the Aboveground and Below-ground Components. Princeton University Press, Princeton, NJ.

Wardle, D.A., Bardgett, R.D., Klironomos, J.N., Setälä, H., Van der Putten, W.H., Wall, D.H., 2004. Ecological linkages between aboveground and belowground biota. Science 304, 1629–1633.

Wardle, D.A., Williamson, W.M., Yeates, G.W., Bonner, K.I., 2005. Trickle-down effects of aboveground trophic cascades on the soil food web. Oikos 111, 348–358.

Wardle, D.A., Bardgett, R.D., Callaway, R.M., Van der Putten, W.H., 2011. Terrestrial ecosystem responses to species gains and losses. Science 332, 1273–1277.

Whalen, M.A., Duffy, J.E., Grace, J.B., 2013. Temporal shifts in top-down vs. bottom-up control of epiphytic algae in a seagrass ecosystem. Ecology 94, 510–520.

Williams, R.J., Martinez, N.D., 2000. Simple rules yield complex food webs. Nature 404, 180–183.

Wolters, V., Silver, W.L., Bignell, D.E., Coleman, D.C., Lavelle, P., Van der Putten, W.H., De Ruiter, P.C., Rusek, J., Wall, D.H., Wardle, D.A., Brussaard, L., Dangerfield, J.M., et al., 2000. Effects of global changes on above- and belowground biodiversity in terrestrial ecosystems: implications for ecosystem functioning. Bioscience 50, 1089–1098.

Woodward, G., Ebenman, B., Emmerson, M., Montoya, J.M., Olesen, J.M., Valido, A., Warren, P.H., 2005. Body size in ecological networks. Trends Ecol. Evol. 20, 402–409.

Wootton, T.J., 1994. The nature and consequences of indirect effects in ecological communities. Annu. Rev. Ecol. Syst. 25, 443–466.

Worm, B., Lotze, H.K., Myers, R.A., 2003. Predator diversity hotspots in the blue ocean. Proc. Natl. Acad. Sci. U.S.A. 100, 9884–9888.

Yodzis, P., 1982. The compartmentalization of real and assembled ecosystems. Am. Nat. 120, 551–570.

Zak, D.R., Holmes, W.E., White, D.C., Peacock, A.D., Tilman, D., 2003. Plant diversity, soil microbial communities and ecosystem function: are there any links. Ecology 2003, 2042–2050.

CHAPTER FIVE

Persistence of Plants and Pollinators in the Face of Habitat Loss: Insights from Trait-Based Metacommunity Models

Julia Astegiano*,†,‡,1, Paulo R. Guimarães Jr.*, Pierre-Olivier Cheptou†, Mariana Morais Vidal*, Camila Yumi Mandai*,§, Lorena Ashworth‡, François Massol†,¶

*Departamento de Ecologia, Instituto de Biociências, Universidade de São Paulo, São Paulo, Brazil
†CEFE UMR 5175, CNRS—Université de Montpellier—Université Paul-Valéry Montpellier—EPHE campus CNRS, Montpellier, France
‡Instituto Multidisciplinario de Biología Vegetal, Universidad Nacional de Córdoba–CONICET, Córdoba, Argentina
§Imperial College London, Silwood Park Campus, Ascot, London, United Kingdom
¶Laboratoire EEP, CNRS UMR 8198, Université Lille 1, Villeneuve d'Ascq, Lille, France
[1]Corresponding author: e-mail address: juastegiano@gmail.com

Contents

1. Introduction 202
   1.1 Habitat Loss and Fragmentation Effects on Pollinator Diversity, Pollination Service and Plant Reproduction 202
   1.2 Traits Modulating Wild Plant Response to Habitat Fragmentation 204
   1.3 Linking Plant Breeding System, Dispersal Ability and Pollination Generalization 206
   1.4 Plant–Pollinator Networks Organization and Its Role in Promoting Species Persistence in Fragmented Landscapes 208
   1.5 Predicting Species Persistence in the Face of Habitat Loss Using Trait-Based Metacommunity Models 210
2. A Trait-Based Metacommunity Model to Understand Plant and Pollinator Persistence in the Face of Habitat Loss 212
   2.1 Constructing Theoretical Plant–Pollinator Networks 212
   2.2 Constructing the Trait-Based Metacommunity Model 213
   2.3 Constructing Theoretical Scenarios Linking Plant Traits 214
   2.4 Statistical Analyses 219
3. Results 220
   3.1 Plant–Pollinator Metacommunity Persistence 220
   3.2 Plant–Pollinator Network Organization 227
   3.3 Plant Traits 227
4. Discussion 231

Advances in Ecological Research, Volume 53
ISSN 0065-2504
http://dx.doi.org/10.1016/bs.aecr.2015.09.005

4.1 Harbouring Plant Species with Different Dispersal Ability Matters for Metacommunity Persistence 231
4.2 Habitat Loss May Select for Higher Dispersal and Autonomous Self-Pollination but Not Pollination Generalization 234
4.3 Integrating the Network Approach to the Management of Pollination Services in Human-Dominated Landscapes 236
4.4 Evidencing the Contrasting Effects of Lower Dependence on Pollinators on Species Persistence 238
4.5 Caveats of the Trait-Based Metacommunity Model 239
5. Future Directions: Pollination Services in Human-Dominated Landscapes 241
Acknowledgements 242
Appendix A. Generating Bipartite Incidence Matrices with Determined Degree Sequences 243
Appendix B. Nestedness Depends on the Distribution of Degrees 247
References 248

Abstract

The loss of natural habitats is one of the main causes of the global decline of biodiversity. Understanding how increasing habitat loss affects ecological processes is critical for mitigating the effects of environmental changes on biodiversity and thus on the supply of ecosystem services by natural habitats. Habitat loss negatively affects pollinator diversity and the pollination service provided by insects, a key ecosystem service supporting the quantity, quality and diversity of crops directly consumed by humans and the sexual reproduction of most flowering plants. By integrating evolutionary relationships among traits that may modulate plant response to habitat loss, the structure of plant–pollinator interaction networks and metacommunity models, we examine how plant–pollinator metacommunities might respond to habitat loss. The main predictions of our trait-based metacommunity model are that (1) variation on dispersal ability among plant species may prevent full metacommunity collapse under pollinator loss associated with increasing habitat loss; (2) habitat loss may select for plants with higher dispersal ability and higher autogamous self-pollination, and will typically decrease the incidence of pollination generalist plants; (3) metacommunities that comprise plants with high autonomous self-pollination ability may harbour higher richness of rare plant species when pollinator diversity declines with increased habitat loss. We discuss the implications of our results for the vulnerability of pollination services for biotically pollinated wild plants and crops co-occurring in human-dominated landscapes.

1. INTRODUCTION

1.1 Habitat Loss and Fragmentation Effects on Pollinator Diversity, Pollination Service and Plant Reproduction

It is expected that anthropogenic land-use change will have the largest impact on global biodiversity for the foreseeable future (Haddad et al.,

2015; Krauss et al., 2010; Sala et al., 2000). Habitat loss and habitat fragmentation *per se* are among the main consequences of land-use change (Fahrig, 2003; Fisher and Lindemayer, 2007). These two consequences of anthropogenic land-use change have not only detrimental effects on biodiversity but also on the supply of multiple ecosystem services (e.g. pollination and flood and pest regulation) by remnant fragments of natural habitat to the surrounding human-dominated landscapes (Mitchell et al., 2015a).

Conceptually, habitat loss implies the removal of natural habitat from a landscape, while habitat fragmentation *per se* implies the "breaking apart" of continuous habitat (Fahrig, 2003). Although habitat loss has greater negative effect on biodiversity than habitat fragmentation (Fahrig, 2003), few empirical studies have evaluated separately the effects of these two different consequences of land-use change (Fahrig, 2003; Hadley and Betts, 2012; Tscharntke et al., 2012) mainly because habitat loss and fragmentation typically occur together in human-dominated landscapes (Fahrig, 2003). Thus, even most of the available empirical studies are presented as evaluations of the effects of habitat fragmentation they generally should reflect the confounding effects of both processes (Fahrig, 2003).

Habitat loss and fragmentation have negative effects on the population size of plants and animals and on the ecological interactions between them (Biesmeijer et al., 2006; Fahrig, 2003). These negative effects seem to be more consistent for mutualistic (pollination and seed dispersal) than for antagonistic interactions (predation and herbivory; Magrach et al. 2014) and are considered one of the chief causes of the decline of pollinator richness and abundance (Potts et al., 2010; Steffan-Dewenter et al., 2005; Winfree et al., 2009). Interestingly, pollinator diversity seems to be rather resilient to intermediate levels of habitat loss, with noticeable negative effects on diversity only at high levels of habitat loss (Ekroos et al., 2010; Winfree et al., 2009). However, even without species extinctions, important reductions in population size may decrease species encounter probability and thus can lead to the loss of ecological interactions well before species disappearance in human-dominated landscapes (Aizen et al., 2012; Sabatino, et al. 2010; Valiente-Banuet et al., 2015).

Plant–pollinator interactions are essential for generating and maintaining biodiversity and ecosystem services and functions (Bascompte et al., 2006; Fontaine et al., 2006; MEA, 2005; Potts et al., 2010; van der Niet and Johnson, 2012). Most flowering plants (87.5%) and most crops directly consumed by humans depend to some degree on the pollination service provided by animals to produce fruits and seeds (Klein et al., 2007; Ollerton et al., 2011). Thus, animal pollination contributes not only to the productivity of crops but also to the sexual reproduction of wild plants that either provide other services (e.g. medicinal plants) or serve as food sources for other organisms that provide other ecosystems services (e.g. natural enemies; Kremen et al., 2007). The amount and the performance of the seeds produced by wild plants and crops are important demographic and agricultural yield components. Seed quantity and quality determine the maximum population recruitment potential for the next generation (Ashworth et al., 2015a; González-Varo et al., 2010; Mathiasen et al., 2007; Wilcock and Neiland, 2002) and the productivity and nutritional quality of crops (Ashworth et al., 2009; Eliers et al., 2011). Changes induced by habitat loss and fragmentation on pollinator diversity and behaviour may therefore affect directly plant species diversity (Aguilar et al., 2008; Anderson et al., 2011; Fontaine et al., 2006; Potts et al., 2010) and crop yield, quality and the diversity of production (Ashworth et al., 2009; Eliers et al., 2011; Garibaldi et al., 2013, 2015; Klein et al., 2007).

The proximate and ultimate causes of wild plant diversity loss due to habitat fragmentation have been recently reviewed. A quantitative global synthesis showed negative effects of fragmentation on the pollination of wild animal-pollinated plants with concomitant reductions in plant reproductive success (Aguilar et al., 2006). Moreover, wild plant progeny generated in fragmented habitats can have lower performance, e.g., lower capacity of seed germination and seedling growth (e.g. Aguilar et al., 2012; Breed et al., 2012; González-Varo et al., 2010) as they had higher inbreeding coefficients compared to progeny from continuous habitats (Aguilar et al., 2008; Eckert et al., 2010). Thus, wild plant species in fragmented habitats produce not only a lower quantity but also a lower quality of progeny compared to those in continuous habitats (Aguilar et al., 2006, 2008).

However, we expect that plant species persistence in fragmented landscapes may depend on biological traits associated with plant sensitivity to pollinator loss such as breeding system (i.e. plant reproductive dependence on pollinators), pollination generalization and plant dispersal ability (Aguilar et al., 2006; Eckert et al., 2010; Hagen et al., 2012; Magrach et al., 2014). Here, we review evidence supporting the idea that these biological traits may determine differential responses of plants to habitat fragmentation.

1.2 Traits Modulating Wild Plant Response to Habitat Fragmentation

To produce seeds sexually, flowering plants range from complete dependence on animal pollination up to complete autonomy from pollinators either via spontaneous self-pollination (Lloyd, 1992; Richards, 1997; Vogler and Kalisz, 2001) or via wind pollination (Fægri and van der Pijl, 1979; Mulder et al., 2005). Dioecious, monoecious and hermaphrodite self-incompatible plant species are obligate outbreeders that completely depend on pollinator agents to exchange pollen among plants and to sexually reproduce with success. Conversely, self-compatible plant species may be considered facultative outbreeders that partially depended on animal pollination. Although animal pollinators are needed to transport pollen, a single visit of a pollinator to each individual flower may allow seed production. Moreover, some self-compatible species may have the ability to reproduce sexually via autonomous self-pollination, without the intervention of pollinators (Richards, 1997). As expected, results from a meta-analysis on the effects of habitat fragmentation on plant pollination and reproduction show that the reproductive success of plant species with higher dependence on animal pollination (i.e. self-incompatible plants) was more negatively and strongly affected than that of less dependent ones (i.e. self-compatible species; Aguilar et al., 2006). Moreover, habitat fragmentation can decrease the incidence of species highly dependent on animal pollination, as reported for tropical trees of a fragmented landscape of the Brazilian Atlantic Forest (Girão et al., 2007).

The sensitivity of plants to habitat fragmentation may also be determined by their degree of pollination generalization (Bond, 1994; Johnson and Steiner, 2000; Renner, 1998). Plant species range from "super-generalists" that interact with hundreds of pollinator species to "extreme specialists" interacting with just a single pollinator species (Fægri and van der Pijl, 1979; Waser et al., 1996). Conventionally, the expectation has been that the sexual reproduction of specialist plants should be more affected by habitat fragmentation than that of generalists because losing a few pollinator species locally is more likely than losing all the pollinators associated with a generalist plant species. This prediction was grounded in the idea that any change imposed by fragmentation in pollinator assemblages is more likely to cause reproductive failure in plants interacting with pollinator assemblages of lower richness (Aizen et al., 2002; Bond, 1994; Waser et al., 1996). Conversely, generalist plants are expected to be more resilient to the changes imposed by fragmentation on their pollinator assemblages because of the functional redundancy among their pollinators (Fægri and van der Pijl, 1979; Morris, 2003). For both self-compatible and self-incompatible species, however, the negative effect of habitat fragmentation on plant reproductive success seems to be independent of plant pollination generalization (Aguilar et al., 2006).

The number of seeds produced by plants and their dispersal mode are the main traits determining species dispersal success (Willson and Traveset, 2000). Habitat fragmentation may modify seed dispersal success by affecting seed size and quantity (e.g. Aguilar et al., 2006; Fakheran et al., 2010; Galetti et al., 2013), plant and inflorescence height (Fakheran et al., 2010; Lobo et al., 2011) and the diversity and behaviour of dispersal vectors (Cordeiro et al., 2009; Galetti et al., 2013). Overall, increased dispersal ability would appear to be favoured in fragmented landscapes (Hagen et al., 2012; but see Cheptou et al., 2008). It has been reported, for instance, that habitat fragmentation affects more negatively the proportion of seeds of plant species with larger seeds and of animal-dispersed plants arriving in habitat fragments (Magrach et al., 2014; McEuen and Curran, 2004). The negative relationship between seed size and fragment occupancy (Ehrlén and Eriksson, 2000) and the lower diversity of animal-dispersed plant species in forest fragments (Tabarelli et al., 1999) also suggest that fragmentation may select for smaller seed size and abiotically dispersed species (Fakheran et al., 2010; Galetti et al., 2013; Lobo et al., 2011; Magrach et al., 2014; Melo et al., 2010). Moreover, as seed production may be positively related to the probability of plant species occurrence in isolated habitat fragments (Evju et al., 2015), more fecund plant species will have higher probabilities of persistence in fragmented landscapes (McEuen and Curran, 2004). Finally, it has been recently suggested that when a landscape becomes more fragmented over evolutionary relevant time scales, increased (mean and long-distance) dispersal rates will be selected (Aparicio et al., 2008; Koh et al., 2015; but see Cheptou et al., 2008). This prediction seems to be supported by empirical evidence showing that increased isolation among patches leads to increased richness of species with long-distance dispersal and to decreased richness of species with short-distance dispersal (Aparicio et al., 2008; Koh et al., 2015).

1.3 Linking Plant Breeding System, Dispersal Ability and Pollination Generalization

As species are characterized by sets of traits, associations among these plant traits may ultimately determine plant response to fragmentation. The question of how breeding systems and dispersal traits interact in plants has been discussed widely in the literature. As sexual reproduction typically requires more than one partner, it is expected a link between the traits of movement (dispersal) and those associated with the breeding system. Moreover, seeds are mostly the product of sexual reproduction across plant species, which suggests *a priori* that breeding and dispersal may be functionally constrained. In this regard, a very influential line of argument has been inspired by island studies in the 1960s. Baker (1955) hypothesized that uniparental reproduction should be advantageous in recently colonized areas where pollinators or mating partners are scarce (this hypothesis is often referred to as "Baker's Law"). The ability of species to frequently colonize new areas is expected to be correlated with high dispersal ability. As a consequence, outbreeding strategies such as full outcrossers or dioecious species are expected to be associated with low colonization (dispersal) ability, while selfers are expected to be associated with higher colonization (dispersal) ability. The data for associations between these traits are, however, inconclusive (Auld and de Casas, 2013; Martén-Rodríguez et al., 2015) and equivocal (see Cheptou, 2012 for a review).

Historically, the high proportion of dioecious plants on islands was considered a problem for Baker's law (Carlquist, 1966). Remote island floras are, however, difficult to interpret because post-colonization evolution may obscure effects consistent with Baker's expectations. Using historical data on forest colonization, Réjou-Méchain and Cheptou (2015) were able to show unambiguously that recently colonized areas exhibit a higher proportion of dioecious species than the mature forests close by. Thus, in contrast to the expectations of Baker's law, Réjou-Méchain and Cheptou (2015) data suggest a positive association between outcrossing levels and plant dispersal ability that is also predicted by some theoretical models.

In agreement with this last empirical finding, a recent metapopulation model examining the joint evolution of self-fertilization and seed dispersal with locally variable pollination environment over time showed that outcrossing and dispersal jointly evolve (Cheptou and Massol, 2009). The outcrossing–disperser syndrome emerges because the temporal variability in the deposition of outcross pollen into stigmas creates fitness heterogeneity for outcrossers but not for self-pollinated plants. This temporal heterogeneity encountered by outcrossers selects for good dispersal, as already demonstrated in evolutionary models of dispersal (Comins et al., 1980; see also Massol and Débarre, 2015). As fluctuations in pollinator service may limit the deposition of outcross pollen on stigmas, Cheptou and Massol's (2009) model highlights the importance of these fluctuations for the evolution of plant dispersal ability. As a consequence of buffering fluctuations in pollination service, plant pollination generalization may reduce selection for good dispersal, and it can be hypothesized that low dispersing plants with high dependence on pollinators to reproduce should be generalists

(Astegiano et al., 2015). A positive association between plant dependence on pollinators and generalization was recently reported for a dune marshland plant community in the Balearic Islands (Tur et al., 2013). However, a study comparing plant species with different dependence on pollinators and dispersal ability in 10 plant–pollinator communities around the world found that plants highly dependent on pollinators or with low dispersal ability may not be more generalist (Astegiano et al, 2015).

The detailed study of interactions between individuals of co-occurring plant and pollinator species shows that these interactions are immersed in complex networks, with highly regular patterns of organization (Bascompte et al., 2003; Jordano, 1987; Jordano et al., 2003; Olesen et al., 2007). The topological properties of networks have different consequences for the ecology and the evolution of species (Bascompte and Jordano, 2007; Bascompte et al., 2006; Guimarães et al., 2011). Thus, our understanding of the way metacommunities of plants and pollinators may persist in fragmented landscapes may be improved by studies integrating not only relationships among plant traits determining plant sensitivity to pollinator loss but also the topological properties of plant–pollinator interaction networks.

1.4 Plant–Pollinator Networks Organization and Its Role in Promoting Species Persistence in Fragmented Landscapes

In plant–pollinator networks, most species interact with a small proportion of possible partners, whereas few species are "super-generalists" (Jordano et al., 2003; Vázquez, 2005). Indeed, interactions are mainly organized in a nested way (Bascompte et al., 2003; but see Blüthgen, 2010; Dorman et al., 2009), which means that specialist plants interact with subsets of pollinators interacting with more generalist plants. Moreover, there is a high incidence of asymmetric plant–pollinator interactions, with specialist plants and pollinators interacting, respectively, with generalist pollinators and plants (Vázquez and Aizen, 2004). These network features imply that a high proportion of plants and pollinators may persist, while generalist species persist and that plant–pollinator networks may be highly stable (Astegiano et al., 2015; Bastolla et al., 2009; Kaiser-Bunbury et al., 2010; Memmott et al., 2004; Okuyama and Holland, 2008; Rezende et al., 2007; Rohr et al., 2014; Suweis et al., 2013; Thébault and Fontaine, 2010; but see Allesina and Tang, 2012; James et al., 2012; Vieira and Almeida-Neto, 2015).

Consequently, it has been proposed that the nested and asymmetric nature of the interactions among plants and pollinators could explain the fact that pollination generalist and specialist plants show similar decrease in their reproductive success in fragmented landscapes (Aizen et al., 2002; Ashworth et al., 2004). While specialist pollinators are typically those most affected by habitat fragmentation (Bommarco et al., 2010), their decline should only affect generalist plant species (Ashworth et al., 2004). Generalist plants should maintain their pollination service by interacting with generalist (redundant) pollinators and specialist plants may interact mostly with generalist pollinators, thus generalist and specialist plants should have their reproduction equally affected by fragmentation (Ashworth et al., 2004). In support of this hypothesis, recent theoretical work predicts that specialist species may have lower probability of extinction in networks with a higher incidence of asymmetric interactions (Abramson et al., 2011). However, differences in the probability of specialist species dying out among these differentially structured networks decline progressively with increasing levels of habitat loss (Abramson et al., 2011). In the same vein, a theoretical study explicitly evaluating plant–pollinator network robustness to habitat loss suggested that the nested organization of interactions may increase the persistence of mutualistic species in the face of habitat loss (Fortuna and Bascompte, 2006). In this study, mutualistic networks with nested structures persisted at higher levels of habitat loss than randomly structured networks (Fortuna and Bascompte, 2006). Nestedness implies both a redundancy of mutualistic partners and an indirect facilitation effect among species sharing interaction partners (Lever et al., 2014), which may increase species persistence in the face of species loss.

Although we can identify some general patterns for the consequences of habitat fragmentation on species persistence (see Sections 1.1. and 1.2), our understanding of how the organization of ecological networks of mutualistic species at local and regional scales is affected by habitat fragmentation is still in its infancy (Gonzalez et al., 2011; Hagen et al., 2012). As the disruption of mutualistic interactions may predict future species extinctions and network collapse (Aizen et al., 2012; Fortuna et al., 2013; Valiente-Banuet et al., 2015), interactions, and not species, should be the focus of studies aiming to understand plant and pollinator species persistence in fragmented landscapes. Theoretical work suggests that plant–pollinator interactions are highly sensitive to habitat loss and that the structure of networks can change abruptly once a critical fraction of interactions have been lost this critical fraction being positively associated with the number of interactions of the network and to the number of possible interactions that are actually realized (network connectance; Fortuna et al, 2013). Moreover, the distribution of the number of interactions in each fragment may change from homogeneous when the habitat is continuous to a very skewed distribution when habitat loss reaches levels close to the global extinction of interactions (Fortuna et al., 2013).

To our knowledge, only three empirical studies have explicitly evaluated the effects of habitat fragmentation on plant–pollinator network structure. Sabatino et al. (2010) studied plant pollination webs of 12 isolated hills immersed in an agricultural matrix in the Argentinean pampas. They showed that both species and interaction richness decreased with decreasing habitat area but interactions were lost faster than species (Sabatino et al., 2010). Although it was only a marginal effect, isolation (a fragment metric directly related to habitat loss levels) also diminished interaction richness (Sabatino et al., 2010). In the same study system, Aizen et al. (2012) showed that interactions involving high specialization between interacting partners and occurring at low frequency were more likely to be lost with decreases in habitat area, potentially reflecting lower probability of encounter among specialist species with extremely low abundances in smaller fragments. Finally, Spiesman and Inouye (2013) explicitly studied the effects of sandhill habitat loss on 15 plant–pollinator local webs in North Florida, USA. They found that regional habitat loss contributes directly to species loss and indirectly to the reorganization of plant–pollinator interactions in local communities. Local networks became more connected and modular, and less nested with increasing habitat loss.

Therefore, to bridge the existing gap between theoretical models of habitat loss impact on plant–pollinator networks (Fortuna and Bascompte, 2006) and empirical observations of the impacts of habitat loss on real plant–pollinator communities (Sabatino et al., 2010; Spiesman and Inouye, 2013), the next step is to integrate links between plant species traits and their level of generalization on pollinators (Section 1.3) within a metacommunity model incorporating the effects of habitat loss on species occupancy.

1.5 Predicting Species Persistence in the Face of Habitat Loss Using Trait-Based Metacommunity Models

Predictions on the effects of habitat fragmentation on local species persistence may be substantially altered when the dispersal of individuals among fragments is explicitly considered, as proposed by metacommunity theory (Leibold et al., 2004). Landscapes can be viewed as a set of patches inhabited by communities and connected by the dispersal of individuals (Leibold et al., 2004; Urban et al., 2008). The explicit modelling of dispersal and of other traits associated with the movement of individuals across the landscape can substantially change predictions of the effects of habitat fragmentation on biodiversity and therefore on the supply of ecosystem services from remnants fragments to the surrounding matrix (Keitt, 2009; Mitchell et al., 2015a). The existence of trade-offs between colonization and competitive ability may lead to superior competitors being either more negatively affected by habitat loss due to their low dispersal ability (Nee and May, 1992; Tilman et al., 1994) or positively affected by habitat destruction in metacommunities with source–sink spatial structures (Mouquet et al., 2011), for example. In spatially structured food chains, the association of predator presence with different rates of colonization or extinction, through top-down control, leads to different outcomes on the average food chain length at the metacommunity level (Calcagno et al., 2011). Results from metacommunity models also predict that the positive effects of high dispersal ability on metacommunity persistence might depend on the costs associated with the dispersal of individuals throughout the matrix surrounding fragments. In small food webs, high dispersal ability can increase metacommunity persistence by reducing, via rescue effects, the risk of bottom-up extinction cascades (Eklöf et al., 2012). However, when surrounding matrices decrease the probability of survival of species dispersing among fragments, high dispersal decreases metacommunity persistence (Eklöf et al., 2012).

The few metacommunity models that have studied the persistence of mutualistic species in fragmented landscapes have focused on the effects of network structure rather than the influence of dispersal (Fortuna and Bascompte, 2006; Fortuna et al., 2013). An interesting result is that, even when there is no habitat loss and plants are allowed to persist without pollinators (i.e. plants only depend on pollinators to colonize new fragments), metacommunity collapse occurs when the pollinator extinction rates approach colonization rates (Fortuna and Bascompte, 2006). Thus, habitat loss may lead to the collapse of plant–pollinator metacommunities because of a reduction of both the habitat available to persist and the species colonization ability via decreasing the availability of mutualistic partners. Moreover, a metacommunity model that explicitly considered the interaction of pollinators with animal-pollinated crops (but not network structure) showed that allowing pollinators to use crops as food sources might prevent the total collapse of pollinators but not the extinction of wild plants depending on pollinators to reproduce, if habitat loss is high (>50%; Keitt, 2009).

Results from recent studies provide some indication of the importance of modelling, explicitly, dispersal and other biological traits in metacommunity models in order to understand the persistence of mutualistic assemblages in fragmented landscapes. The analysis of the association between species interaction patterns and species sensitivity to partner loss in empirical plant–pollinator networks suggests that the persistence of plants with both low dispersal ability (e.g. dispersed by gravity) and high reproductive dependence on pollinators (e.g. dioecious species) may be highly compromised if their pollinators disappear (Astegiano et al., 2015). These plants share a low proportion of their interaction partners with other plants of the community (Astegiano et al., 2015), and thus their pollinators may only be maintained by the persistence of other less sensitive plants (Lever et al., 2014). Moreover, in communities where plants show a lower mean ability to self-pollinate, a higher number of co-extinctions per extinction event may occur, which may increase network fragility to the loss of generalists (Vieira and Almeida-Neto, 2015).

Although dispersal ability and breeding systems may be key traits determining plant persistence in fragmented landscapes (see Sections 1.1 and 1.2), we still poorly understand how these biological traits may influence the robustness of plant–pollinator networks to habitat loss and thus species persistence. By unifying the available theory on how breeding system, dispersal ability and the structure of complex networks may modulate the response of species assemblages to species loss, the present study aims at improving our understanding of how plant–pollinator webs may persist in the face of habitat loss. We propose a trait-based metacommunity model to investigate the persistence of plant–pollinator networks under different levels of habitat loss. We hypothesize different scenarios of evolutionary associations between biological traits formerly associated to species response to habitat fragmentation (i.e. plant dispersal ability, breeding system and pollination generalization) to understand (i) how plant and pollinator species and interactions between them persist in the landscape, (ii) how regional network structure changes and (iii) how biological traits in the metacommunity vary, with increasing levels of habitat loss.

2. A TRAIT-BASED METACOMMUNITY MODEL TO UNDERSTAND PLANT AND POLLINATOR PERSISTENCE IN THE FACE OF HABITAT LOSS

2.1 Constructing Theoretical Plant–Pollinator Networks

We constructed 800 interaction networks between 60 plant species and 120 pollinator species, in which 20% of the possible interactions among plants and pollinators were actually realized (i.e. connectance = 0.2). These 800 networks represented a subsample of the possible network configurations that can be achieved by considering that the degree distribution of plants and pollinators follows a power law function with an exponent ranging from 2.2 to 2.9, as described in Appendix A. By varying the exponent of the power law degree distributions (i.e. varying the heterogeneity in degrees among species of the same trophic level), we were able to generate a gradient of nestedness (Almeida-Neto et al., 2008; Podani and Schemera, 2012; see Box A1 in Appendix A). We decided to use different power law exponents for network construction because of the small range of nestedness that can be achieved by using the same power law exponent to generate a given number of networks (see Appendix B). Thus, in agreement with results reported by Dorman et al. (2009), we found that variation in nestedness among networks is highly explained by variation in degree distribution, with lower nestedness being associated with higher power law exponents (see Appendix B).

2.2 Constructing the Trait-Based Metacommunity Model

We modified the metacommunity model proposed by Fortuna and Bascompte (2006) to study the persistence of mutualistic species in fragmented landscapes, in order to evaluate how different associations among plant traits (i.e. autonomous self-pollination, dispersal ability and plant pollination generalization) may influence plant–pollinator network persistence under different levels of habitat loss. The original model considers plants and pollinators inhabiting a landscape consisting of an infinite number of identical, well-mixed fragments (Fortuna and Bascompte, 2006). Thus, it represents an n-species version of the classic metapopulation model proposed by Levins (1969). In Fortuna and Bascompte's (2006) model, the interaction is obligate for animals, but plant species are able to survive in the absence of pollinators. However, plants cannot colonize new fragments without the presence of pollinators (Fortuna and Bascompte, 2006). Thus, the fractions of patches occupied by plants (Eq. 1) and pollinators (Eq. 2) in species-rich mutualistic metacommunities are described by the following differential equations:

$$\frac{dp_i}{dt} = \sum_{j=1}^{A} \left(s_{ij} \frac{p_i a_j}{M_j} \right) (1 - d - p_i) - e_i p_i \tag{1}$$

$$\frac{da_j}{dt} = c_j a_j (M_j - a_j) - e_j a_j \tag{2}$$

where A is the number of pollinator species, p_i and a_j are the fractions of fragments that plant i and animal j inhabit, respectively, s_{ij} is the colonization rate of new fragments by plant i through seeds produced by the pollination service performed by animal j, c_j is the colonization rate of new fragments by animal j, M_j is the fraction of fragments inhabited by plants used by animal j, d is the fraction of habitat fragments that is lost due to human activity and e_i and e_j are the local extinction rates of plant species i and animal species j, respectively. The colonization rate of a given plant encompasses both reproduction and subsequent establishment of new populations via random dispersal. The extinction rate summarizes all forms of extinction sources for plants and animals (Fortuna and Bascompte, 2006). Finally, we assumed in this model that plant extinction causes the subsequent extinction of animals depending exclusively on that plant,i.e., there is no rewiring (Fortuna and Bascompte, 2006).

In the model proposed here, we have assumed that plants can colonize new sites without pollinators by producing seeds through autonomous self-pollination. Moreover, colonization rate also depends on plant dispersal ability, which is explicitly considered in the model. Thus, in our modified version of the model, the dynamics of the fraction of patches occupied by plant species i, p_i, is described by the following differential equation:

$$\frac{\mathrm{d}p_i}{\mathrm{d}t} = \alpha_i\left[(1-\delta)b_i + (1-b_i)\sum_{j=1}^{A} s_{ij}\frac{a_j}{M_j}\right]p_i(1-d-p_i) - e_i p_i \qquad (3)$$

where α_i is the dispersal rate of plant i, b_i is the proportion of seeds produced by autonomous self-pollination by plant i, δ is the inbreeding depression rate endured by seeds produced by autogamy, $(1-b_i)$ represents the fraction of the progeny produced by pollination due to pollinators visits to plant i and s_{ij} measures the effects of pollination by a given pollinator on total seed production. All other parameters and variables are the same as described previously for the original model (Fortuna and Bascompte, 2006). For simplicity, we assumed that δ and s_{ij} are equal for all species in all the scenarios. In the following, all pollinator extinction rates were set at e_A, all plant extinction rates were set at e_P and all pollinator colonization rates were set at c_A.

2.3 Constructing Theoretical Scenarios Linking Plant Traits

In order to evaluate how different associations among plant traits may modulate plant–pollinator metacommunity persistence, we constructed metacommunity models with different covariance structures among autonomous self-pollination and dispersal rates, and plant degree, i.e., the number of pollinator species interacting with a given plant species. Autonomous selfing and dispersal rates of each plant species were sampled from a multinormal distribution, which linked normalized versions of these plant traits. Dispersal rate, α, took values between zero and ∞. Thus, we set the normalized linearized version of this variable as $\beta = \log\alpha$. The proportion of seeds produced by autonomous self-pollination, b, took values between zero and one. Thus, we set the normalized linearized version of this variable as $\xi = \log\frac{b}{1-b}$.

We tested metacommunity persistence under the following eight scenarios, summarized below (Fig. 1):

(a) *Neutral*, in which all species received the same value of β and ξ;

(b) *Random*, in which β and ξ were randomly assigned to plant species following two independent normal distributions (with variances set at 0.1);

(c) *AGnegDrdm*, in which ξ was negatively correlated with species degree (i.e. its mean was determined by species degree), and β was randomly assigned, following two independent normal distributions (with variances set at 0.1);

(d) *AGnegDneutral*, in which ξ was negatively correlated with species degree (i.e. its mean was determined by species degree) and assigned following a normal distribution (with variance set at 0.1), and all species received the same value of β;

(e) *DGnegArdm*, in which β was negatively correlated with species degree (i.e. its mean was determined by species degree), and ξ was randomly assigned, following two independent normal distributions (with variances set at 0.1);

(f) *DGnegAneutral*, in which β was negatively correlated with species degree (i.e. its mean was determined by species degree) and assigned following a normal distribution (with variance set at 0.1), and all species received the same value of ξ;

(g) *DAnegGrdm*, in which β was negatively correlated with ξ, and the values of these traits were randomly assigned to plants independently of their degree, following a correlated multinormal distribution (with variances set at 0.1 and correlation set at −0.5);

(h) *DAposGrdm*, where β was positively correlated with ξ, and the values of these traits were randomly assigned to plants independently of their degree, following a correlated multinormal distribution (with variances set at 0.1 and correlation set at 0.5).

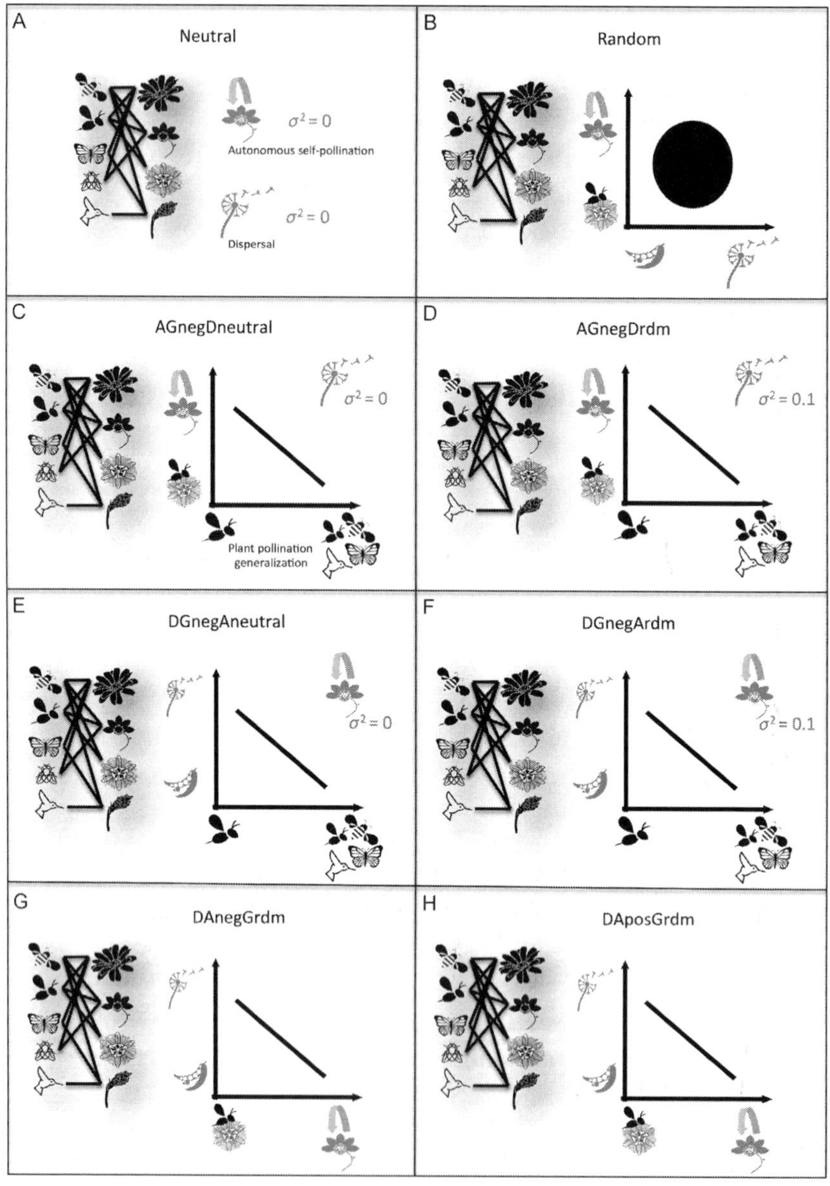

Figure 1 Constructing theoretical scenarios linking plant traits. Schematic representation of the eight scenarios in which metacommunity persistence was explored, as described in Section 2.3. Scenarios represent different associations between plant pollination generalization, autonomous self-pollination and dispersal rates that have been theoretically and empirically explored in previous studies. Designs created by Peter Silk, Galo Chiriboga, Cassie McKnown, Cherish Watson, Tom Ingebretsen, Alec Dhuse, Gabi McKensie, Lane F. Kinkade, Matt Brooks, Anbileru Adaleru and Karen Ardila Olmos for the Noun Project (https://thenounproject.com).

We explored how changing the mean proportion of seeds produced by autonomous self-pollination (b_i=0.25, 0.5, 0.9), the ratio between plant extinction and mean dispersal rate (e_P/d_P=0.25, 0.5, 0.75 and 0.95) and the ratio between pollinator extinction and colonization rates (e_A/c_A=0.25, 0.5, 0.75 and 0.95), affected the persistence of metacommunities under the eight scenarios when all species can colonize all fragments (d=0; Fig. 2). We simulated the dynamics of 96 metacommunities sampled from the initial 800 networks. These 96 metacommunities were obtained by taking 12 networks that encompass the range of nestedness observed in 100 networks from a given power law distribution. Species were considered to have maximum occupancy (100% of fragments of natural habitat) at the beginning of each simulation. Once we identified the combinations of mean autonomous self-pollination rate, plant extinction/dispersal and pollinator extinction/colonization ratios that allowed full plant and pollinator species persistence under the eight scenarios of association among plant traits, we explored how these different scenarios influenced metacommunity persistence under different levels of habitat loss (d=0.3 and 0.6; Fig. 2).

We obtained the final proportion of plant and pollinator species and plant–pollinator interactions that persisted under different levels of habitat loss at the end of each simulation (Fig. 2). We also explored how the distribution of occupancies of (1) plants species, (2) pollinators species and (3) plant–pollinator interactions varied among scenarios, in the absence of habitat loss and 30% and 60% of habitat loss, for all initial combinations of model parameters. We first described the distribution of occupancies in each metacommunity at the end of each simulation and then we obtained the proportion of networks leading to a given distribution per scenario/habitat loss level/initial combination of parameters. To describe the distribution of occupancies of plant and pollinator species within each metacommunity, we constructed rank-occupancy curves, following the method proposed by Jenkins (2011) for species occupancy. We built the rank-occupancy curve of plants or pollinator species within each metacommunity, by plotting species in order of decreasing occupancy. The shape of the decaying curve of this rank-occupancy relationship describes the degree of variation of occupancy among species within metacommunities and the degree of dominance of species with the highest occupancies.

To describe the shape of the rank-occupancy curve within a given metacommunity, we used linear and non-linear regression models. Among the non-linear regressions, we chose two equations of the exponential family—the same used by Jenkins (2011) to fit empirical rank-occupancy distributions. The first equation describes a convex exponential curve, i.e., high dominance of a few species. The second equation describes a concave exponential curve, i.e., low species dominance with species having similar proportions of occupancies. We also decided to fit the linear model because it allowed us to describe an intermediate curve between the convex and concave exponential curves, i.e., an intermediate case of dominance. For all the regression models, we assumed normal errors and homogeneity of variance. We fitted the models only for the data sets (metacommunities) with a minimum variation of species occupancy. When the difference in occupancy between the most and the least frequent species was smaller than 0.1, we considered the variation among species to be low and thus dominance to be null, i.e., that the curve for this distribution was approximately constant. This lack of dominance with a constant curve includes metacommunity dynamics leading to either full species extinction or to species (interactions) with similar frequencies of occupancy. We divided the null dominance into two categories: (i) null dominance associated with full species extinction and (ii) null dominance associated with either all species or some species persisted. Thus, we had five possible descriptions of curves, i.e., (i) constant with total extinction of species (complete colapse), (ii) constant with persistence of all or some species (no dominance with persistence of species), (iii) concave (low dominance), (iv) linear (intermediate dominance) and (v) convex (high dominance). All models were fitted to the data by using maximum likelihood, with the values of maximum likelihood also used to select the best model through a selection procedure that did not penalize by the number of parameters. The highest value of likelihood, then, indicated the best model. We obtained the percentage of occupancy distributions that were better described by each rank-occupancy model for each initial combination of parameters, scenario and level of habitat loss (i.e. absence of habitat loss, 30% and 60% of habitat loss).

To describe the distribution of occupancy of plant–pollinator interactions within metacommunities, we followed the same procedure as for species. The occupancy of each plant–pollinator interaction (I_{ij}) was obtained as described in Eq. (3) in Section 2.2:

$$I_{ij} = \frac{p_i a_j}{M_j} \tag{4}$$

where p_i and a_j are the fractions of fragments that plant i and animal j inhabit, respectively, and M_j is the fraction of fragments inhabited by plants used by animal j.

We also measured the relative change in nestedness $(\mathrm{NODF}_{\mathrm{max}_{\mathrm{final}}} - \mathrm{NODF}_{\mathrm{max}_{\mathrm{initial}}})/\mathrm{NODFmax}_{\mathrm{intial}}$ and in connectance ($\mathrm{connectance}_{\mathrm{final}} - \mathrm{connectance}_{\mathrm{initial}}$) after each simulation (Fig. 2). Nestedness was measured as $\mathrm{NODF}_{\mathrm{max}}$ and PRSN (Podani and Schemera, 2012; Appendix A, Box A1). As both measures showed similar behaviours, we only show results from $\mathrm{NODF}_{\mathrm{max}}$.

Finally, in order to evaluate if particular sets of traits favoured the persistence of plants under the different scenarios and at different levels of habitat loss, we compared the relative change in the mean and the coefficient of variation of plant dispersal, autonomous self-pollination and degree (i.e. plant pollination generalization), among scenarios within combinations of parameters that led to the extinction of at least 10% of the plant species and to the persistence of at least one plant species (Fig. 1). Comparisons were performed within those combinations of parameters that met these criteria in most scenarios. The relative changes in the mean and the coefficient of variation of the three plant traits were measured as (final value − initial value)/initial value.

2.4 Statistical Analyses

We compared the parameter range of plant and pollinator species persistence among all scenarios, in the absence of habitat loss and with 30% and 60% of habitat loss. These two levels of habitat loss were chose arbitrarily, but they reflect mean and maximum estimations of the percentage of natural habitats that have been converted across different biomes (Hoekstra et al., 2005). We also evaluated the statistical significance of differences in the final proportion of plants and pollinators species, and in the relative change in nestedness and connectance, among *a priori* planned pair-wise comparisons between the different scenarios, within situations of absence of habitat loss and 30% and 60% of habitat loss within combinations of parameters. The *Random* and *Neutral* scenarios were compared with all scenarios. We also performed pair-wise comparisons within scenarios describing negative associations between species degree and autonomous self-pollination or dispersal rates, by contrasting the results obtained when values of dispersal rates or autonomous self-pollination, respectively, were randomly assigned or were the same for all species. Finally, the two scenarios describing negative and positive associations between plant autonomous self-pollination and dispersal rates were also compared. Differences between means among pairs of scenarios were considered significant when the 95% confidence interval of the difference did not breach zero. All simulations were performed in Matlab, 2011.

We evaluated if the mean relative change in the mean and the coefficient of variation of the dispersal rate, autonomous self-pollination and degree of species within metacommunities, within habitat loss level, scenarios and parameter combinations, were significantly different from zero, by calculating the 95% confidence interval of each mean. Then we evaluated the statistical significance of differences among *a priori* pair-wise comparisons between the different scenarios, within situations of no habitat loss and 30% and 60% of habitat loss within combinations of parameters as described before.

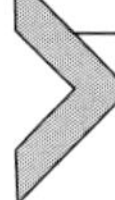

3. RESULTS

3.1 Plant–Pollinator Metacommunity Persistence

We first evaluated the dynamics of the 96 metacommunities in the absence of habitat loss, for different initial combinations of three parameters: autonomous self-pollination rate, plant extinction-to-dispersal ratio and animal extinction-to-dispersal ratio (Fig. 2). For most combinations of these parameters, full metacommunity persistence (i.e. no species went extinct) occurred (Fig. 3). However, the prevalence of full metacommunity persistence varies among scenarios (Fig. 3). Full metacommunity persistence was observed for a wider combination of parameters under the assumption of neutrality in both dispersal ability and in autonomous self-pollination rate, and of neutrality in dispersal ability and a negative association between plant pollination generalization and autonomous self-pollination (Neutral and AGnegDneutral scenarios, respectively; Fig. 3A and C). Lower plant and pollinator diversity and high plant dominance were observed in simulations with the highest levels of plant autonomous self-pollination, plant extinction-to-dispersal ratios (e_P/d_P) and animal extinction-to-colonization ratios (e_A/c_A) for most scenarios, while under all other initial combinations of these parameters, the most prevalent pattern was the absence of dominance, in all scenarios (Fig. 3 and Table S1 (http://dx.doi.org/10.1016/bs.aecr.2015.09.005)). Under the assumption of neutrality in dispersal ability and autonomous self-pollination (Neutral scenario), metacommunities always showed no plant dominance, with mean plant occupancy decreasing with increases of plant extinction-to-dispersal ratios and of animal extinction-to-colonization ratios (Table S1 and Fig. S2A (http://dx.doi.org/10.1016/bs.aecr.2015.09.005)). Metacommunities showed no dominance of pollinator species under all scenarios and combinations of parameter values, with mean pollinator occupancy decreasing with increases in the

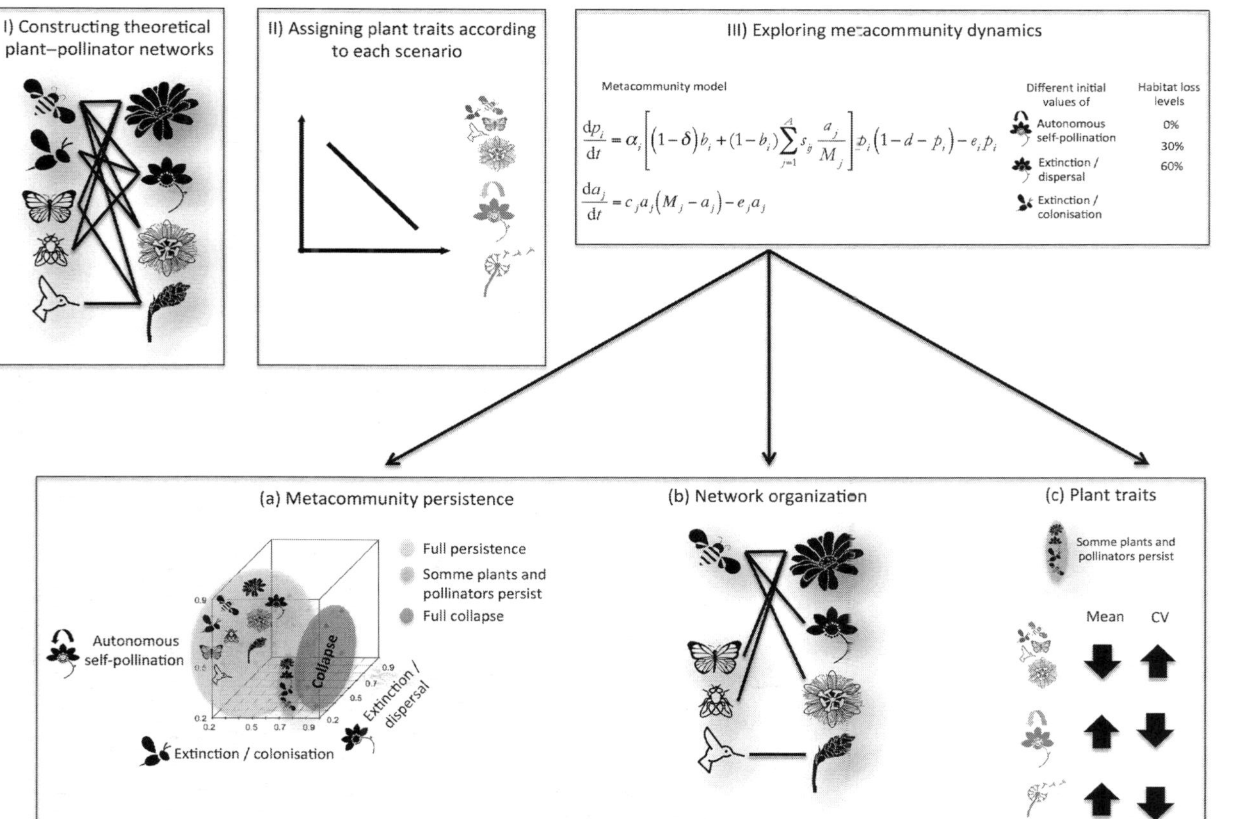

Figure 2 Schematic summary of the general theoretical and methodological framework used in this study. Each box summarizes the main steps described in Section 2. References: CV, coefficient of variation. Designs are as in Fig. 1.

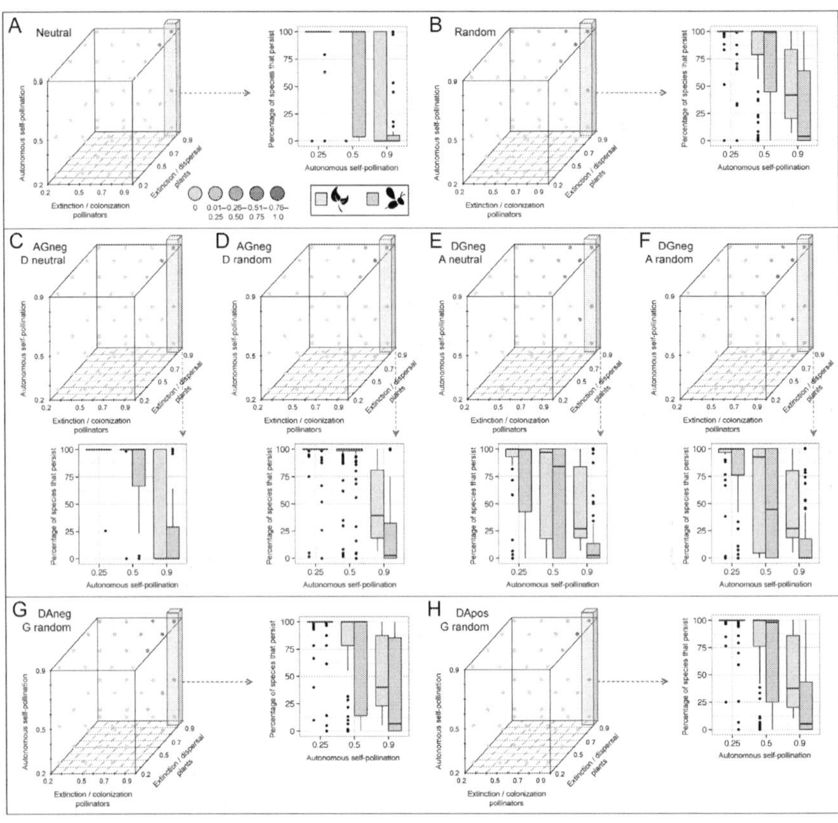

Figure 3 Metacommunity persistence in non-fragmented landscapes. The percentage of network replicates ($n = 96$) leading to at least one species extinction for each combination of parameter values (initial values of pollinator extinction/colonization ratios, mean plant extinction/dispersal ratios and mean autonomous self-pollination rates) are shown inside each cube, under each scenario. This percentage ranges from 0% of network replicates leading to species extinction (yellow (light grey in the print version) dots) to most or all network replicates leading to species extinction (75–100%, red (dark grey in the print version) dots). Box-plots show final plant and pollinator richness for initial conditions set at the maximum e_A/c_A and e_P/d_P considered in this study (0.95), at different initial values of mean autonomous self-pollination rate (abscissas). Green (light grey in the print version) and purple (grey in the print version) boxes represent plant and pollinator species richness, respectively. Black lines within boxes represent median values. Upper and lower limits of boxes represent first and third quartiles, respectively. Black dots represent outliers. Scenarios are those described in Section 2.3 and Fig. 1. Plant and pollinator designs created by Guillaume Bahri and Peter Silk for the Noun Project (https:/thenounproject.com).

extinction-to-colonization ratio of animal species (Table S1; Fig. S5 (http://dx.doi.org/10.1016/bs.aecr.2015.09.005)). Likewise, metacommunities showed no dominance of plant–pollinator interactions when full metacommunities persisted, under all scenarios, with interactions occurring in a decreasing fraction of the landscape with increasing animal extinction-to-colonization ratios and plant extinction-to-dispersal ratios (Table S1; Fig. S8 (http://dx.doi.org/10.1016/bs.aecr.2015.09.005)). In the absence of habitat loss, extinction was limited to the combination of the highest values of autonomous self-pollination (0.9), the highest extinction-to-colonization ratio for animals (0.95) and the two highest extinction-to-dispersal ratios considered for plants (0.75 and 0.95; Fig. 3). For this combination of parameters, total neutrality in dispersal ability and autonomous self-pollination rate (Neutral scenario), and neutrality in dispersal associated with a negative relationship between autonomous self-pollination and plant pollination generalization (AGnegDneutral scenario) led to the extinction of all plant and pollinator species (i.e. complete metacommunity collapse occurred), whereas some plants and pollinators persisted in a small fraction of the landscape under the other scenarios (Figs. 3, S2 and S5 (http://dx.doi.org/10.1016/bs.aecr.2015.09.005)).

We now describe the effects of habitat loss. No matter the scenario, the complete collapse of the metacommunities—or at least of all pollinator species—occurred when 30% of the original natural habitat was removed, for some of the initial combinations of parameters that allowed full metacommunity persistence in absence of habitat loss (Fig. 4). These collapses occurred when animal extinction-to-colonization ratio was high (0.75 or 0.95), for all values of extinction-to-dispersal ratio for plants (Fig. 4). As occurred in the absence of habitat loss, the collapse of metacommunities was the most prevalent catastrophic outcome under both the scenario assuming neutrality in plant dispersal ability and autonomous self-pollination rate, and that assuming neutrality in plant dispersal ability and a negative relationship between autonomous self-pollination and plant pollination generalization (Neutral and AGnegDneutral scenarios; Fig. 4A and C). Pollinator collapse prevailed as the most likely catastrophic outcome under the other scenarios (Fig. 4B, D–H). The percentage of plant and pollinator species surviving with extinction-to-colonization ratio = 0.75 varied with mean autonomous self-pollination rate and among scenarios (Fig. 4). For instance, when the mean autonomous self-pollination rate varied from 0.5 to 0.9 and plant extinction-to-dispersal ratio was set to 0.25, the percentage of plant species surviving increased with autonomous self-pollination rate in five

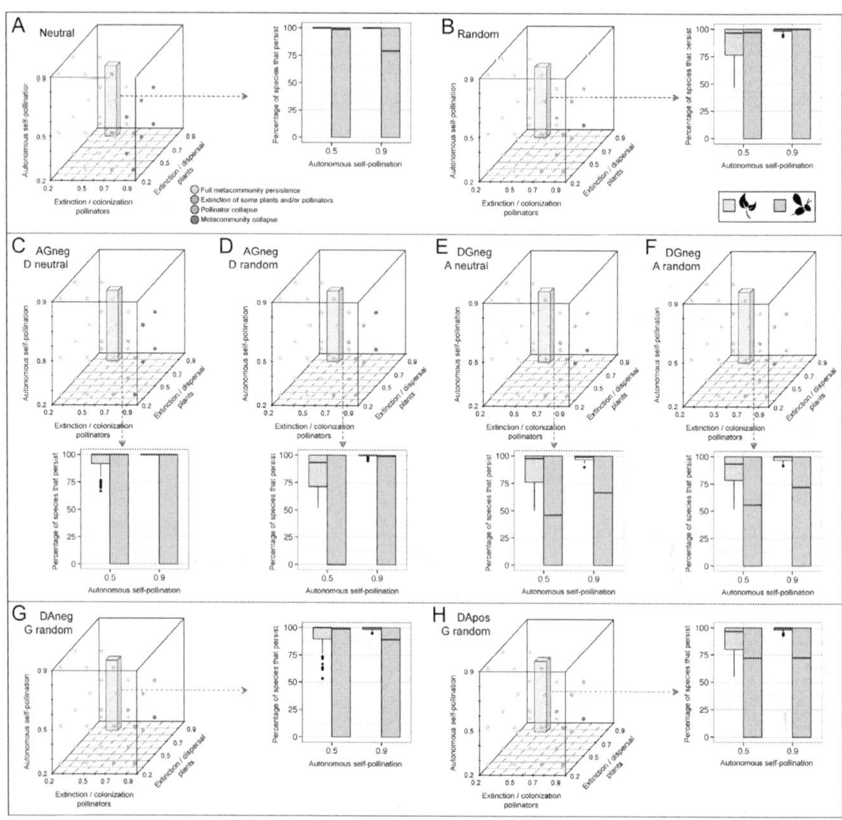

Figure 4 Metacommunity persistence with 30% of habitat loss. Combination of initial values for three parameters (pollinator extinction/colonization ratios, mean plant extinction/dispersal ratios and mean autonomous self-pollination rates) leading to full metacommunity persistence (yellow (light grey in the print version) dots), the extinction of some plants and/or some pollinators (orange (grey in the print version) dots), complete pollinator collapse (i.e. only plant species persist; green (grey in the print version) dots) and complete metacommunity collapse (red (dark grey in the print version) dots), is shown inside each cube for each scenario. Box-plots show variation in final plant and pollinator richness for initial conditions set at $e_A/c_A = 0.75$ and $e_P/d_P = 0.25$, at different initial values of mean autonomous self-pollination rate (abscissas). Green (light grey in the print version) and purple (grey in the print version) boxes represent plant and pollinator species richness, respectively. Box-plot interpretation is as in Fig. 3. Scenarios are those described in Section 2.3 and Fig. 1. Plant and pollinator designs are as in Fig. 3.

scenarios (Fig. 4B, D–F, H), whereas the percentage of pollinators surviving increased in four scenarios (Fig. 4B, D–F) and decreased in two (Fig. 4A and G).

With the loss of 30% of the original natural habitat, a combination of high frequency of autonomous self-pollination (0.9), low extinction-to-dispersal ratios for plants ($e_P/d_P = 0.25$) and high extinction-to-colonization ratios for animals ($e_A/c_A = 0.75$) led to metacommunities highly dominated by some plant species under most scenarios (Table S1 (http://dx.doi.org/10.1016/bs.aecr.2015.09.005)). Lower levels of selfing (0.5) lead to metacommunities showing no dominance or intermediate dominance of plants (Table S1 (http://dx.doi.org/10.1016/bs.aecr.2015.09.005)). Metacommunities showed no dominance of pollinator species under all combinations of parameters (Table S1 (http://dx.doi.org/10.1016/bs.aecr.2015.09.005)). Nevertheless, the mean occupancy of pollinators decreased with increased extinction-to-colonization ratios (Table S1; Fig. S6 (http://dx.doi.org/10.1016/bs.aecr.2015.09.005)). Accordingly, the patterns of interaction occupancy showed no dominance under most parameter combinations, with decreasing interaction occupancy with both increasing animal extinction-to-colonization ratio and plant extinction-to-dispersal ratio (Table S1; Fig. S9 (http://dx.doi.org/10.1016/bs.aecr.2015.09.005)).

With the loss of 60% of the original natural habitat, no matter the combination of parameters or the scenario, complete metacommunity collapse was the prevalent outcome of metacommunity dynamics (Fig. 5). Full metacommunity persistence only occurred when pollinators had the lowest extinction-to-colonization ratio (0.25, Fig. 5). All plant species showed maximum occupancy when the complete metacommunity persisted, under most scenarios (Table S1; Fig. S4 (http://dx.doi.org/10.1016/bs.aecr.2015.09.005)). With full metacommunity persistence, pollinators showed similar occupancies (i.e. no dominance), occurring in less than 20% of the landscape under all scenarios (Table S1; Fig. S7 (http://dx.doi.org/10.1016/bs.aecr.2015.09.005)). Interactions among plants and pollinators also occurred at a similar proportion of fragments, which was lower than 0.1 under all scenarios, when there was full metacommunity persistence (Table S1; Fig. S10 (http://dx.doi.org/10.1016/bs.aecr.2015.09.005)). When the initial ratio of extinction-to-colonization for animals was higher than 0.25, the whole set of pollinator species died out, under all scenarios (Fig. 5). Assuming neutrality in plant dispersal and autonomous self-pollination (Neutral scenario), initial values of extinction-to-colonization ratio for animals higher than 0.25 also led to the extinction of the whole set of plant species in almost all combinations of extinction-to-dispersal ratio for plants (Fig. 5A). Plant persistence without pollinators was observed for a similar range of parameters under most of the other scenarios, except in the scenario assuming neutrality in plant dispersal and a negative association between autonomous self-pollination and plant pollination generalization (AGnegDneutral scenario)

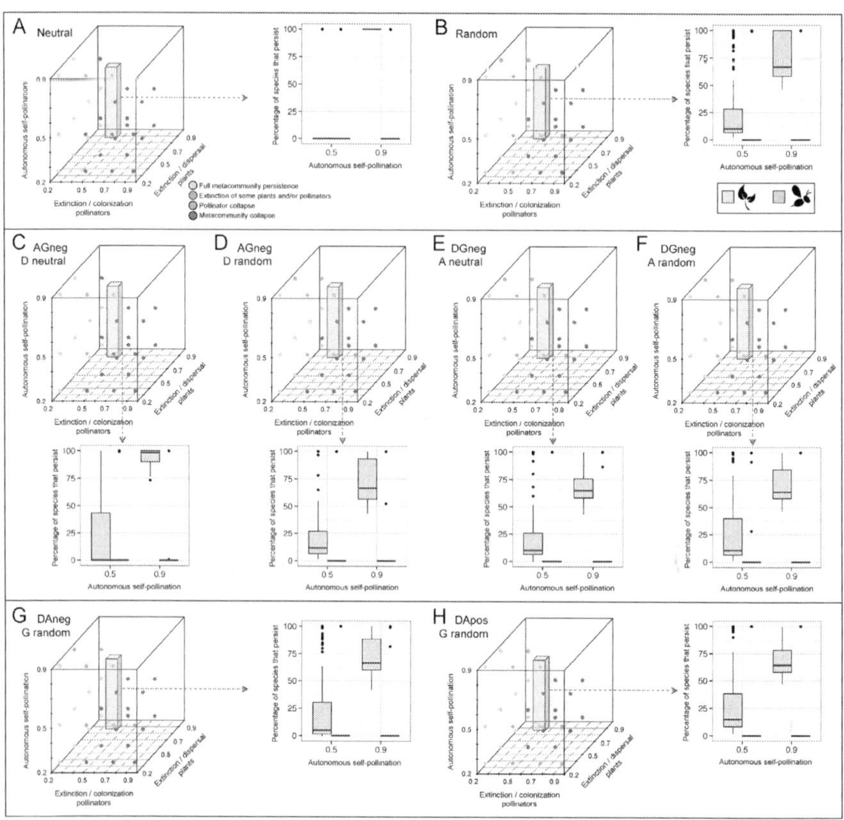

Figure 5 Metacommunity persistence with 60% of habitat loss. Combination of initial values for three parameters (pollinator extinction/colonization ratios, mean plant extinction/dispersal ratios and mean autonomous self-pollination rates) leading to full metacommunity persistence (yellow (light grey in the print version) dots), the extinction of some plants and/or some pollinators (orange (grey in the print version) dots), complete pollinator collapse (i.e. only plant species persist; green (grey in the print version) dots) and complete metacommunity collapse (red (dark grey in the print version) dots), is shown inside each cube for each scenario. Box-plots show variation in final plant and pollinator richness for initial conditions set at $e_A/c_A = 0.75$ and $e_P/d_P = 0.25$, at different initial values of mean autonomous self-pollination rate (abscissas). Green (light grey in the print version) and purple (grey in the print version) boxes represent plant and pollinator species richness, respectively. Box-plot interpretation is as in Fig. 3. Scenarios are those described in Section 2.3 and Fig. 1. Plant and pollinator designs are as in Fig. 3.

for which this outcome was observed under a narrower combination of parameter values (Fig. 5C). Increases from intermediate (0.5) to high (0.9) values of autonomous self-pollination rate increased the percentage of plant species surviving and led to the prevalence of high plant dominance when there was complete pollinator collapse, but only at intermediate values of plant extinction-to-dispersal ratios ($e_P/d_P = 0.5$) (Table S1 (http://dx.doi.org/10.1016/bs.aecr.2015.09.005); Fig. 5).

3.2 Plant–Pollinator Network Organization

Relative changes in network connectance and nestedness were only analyzed for 30% of habitat loss. With 60% of habitat loss, most metacommunities either fully persisted or completely collapsed, or only plants persisted, and thus in all cases, there was either no change in connectance and nestedness or both dropped to zero. When 30% of the habitat was lost, most scenarios showed either no change in connectance or nestedness, or these metrics dropped to zero (relative change $= -1$; Figs. S11 and S12 (http://dx.doi.org/10.1016/bs.aecr.2015.09.005)). When there was a change, both connectance (i.e. the proportion of realized interactions) and the overlap of interactions among species of the same trophic level (i.e. nestedness) barely decreased (e.g. Figs. S11E and F, and S12E and F (http://dx.doi.org/10.1016/bs.aecr.2015.09.005)).

3.3 Plant Traits

Mean dispersal rate increased and its variation decreased, across all habitat loss levels, initial parameter combinations and under most scenarios (Fig. 6; Table S2 (http://dx.doi.org/10.1016/bs.aecr.2015.09.005)). In contrast, the mean autonomous self-pollination rate and its variation showed no or small variation after some plant species were lost (Fig. 7; Table S3 (http://dx.doi.org/10.1016/bs.aecr.2015.09.005)). Among the different combinations of parameters, the highest increases in mean autonomous self-pollination rates were observed in the scenario assuming neutrality in dispersal rates and a negative association between autonomous self-pollination ability and plant pollination generalization (AGnegDneutral), and the scenario assuming a positive association between dispersal and autonomous self-pollination rate (DAposGrdm scenario), under both habitat loss levels (Fig. 7B and D; Table S3 (http://dx.doi.org/10.1016/bs.aecr.2015.09.005)). These two scenarios also showed the highest relative decreases in the coefficient of variation of autonomous self-pollination with 60% of habitat loss (Fig. 7D–F; Table S3 (http://dx.doi.org/10.1016/bs.aecr.2015.09.005)).

Mean plant pollination generalization (i.e. plant degree) decreased and its variation increased, significantly, with 30% and 60% of habitat loss, across most scenarios and initial parameter combinations (Fig. 8; Table S4 (http://dx.doi.org/10.1016/bs.aecr.2015.09.005)). Under the Random

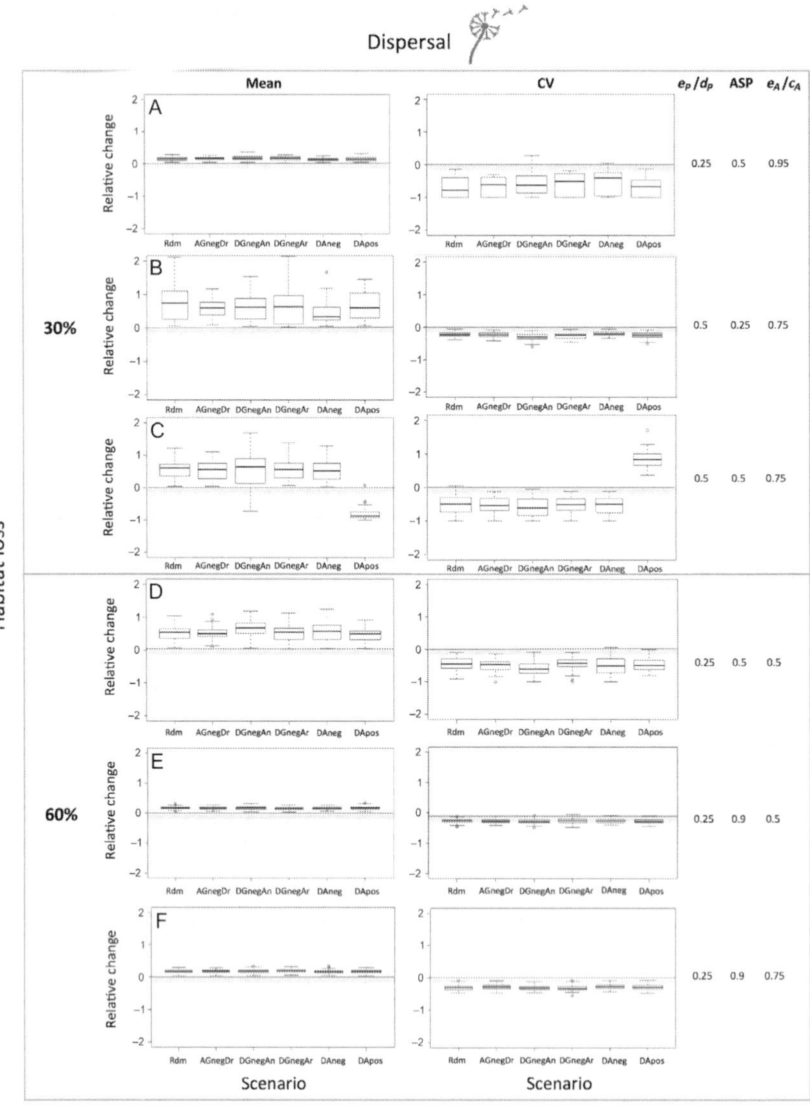

Figure 6 Relative change of the mean and the coefficient of variation (CV) of plant dispersal within metacommunities when >10% of plant species went extinct. Results for initial combinations of parameters allowing the persistence of plants in most scenarios at each habitat loss level are shown (A–C for $d=0.3$; D–F for $d=0.6$). Scenarios in which species had the same value of dispersal (Neutral and AGnegDneutral) are not shown. Relative change was obtained as (final value − initial value)/initial value, as described in *Methods*. Initial values for each parameter are shown on the right side of the figure: ASP, initial mean autonomous self-pollination rate; e_A/c_A, initial ratio between the extinction and the colonization rate of pollinators; e_P/d_P, initial ratio between the extinction and dispersal rate of plants. Box-plot interpretation is as in Fig. 3. Boxes were drawn with widths proportional to the number of networks in which the extinction of >10% of plant species occurred. Scenarios are those described in Section 2.3 and Fig. 1. References: *d*, proportion of habitat that was lost. Designs are as in Fig. 1.

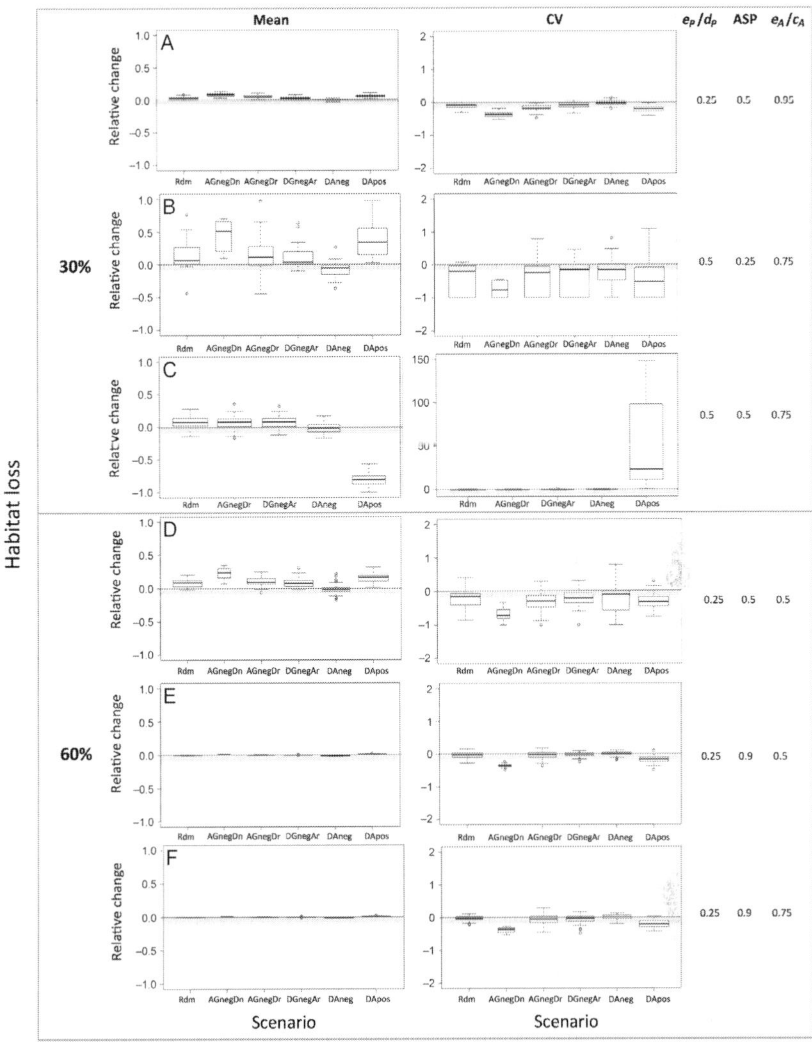

Figure 7 Relative change of the mean and the coefficient of variation (CV) of autonomous self-pollination rate within metacommunities when >10% of plant species went extinct. Results for initial combinations of parameters allowing the persistence of plants in most scenarios at each habitat loss level are shown (A–C for $d=0.3$; D–F for $d=0.6$). Scenarios in which species had the same value of autonomous self-pollination rate (Neutral and DGnegAneutral) are not shown. Relative change was obtained as (final value − initial value)/initial value, as described in Section 2.3. Initial values for each parameter are shown on the right side of the figure: ASP, initial mean autonomous self-pollination rate; e_A/c_A, initial ratio between the extinction and the colonization rate of pollinators; e_P/d_P, initial ratio between the extinction and dispersal rate of plants. Boxplot interpretation is as in Fig. 3. Boxes were drawn with widths proportional to the number of networks in which the extinction of >10% of plant species occurred. Scenarios are those described in Section 2.3 and Fig. 1. References: d, proportion of habitat that was lost. Designs are as in Fig. 1.

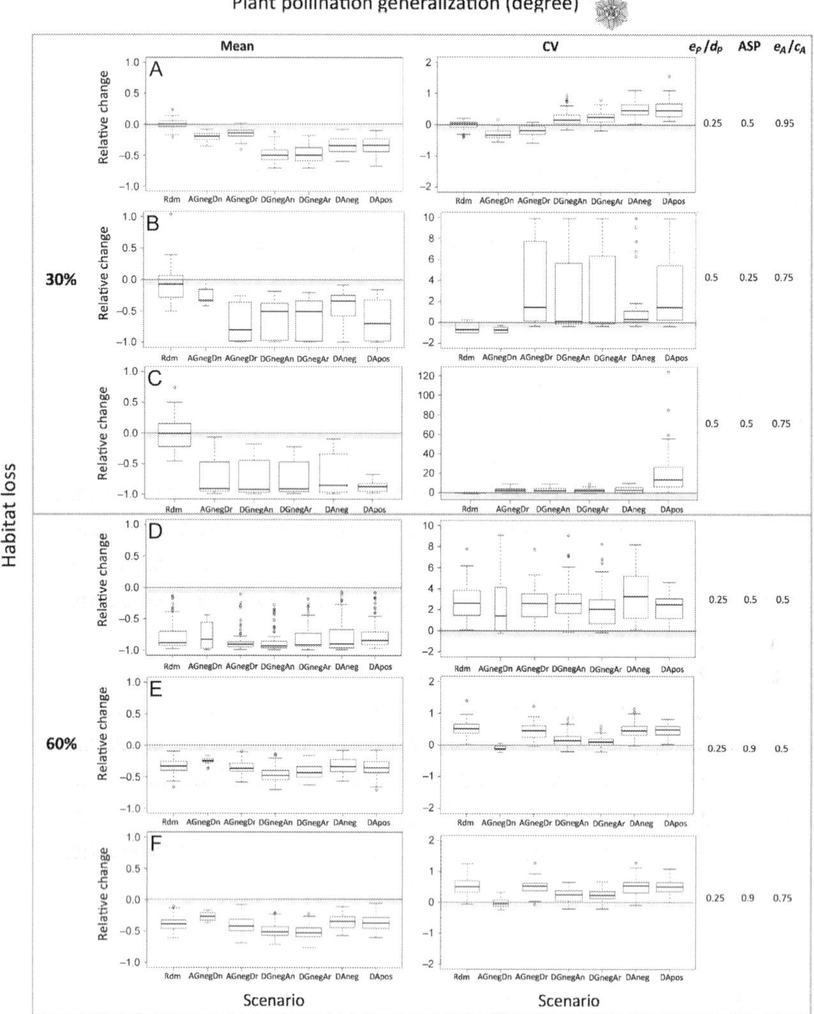

Figure 8 Relative change of the mean and the coefficient of variation (CV) of plant pollination generalization (degree) within metacommunities when >10% of plant species went extinct. Results for initial combinations of parameters allowing the persistence of plants in most scenarios at each habitat loss level are shown (A–C for $d=0.3$; D–F for $d=0.6$). Relative change was obtained as (final value − initial value)/initial value, as described in *Methods*. Initial values for each parameter are shown on the right side of the figure: ASP, initial mean autonomous self-pollination rate; e_A/c_A, initial ratio between the extinction and the colonization rate of pollinators; e_P/d_P, initial ratio between the extinction and dispersal rate of plants. Box-plot interpretation is as in Fig. 3. Boxes were drawn with widths proportional to the number of networks in which the extinction of >10% of plant species occurred. Scenarios are those described in Section 2.3 and Fig. 1. References: d, proportion of habitat that was lost. Designs are as in Fig. 1.

dispersal ability), ultimately reducing the resilience of ecosystem functions (Elmqvist et al., 2003; Laliberte et al., 2010).

"Landscape-moderated" filtering (*sensu* Tscharntke et al., 2012) of biotically pollinated plant species with low dependence on pollinators can be expected under habitat loss (Aguilar et al., 2006; Eckert et al., 2010). Unlike dispersal, our model predicts that the relative incidence of species with higher autonomous self-pollination may barely increase with habitat loss, as relative changes in mean autogamous selfing and its variation are comparatively small to those observed for dispersal. It is important to point out that, with 30% of habitat loss, our model predicts that several pollinator species may persist even when some plants went extinct. Thus, with low habitat loss, plants with high reproductive dependence on pollinators may persist by being generalists (Astegiano et al., 2015; but see Aguilar et al., 2006; Aizen et al., 2002). However, since complete pollinator collapse is predicted to occur with both 30% and 60% of habitat loss, high dispersal ability seems to be crucial for the persistence of plant species with low autonomous self-pollination ability.

Surprisingly, the relative incidence of generalist plants was not favoured by habitat loss compared with the original set of plant species. The subset of surviving species had lower mean and higher variation in pollination generalization than plant species in non-fragmented landscapes. Thus, under most scenarios, generalist plant species were relatively more affected by habitat loss than specialist species. These results contrasted with predictions based on the pervasiveness of asymmetric interactions in mutualistic networks (Ashworth et al., 2004; Vázquez and Aizen, 2004). It has been proposed that specialist and generalist plants may be equally affected by habitat fragmentation because the higher decline of specialist pollinators (Bommarco et al., 2010; Steffan-Dewenter and Tscharntke, 2002) may mainly affect generalist plants (Ashworth et al., 2004). Specialist plants may decrease their extinction probability with increasing habitat loss by interacting with generalist pollinators (Abramson et al., 2011). In our model, persistence ability may differ among pollinator species interacting with different number of plant species. However, we assumed that pollinator colonization and extinction rates did not differ among pollinator species. The pollinator extinction-to-colonization ratio seems to govern the dynamics of the occupancy of pollinators, which may lead to specialist and generalist pollinators being similarly affected by habitat loss when extinction is approximately equal to colonization, even if the feed on a different number of resources. Given that generalist plants had lower autonomous self-pollination or dispersal rate under most scenarios, pollinator extinction may negatively affect more the occupancy of generalist plants than that of specialist ones. In this regard, when pollination generalization was not negatively associated with dispersal or autogamous selfing (i.e. under the *Random* scenario) and habitat loss was low (30%), surviving plants were as generalists as the initial set of plants in non-fragmented landscapes.

4.3 Integrating the Network Approach to the Management of Pollination Services in Human-Dominated Landscapes

Understanding how landscape fragmentation may affect ecological interactions has largely been improved by network approaches (Aizen et al., 2012; Cagnolo et al. 2009; Ebeling et al., 2011; Fabian et al., 2013; Fortuna and Bascompte, 2006; Hagen et al., 2012; Massol and Petit, 2013; Melián and Bascompte, 2002; Sabatino et al., 2010; Spiesman and Inouye, 2013; Tylianakis et al. 2010; Valiente-Banuet et al., 2014). In this regard, it has been proposed that conservation of interaction networks should involve the monitoring of network structural characteristics, such as connectance or nestedness (Tylianakis et al., 2010; but see Kaiser-Bunbury and Blüthgen, 2015). In principle, changes in connectance and nestedness should alert about changes in functional redundancy of species within networks and thus may be good indicators of the fragility of ecological networks in the face of species loss (Tylianakis et al., 2010; Vieira and Almeida-Neto, 2015). However, for plant–pollinator webs with nested structures, our model predicts that network connectance and nestedness may either barely change or drop to zero with increased habitat loss. This lack of change in network connectance and nestedness when some plant and pollinator species went extinct indeed show the lack of sensitivity of these structural network measures to the loss of a few species (Kaiser-Bunbury and Blüthgen, 2015; Nielsen and Bascompte, 2007).

Scale-free networks, such as the ones modelled here, are intrinsically robust (i.e. connectance does not change, and the general shape of the distribution of degrees does not change either) to random removals of nodes (i.e. species in our networks) but are particularly fragile in the face of removals targeted at hubs (i.e. the "super-generalists" in our networks; Barabási and Albert, 1999; Cohen et al., 2000, 2001). This robustness to random extinction of species has been showed for several plant–pollinator networks in studies simulating the random removal of either plants or pollinators (Astegiano et al., 2015; Kaiser-Bunbury et al., 2010; Memmott et al., 2004; but see Vieira and Almeida-Neto, 2015). Secondary extinctions are generally low until the core of the network (symmetric generalist–generalist interactions) has been highly eroded. However, as discussed in the previous sections, generalist plant species seem to be more affected than specialist plants by habitat loss when pollination generalization is negatively associated with either dispersal or autonomous self-pollination rate. Thus, the small change on network connectance even when some species are lost suggests that habitat loss is a perturbation that might not be intensively targeting high-degree species, but rather more random. As far as nestedness is concerned, the same reasoning can apply since we have shown (Appendix B) that there is a strong association between the degree of nestedness of a network ($NODF_{max}$ scores) and the shape parameter of the power law distribution determining the distribution of degree. If habitat loss does not select against "super-generalist" species (hubs) and maintains the shape of the degree distribution, the robustness of nestedness to habitat loss may be expected. In this sense, changes in connectance and nestedness should occur when networks are near to global collapse, i.e., when the distribution of interactions becomes very skewed as showed recently by Fortuna et al. (2013).

It has been proposed that integrating the network approach in studies aiming to conserve or restore natural biodiversity and to manage ecosystem services in agricultural landscapes should improve management results and also advance ecological network theory (Bohan et al., 2013; Kaiser-Bunbury and Blüthgen, 2015; Tylianakis et al., 2010). In the face of our results, which metrics will be useful to track in human-dominated? As stated by Kaiser-Bunbury and Blüthgen (2015), the first step will be to establish monitoring goals. For example, as natural areas are usually converted to expand cultivable lands, monitoring goals may be associated with the preservation of the pollination service of both wild plants providing several ecosystem services and insect-pollinated crops. This double goal may impose conflicting interests to landscape design (Keitt, 2009; Kremen and Tscharntke, 2007; Mitchell et al. 2015a) because the provision of pollination services to wild plants and the supply of pollinators to crop pollination may be maximized by different configurations of landscapes (Mitchell et al., 2015a). In this sense, the provision of pollination services generally is focused on maximizing crop production (Garibaldi et al., 2014; Kremen et al., 2004; Scheper et al., 2013; but see Gill et al., 2016; Kennedy et al., 2013). However, if the goal is to maximize crop pollination services but also assure the reproduction of wild plants in natural areas to maintain the provision of the pollination service in the long term, monitoring the size and the functional redundancy of the core of plant–pollinator networks in natural areas might represent a more simplified strategy than monitoring the structure of the network. Detecting decreases in the interaction diversity and evenness (Kaiser-Bunbury and Blüthgen, 2015) of species representing the core of plant–pollinator networks (those "super-generalists" symmetrically interacting among them) should alert about the capacity of natural areas to supply these services and to support natural metacommunities. The seed production of insect-pollinated crops and wild plants is increased by interacting with richer pollinator assemblages (Fontaine et al., 2006; Garibaldi et al., 2013; Hoehn et al., 2008). On the other hand, maintaining the preferred host plants for generalist pollinators should be a key strategy to assure pollination services (Scheper et al., 2014). Indeed, by maintaining generalist pollinators, we should promote the persistence of high pollinator-dependent and low dispersal wild plants (Astegiano et al., 2015; Tur et al., 2013). Moreover, monitoring not only pollinator richness but also evenness matters for crop production (Garibaldi et al., 2015). However, the morphological matching between crop flowers and different pollinator features affect crop production, thus different crops should benefit from the maintenance of different functional groups of pollinators (Garibaldi et al., 2015). Therefore, as recently proposed, a hierarchical network approach to the conservation of interactions should advance adaptive management in human-dominated landscapes (Kaiser-Bunbury and Blüthgen, 2015).

4.4 Evidencing the Contrasting Effects of Lower Dependence on Pollinators on Species Persistence

Recently, it has been proposed that plant–pollinator networks harbouring plants with high dependence on pollinator service may be less robust to extinctions (Vieira and Almeida-Neto, 2015). Our model predicts that increasing the autonomous self-pollination rate of plants within metacommunities may lead to contrasting results depending on both the level of habitat loss and the association between plant biological traits. In non-fragmented landscapes, when pollinator extinction/colonization ratios and plant extinction/dispersal ratios approached one, increasing plant autogamous selfing led to decreases in plant and pollinator richness under all scenarios of associations among plant traits. Moreover, under scenarios in which only species generalization varied among plants and pollinators (i.e. the *Neutral* scenario) or when autonomous self-pollination was negatively associated with plant pollination generalization, increasing autonomous self-pollination led to full metacommunity collapse. Decreases in species richness or even complete metacommunity collapse may be explained by increasing autogamous selfing decoupling plant and pollinator dynamics. In our model, increasing seed production by autonomous self-pollination decreased the contribution of pollinators to plant reproduction, i.e., pollinator dynamics may affect less plant dynamics. Moreover, if plant and pollinator dynamics are decoupled, the facilitation effect among species that may arise because of high partner overlap in plant–pollinator webs may be less important (Lever et al., 2014). Thus, contrary to common expectations, increasing autonomous selfing may decrease the robustness provided by high interaction overlap in plant–pollinator assemblages, ultimately decreasing metacommunity persistence. This is a surprising result that may shed light on likely effects of climate change. For instance, it has been predicted that climate change may alter species phenology, increasing the mismatch between plants and their pollinators (Hegland et al., 2009; Memmott et al., 2007; Miller-Rushing and Inouye, 2009; but see Bartomeus et al., 2011). This phenological mismatch might trigger a cycle in which higher autogamy may be selected in plants, decreasing floral attraction and reward, and thus also pollinator visitation (Eckert et al., 2010).

Another prediction of our model is that metacommunities originally comprising plants with higher autonomous self-pollination ability may harbour higher plant richness with increased habitat loss than metacommunities with lower autogamous selfing levels. However, pollinator richness may either not change or barely increase with increased mean plant autogamous selfing. These results may be explained by metacommunities originally comprising plants with lower reproductive dependence on pollinators harbouring more rare plants, even with high levels of habitat loss. The higher plant dominance found in these metacommunities may support this last idea. Rare plants occupying remnants of natural habitat may suffer from pollen limitation if pollinators preferentially feed on mass co-flowering crops (the "dilution hypothesis"; Holzschuh et al., 2011; Tscharntke et al., 2012). Interactions between wild plant species and generalist pollinators are more prone to be temporally lost, since these pollinators are the most likely to be attracted by these crops as reported in recent empirical studies (Holzschuh et. al., 2010; Kleijn et al., 2015).

4.5 Caveats of the Trait-Based Metacommunity Model

In our model, it was assumed equal inbreeding depression for all plant species. Inbreeding depression can reduce the performance of the progeny with consequences for populations and species persistence. Higher probability of inbreeding depression is expected when mating system shift from outcrossing or mixed mating to mainly selfing (Goodwillie et al., 2005). Indeed, it has been reported that habitat fragmentation and disturbance can modify mating systems by decreasing outcrossing rates (Aguilar et al., 2008; Eckert et al., 2010). Such change in mating system is expected to differentially affect species seed production, dispersal and survival in fragmented landscapes.

Lower outcrossing rates in strictly self-incompatible species should decrease seed production (Aguilar et al., 2006). How such changes in the mating system of self-compatible species may affect plant reproduction is less predictable (Eckert et al., 2010). Historically selfing species may suffer slight or no inbreeding depression, but outcrossing and mixed mating species may show a reduction in seed production and in progeny performance (Goodwillie et al., 2005). Moreover, the effects of inbreeding depression can be stronger in stressful habitats (Armbruster and Reed, 2005). Thus, habitat fragmentation may select for the persistence of historically selfing species.

Our model also assumes that pollinator colonization and extinction rates do not differ among pollinator species. Differences in occupancy among pollinator species can only arise as a result of generalist species being temporally favoured by interacting with more plant species, in accordance with empirical data showing lower negative impacts of habitat fragmentation in species with wider foraging diets (Bommarco et al., 2010; Öckinger et al., 2010; but see Williams et al., 2010). The sensitivity of animal pollinators to habitat loss and fragmentation also seems to be related to other biological traits. Higher sensitivity has been associated with smaller body size, lower dispersal ability, lower reproductive capacity, species that nest above ground and solitary species (Ferreira et al., 2015; Jauker et al., 2013; Klein et al., 2008; Kotiaho et al., 2005; Öckinger et al., 2010; Williams et al., 2010). Moreover, like plants, pollinator response to fragmentation may be conditioned by relationships among biological traits (Bommarco et al., 2010; Williams et al., 2010). By being able to use food resources and nest sites from different patches and even from the surrounding matrix, generalist species may have higher dispersal ability and may be less affected by habitat loss. However, as showed by Fahrig (2007), whether dispersal is favourable or increases, the extinction probability of pollinators will depend on the suitability of the surrounding matrix. Moreover, identifying links between suites of traits that may determine pollinator sensitivity to habitat loss and the importance of these pollinators to wild plants by characterizing pollinator centrality in interaction networks seems to be crucial (Hagen et al., 2012). For instance, it has recently been proposed that, by interacting with generalist pollinators, low dispersal plants may persist in local communities (Astegiano et al., 2015). Thus, future trait-based models including associations between plant and pollinator traits may certainly improve our understanding of plant–pollinator persistence in the face of habitat loss.

Functional redundancy among pollinator species was assumed in our model. Functional redundancy implies that for a given plant species, different pollinator species are similarly efficient, such that if one pollinator species goes extinct, another pollinator may fulfil its function (see Valiente-Banuet et al., 2014). Although functional redundancy among pollinators has been reported for several plant species (Fleming et al., 2001; Fumero-Cabán and Meléndez-Ackerman, 2007; Larsen et al., 2005), it seems not to be a general feature not for wild plants (Ashworth et al., 2015b) neither for crop species (Garibaldi et al., 2015). For instance, plant species with bat- or fly-syndrome flowers have higher probabilities of having redundant pollinators from different functional groups than bird- and bee-syndrome flowers (Ashworth et al., 2015b). Moreover, bees and butterflies are redundant pollinators in bat-syndrome flowers (Ashworth et al., 2015b). Given that species richness within Hymenoptera and Lepidoptera can be more strongly diminished by habitat loss than other insect groups (Spiesman and Inouye, 2013), habitat loss might decrease both pollination levels of bee- and butterfly-syndrome flowers and the likelihood of having redundant pollinators in bat-syndrome flowers. Therefore, understanding how pollinator functional redundancy is related to pollinator response to habitat loss may also improve our ability to predict the persistence of species and interactions in fragmented landscapes.

5. FUTURE DIRECTIONS: POLLINATION SERVICES IN HUMAN-DOMINATED LANDSCAPES

Changes induced by habitat loss and fragmentation will modify the taxonomic, genetic and functional diversity of ecosystems in fragmented landscapes over the long term (e.g. Aguilar et al., 2008; Cagnolo et al., 2006; Laurance et al., 2006; Spiesman and Inouye, 2013), decreasing ecosystems resilience and the supply of ecosystem services (Diaz et al., 2006; Haddad et al., 2015; Valiente-Banuet et al., 2014). Impoverished ecosystems will provide low-quality services such as reduced productivity, pollination, pest control and carbon retention (Haddad et al., 2015; Mitchell et al., 2015b). Moreover, natural habitats in fragmented landscapes retain lower diversity of pollinators (e.g. Öckinger et al., 2010; Spiesman and Inouye, 2013; Winfree et al., 2009), which in turn can negatively affect the amount, quality and stability of crop pollination and harvests (Garibaldi et al., 2013; Ricketts et al., 2008). In addition, crop yield is better explained by the trait matching between crop flowers and pollinators and by pollinator evenness, than only by pollinator richness (Garibaldi et al., 2015). Thus, a key question for predicting the vulnerability of ecosystem services faced with changing environmental drivers is how traits determining species' ecosystem-level effects and species responsiveness to drivers also determine species interaction patterns within ecological networks (Díaz et al., 2013; Gill et al., 2016; Lavorel et al., 2013; Mulder et al., 2012). For instance, generalist pollinators form the core of the structure of plant–pollinator networks (Bascompte et al., 2003), providing functional redundancy and complementarity (Blüthgen and Klein, 2011; Tscharntke et al., 2012) and interacting with plants more sensitive to pollinator loss, i.e., specialist, highly pollinator-dependent and low dispersal plants (Astegiano et al., 2015; Tur et al., 2013; Vázquez and Aizen, 2004). Generalist pollinators are also among the main pollinators of several mass-flowering crops (Holzschuh et al., 2011; Kleijn et al., 2015). Thus, it will be crucial to evaluate how functional redundancy among generalist pollinators or pollinator guilds is related to the diversity in the response of pollinators to likely disturbances in human-dominated landscapes (Elmqvist et al., 2003; Kaiser-Bunbury and Blüthgen, 2015; Tylianakis et al., 2010). Pollination service resilience may depend on pollinators response diversity, i.e., how functionally similar pollinators respond differently to disturbance. Generalist pollinators may respond to habitat loss in very different ways (Bommarco et al., 2010; Williams et al., 2010), which may increase the resilience of pollination services to habitat loss. However, how redundancy is distributed in plant–pollinator networks and how it changes with pollinator loss, for instance because of the rewiring process (Kaiser-Bunbury et al., 2010), remain to be understood.

ACKNOWLEDGEMENTS

We thank Pedro Jordano and Carlos Melián for previous discussions on the ideas present in this manuscript. To Jari Oksanen for advice with codes for the construction of theoretical matrices. To Christian Mulder, David Bohan and an anonymous reviewer for useful comments that improved the final version of this manuscript. To David Bohan and Guy Woodward for given us the opportunity to contribute to this special issue. J.A. was supported by Grant 2011/09951-2 and Grant 2012/04941-1 (São Paulo Research Foundation-FAPESP and she is a researcher of CONICET (Argentina) (J.A.)); M.M.V. was supported by Grant 2010/11633-6 (São Paulo Research Foundation-FAPESP);

C.Y.M. was supported by Grant 99999.002978/2014-08 (PDSE-CAPES); F.M. and P.O.C. are researchers from the CNRS; P.R.G. Jr. was supported by Grant 2009/54422-8 (São Paulo Research Foundation-FAPESP) and CNPq and L.A. was supported by Grant PICT 2011–1606 (Agencia Nacional de Promoción Científica y Técnica, ANPCyT) and she is a researcher of CONICET (Argentina).

APPENDIX A. GENERATING BIPARTITE INCIDENCE MATRICES WITH DETERMINED DEGREE SEQUENCES

To generate bipartite incidence matrices (i.e. binary matrices in which rows represented animal species, columns, plant species, and 1's indicated realized interactions), we resorted to the following procedure based on an algorithm proposed by Chung et al. (2003), adapted to the problem of generating incidence matrices with a known number of connections.

Following Chung et al. (2003), a non-increasing sequence of degrees (summing to C) can represent a sample from a power law distribution of parameter γ if the degree of the ith element in the sequence is roughly proportional to $i^{-1/(\gamma-1)}$. In practice, we used sequences of degrees k_i defined by:

$$k_i = \max\left[1, \min\left\{\left\lfloor xi^{-1/(\gamma-1)}\right\rfloor, n\right\}\right] \quad (A.1)$$

where n is the maximal degree (given by the number of species in the other trophic level), $\lfloor\ \rfloor$ is the integer part symbol and x is a constant obtained by numerically solving:

$$C = \sum_i k_i \quad (A.2)$$

The following code can generate these sequences of degrees in R:

```
## function alpha: computes parameter necessary for adjusting the power
law degree sequences
## parameters: s = number of species in the focal group, beta = power law
parameter, c = total number of connections, d = number of species in the
other group
alpha<-function(s,beta,c,d){
        locfun<-function(x){
                ((rep(1,s)%*%sapply(sapply(floor(x*(1:s)^(-(1/
(beta-1)))),function(z) min(z,d)),function(y) max(y,1)))-c)
        }
        uniroot(f=locfun,interval = c(0,2*c))$root
}
```

```
## function powerlawdegreeseq: yields a sequence of non-increasing degrees such that the ensuing distribution follows a power law of parameter beta, following Chung et al. 2003
powerlawdegreeseq<-function(s,beta,c,d){
        a <- alpha(s,beta,c,d)
        sapply(sapply(floor(a*(1:s)^(-(1/(beta-1)))),function
(z) min(z,d)),function(y) max(y,1))
}
```

Once degree sequences are obtained, the rest of the procedure consists in generating random incidence matrices such that summing by row or column generates sequences identical to the desired degree sequences. To do so, we first check that the degree sequences used for animals and plants can effectively generate a bipartite incidence matrix (i.e. are graphical sequences in the mathematical sense). The Gale–Ryser theorem (see Brualdi and Ryser, 1991) manages to check this by recursively comparing sums of degrees to a minimal constraint. This test can be implemented in R as:

```
galerysertest <- function(rowtotal,columntotal){
        row <- length(rowtotal)
        column <- length(columntotal)
        total <- sum(rowtotal)
        test0<-(sum(columntotal)==total)
        delta <- sort(rowtotal, decreasing = T)
        d <- sort(columntotal, decreasing = T)
        left <- cumsum(d)
        right <- sapply(1:column, FUN = function(x) (1/2)*(total +
row*x - sum(abs(rowtotal-x))))
        test1 <- ((right-left)>=0)
        test0 && all(test1)
}
```

If the degree sequences are indeed graphical, we first generate a quantitative incidence matrix (i.e. with integer entries instead of binary entries) that keep row and column sums equal to the matching degree sequences, and then resort to the quasi-swap/sum of squares algorithm of Miklós and Podani (2004) to generate a binary incidence matrix through random swapping of checkerboard patterns that decreased the sum of squares of elements in the matrix. The following R code implements this procedure.

```
library(vegan)
library(igraph)
sumofsquare<-function(rowtotal,columntotal){
        row <- length(rowtotal)
```

```
        column <- length(columntotal)
        m <- r2dtable(1, rowtotal, columntotal)[[1]]
        ssmin <- sum(rowtotal)
        ss <- sum(m*m)
        while(ss>ssmin){
                ik <- igraph.sample(1,row,2)
                jl <- igraph.sample(1,column,2)
        if((m[ik[1],jl[1]]>0)&&(m[ik[2],jl[2]]>0)&&(m[ik[1],jl[1]]
+m[ik[2],jl[2]]-m[ik[1],jl[2]]-m[ik[2],jl[1]]>=2)){
                        ss  <-  ss-2*(m[ik[1],jl[1]]+m[ik[2],jl[2]]-m
[ik[1],jl[2]]-m[ik[2],jl[1]]-2)
                        m[ik[1],jl[1]] <- m[ik[1],jl[1]] - 1
                        m[ik[2],jl[2]] <- m[ik[2],jl[2]] - 1
                        m[ik[1],jl[2]] <- m[ik[1],jl[2]] + 1
                        m[ik[2],jl[1]] <- m[ik[2],jl[1]] + 1
                }
                else {
        if((m[ik[1],jl[2]]>0)&&(m[ik[2],jl[1]]>0)&&(m[ik[1],jl[2]]
+m[ik[2],jl[1]]-m[ik[1],jl[1]]-m[ik[2],jl[2]]>=2)){
                                ss <- ss-2*(m[ik[1],jl[2]]+m[ik[2],jl
[1]]-m[ik[1],jl[1]]-m[ik[2],jl[2]]-2)
                                m[ik[1],jl[2]] <- m[ik[1],jl[2]] - 1
                                m[ik[2],jl[1]] <- m[ik[2],jl[1]] - 1
                                m[ik[1],jl[1]] <- m[ik[1],jl[1]] + 1
                                m[ik[2],jl[2]] <- m[ik[2],jl[2]] + 1
                        }
                }
        }
        m
}
```

Finally, once a single incidence matrix corresponding to the desired degree sequences is generated through the quasi-swap method, we can generate other such matrices by swapping entries in the matrix following the trial swap procedure of Miklós and Podani (2004) and implement in the R package "vegan". The following R code exemplifies how we can use this procedure to generate 100 random bipartite matrices following degree sequences mimicking power laws of parameter 2.5.

```
## function generateZ sums up the various functions defined above
generateZ <- function(sa,sp,betaa,betap,c){
        delta <- powerlawdegreeseq(sa,betaa,c,sp)
```

```
        d <- powerlawdegreeseq(sp,betap,c,sa)
        sumofsquare(delta,d)
}
galerysertest(powerlawdegreeseq(60,2.5,1440,120),
powerlawdegreeseq(120,2.5,1440,60))
initmat <- generateZ(120,60,2.5,2.5,1440)
setofmat<-simulate(nullmodel(initmat,"tswap"),nsim=100,burnin =
1000000, thin = 1000000)
```

BOX A1 Measures of Nestedness

To measure nestedness in a given bipartite network linking plants and pollinators, we used two different measures of nestedness, $NODF_{max}$ and PRSN (Podani and Schmera, 2012). Considering the incidence matrix **M** with element m_{ij} being equal to 1 if pollinator i interacts with plant j, and writing $\delta_i = \sum_j m_{ij}$ the degree of pollinators, $d_j = \sum_i m_{ij}$ the degree of plants, A the number of pollinators and P the number of plants, the formula for $NODF_{max}$ given by Podani and Schmera (2012), averaged over its column-wise and row-wise definitions, can be expanded as:

$$NODF_{max} = \frac{NODF_{max}^{pollinators} + NODF_{max}^{plants}}{2} \tag{A.3}$$

$$NODF_{max}^{pollinators} = \frac{100}{\binom{A}{2}} \sum_{k<l} \left(1 - 0^{|\delta_k - \delta_l|}\right) \frac{\sum_j m_{kj} m_{lj}}{\min(\delta_k, \delta_l)} \tag{A.4}$$

$$NODF_{max}^{plants} = \frac{100}{\binom{P}{2}} \sum_{k<l} \left(1 - 0^{|d_k - d_l|}\right) \frac{\sum_i m_{ik} m_{il}}{\min(d_k, d_l)} \tag{A.5}$$

Essentially, what $NODF_{max}^{plants}$ measures is the percentage of overlap of interacting partners among all pairs of plants, the denominator being given by the plant with the lowest degree. The $1 - 0^{|d_k - d_l|}$ factor indicates that, in case of ties (i.e. when both plant species have the same degree), the comparison always results in the addition of zero to the sum of overlaps. $NODF_{max}^{pollinators}$ measures the same quantity, but from the viewpoint of pairs of pollinator species sharing a more or less high proportion of the interacting plant partners.

The second measure of nestedness used is PRSN (for percentage relativized strict nestedness) can be computed in the same way:

$$PRSN = \frac{PRSN^{pollinators} + PRSN^{plants}}{2} \tag{A.6}$$

BOX A1 Measures of Nestedness—cont'd

$$\mathrm{PRSN}^{\mathrm{pollinators}} = \frac{100}{\binom{A}{2}} \sum_{k<l} \left(1 - 0^{|\delta_k - \delta_l|}\right) \left(1 - 0^{\sum_j m_{kj} m_{lj}}\right) \frac{\sum_j m_{kj} m_{lj} + |\delta_k - \delta_l|}{\delta_k + \delta_l - \sum_j m_{kj} m_{lj}} \tag{A.7}$$

$$\mathrm{PRSN}^{\mathrm{plants}} = \frac{100}{\binom{P}{2}} \sum_{k<l} \left(1 - 0^{|d_k - d_l|}\right) \left(1 - 0^{\sum_i m_{ik} m_{il}}\right) \frac{\sum_i m_{ik} m_{il} + |d_k - d_l|}{d_k + d_l - \sum_i m_{ik} m_{il}} \tag{A.8}$$

APPENDIX B. NESTEDNESS DEPENDS ON THE DISTRIBUTION OF DEGREES

Here, we show that one can cover a rather wide span of nestedness values for a given bipartite network with a given number of connections by simply adjusting the distribution of degrees among the nodes. Or, in other words, that there exists a very strong dependence of the expected values of nestedness indices on the distribution of node degree.

A first observation simply comes from rewriting the elements of Eq. (A.4), which gives $\mathrm{NODF}_{\max}^{\mathrm{pollinators}}$ for a given bipartite network of incidence matrix $\mathbf{M}$. In addition to the degrees of the pollinator species, this quantity depends on the elements of $\mathbf{M}.\mathbf{M}^T$ through the sums $\sum_j m_{kj} m_{lj}$. One can naturally rewrite elements $(\mathbf{M}.\mathbf{M}^T)_{kl}$ as an expectation of random binary interaction variables ($m_{k\bullet}$ and $m_{l\bullet}$):

$$\left(\mathbf{M}.\mathbf{M}^T\right)_{kl} = \sum_j m_{kj} m_{lj} = P \times \mathbb{E}[m_{k\bullet} m_{l\bullet}] \tag{A.9}$$

where P is the number of plant species in the network. If we note $\rho[m_{k\bullet}, m_{l\bullet}]$ the correlation between the random binary variables $m_{k\bullet}$ and $m_{l\bullet}$, the following expression links $\rho[m_{k\bullet}, m_{l\bullet}]$ with $(\mathbf{M}.\mathbf{M}^T)_{kl}$:

$$\left(\mathbf{M}.\mathbf{M}^T\right)_{kl} = P\rho[m_{k\bullet}, m_{l\bullet}] \sqrt{\delta_k \delta_l \left(1 - \frac{\delta_k}{P}\right)\left(1 - \frac{\delta_l}{P}\right)} + \frac{\delta_k \delta_l}{P} \tag{A.10}$$

Thus, in the absence of an explicit correlation between $m_{k\bullet}$ and $m_{l\bullet}$, we still expect $(\mathbf{M}.\mathbf{M}^T)_{kl}$ to take values of $\delta_k\delta_l/P$.

Now, assuming that rows (i.e. pollinators) are sorted in decreasing order of degrees (i.e. $\delta_1 \geq \delta_2 \geq \cdots \geq \delta_A$) and ignoring the $1-0^{|\delta_k-\delta_l|}$ factors in Eq. (A.4), we have the following expectation for $\mathrm{NODF}_{\max}^{\text{pollinators}}$ under the assumption of no correlations between the $m_{k\bullet}$ and $m_{l\bullet}$ of any pair of pollinators:

$$\mathrm{NODF}_{\max}^{\text{pollinators}} \approx \frac{100}{\binom{A}{2}} \sum_{k=1}^{A-1} \sum_{l=k+1}^{A} \frac{\delta_k}{P} = \frac{100}{P\binom{A}{2}} \sum_{k=1}^{A-1} (A-k)\delta_k \qquad (A.11)$$

The expression (A.11) clearly depends on the degree distribution as the quantity $\sum_{k=1}^{A-1} k\delta_k$ will vary depending on the realized sequence of (δ_k).

A practical example of the dependence of $\mathrm{NODF}_{\max}$ on the degree distribution can be found in Fig. S1 (http://dx.doi.org/10.1016/bs.aecr.2015.09.005). For the same connectance and species richness values, the shape of the degree distribution determines the value of $\mathrm{NODF}_{\max}$ unequivocally, with low-power degree distributions leading to higher $\mathrm{NODF}_{\max}$ values than high-power degree distributions (from >70 to <40 $\mathrm{NODF}_{\max}$ values between powers of 2.2 and 2.9).

REFERENCES

Abramson, G., Trejo Soto, C.A., Oña, L., 2011. The role of asymmetric interactions on the effect of habitat destruction in mutualistic networks. PLoS One 6, e21028.

Aguilar, R., Ashworth, L., Galetto, L., Aizen, M.A., 2006. Determinants of plant reproductive susceptibility to habitat fragmentation: review and synthesis through a meta-analysis. Ecol. Lett. 9, 968–980.

Aguilar, R., Quesada, M., Ashworth, L., Herrerias-Diego, Y., Lobo, J., 2008. Genetic consequences of habitat fragmentation in plant populations: susceptible signals in plant traits and methodological approaches. Mol. Ecol. 17, 5177–5188.

Aguilar, R., Ashworth, L., Calviño, A., Quesada, M., 2012. What is left after sex in fragmented forests: assessing the quantity and quality of progeny of *Prosopis caldenia* (Fabaceae) an endemic tree from central Argentina. Biol. Conserv. 152, 81–89.

Aizen, M.A., Ashworth, L., Galetto, L., 2002. Reproductive success in fragmented habitats: do compatibility systems and pollination specialization matter? J. Veg. Sci. 13, 885–892.

Aizen, M.A., Sabatino, M., Tylianakis, J.M., 2012. Specialization and rarity predict non-random loss of interactions from mutualist networks. Science 335, 1486–1489.

Allesina, S., Tang, S., 2012. Stability criteria for complex ecosystems. Nature 483, 205–208.

Almeida-Neto, M., Guimaraes, P., Guimarães, P.R., Loyola, R.D., Ulrich, W., 2008. A consistent metric for nestedness analysis in ecological systems: reconciling concept and measurement. Oikos 117, 1227–1239.

Anderson, S.H., Kelly, D., Ladley, J.J., Molloy, S., Terry, J., 2011. Cascading effects of bird functional extinction reduce pollination and plant density. Science 331, 1068–1070.

Aparicio, A., Albaladejo, R.G., Olalla-Tárraga, M.A., Carrillo, L.F., Rodríguez, M.Á., 2008. Dispersal potentials determine responses of woody plant species richness to environmental factors in fragmented Mediterranean landscapes. For. Ecol. Manag. 255, 2894–2906.

Armbruster, P., Reed, D.H., 2005. Inbreeding depression in benign and stressful environments. Heredity 95, 235–242.

Ashworth, L., Aguilar, R., Galetto, L., Aizen, M.A., 2004. Why do pollination generalist and specialist plant species show similar reproductive susceptibility to habitat fragmentation? J. Ecol. 92, 717–719.

Ashworth, L., Quesada, M., Casas, A., Aguilar, R., Oyama, K., 2009. Pollinator-dependent food production in Mexico. Biol. Conserv. 142, 1050–1057.

Ashworth, L., Calviño, A., Martí, L., Aguilar, R., 2015a. Offspring performance and recruitment of the pioneer tree Acacia caven (Fabaceae) in a fragmented subtropical dry forest. Austral Ecol. 40, 634–641.

Ashworth, L., Aguilar, R., Martén-Rodríguez, S., Lopezaraiza-Mikel, M., Avila-Sakar, G., Rosas-Guerrero, V., Quesada, M., 2015b. Pollination syndromes: a global pattern of convergent evolution driven by the most effective pollinator. In: Pontarotti, P. (Ed.), Evolutionary Biology: Biodiversification from Genotype to Phenotype. Springer International Publishing, Switzerland, pp. 203–224.

Astegiano, J., Massol, F., Vidal, M.M., Cheptou, P.O., Guimarães Jr., P.R., 2015. The robustness of plant-pollinator assemblages: linking plant interaction patterns and sensitivity to pollinator loss. PLoS One 10, e0117243.

Auld, J.R., de Casas, R.R., 2013. The correlated evolution of dispersal and mating-system traits. Evol. Biol. 40, 185–193.

Baker, H.G., 1955. Self-compatibility and establishment after "long-distance" dispersal. Evolution 9, 347–349.

Barabási, A.L., Albert, R., 1999. Emergence of scaling in random networks. Science 286, 509–512.

Bartomeus, I., Ascher, J.S., Wagner, D., Danforth, B.N., Colla, S., Kornbluth, S., Winfree, R., 2011. Climate-associated phenological advances in bee pollinators and bee-pollinated plants. Proc. Natl. Acad. Sci. U.S.A. 108, 20645–20649.

Bascompte, J., Jordano, P., 2007. Plant–animal mutualistic networks: the architecture of biodiversity. Annu. Rev. Ecol. Evol. Syst. 38, 567–593.

Bascompte, J., Jordano, P., Melián, C.J., Olesen, J.M., 2003. The nested assembly of plant–animal mutualistic networks. Proc. Natl. Acad. Sci. U.S.A. 100, 9383–9387.

Bascompte, J., Jordano, P., Olesen, J.M., 2006. Asymmetric coevolutionary networks facilitate biodiversity maintenance. Science 312, 431–433.

Bastolla, U., Fortuna, M.A., Pascual-García, A., Ferrera, A., Luque, B., Bascompte, J., 2009. The architecture of mutualistic networks minimizes competition and increases biodiversity. Nature 458, 1018–1020.

Biesmeijer, J.C., Roberts, S.P.M., Reemer, M., Ohlemuller, R., Edwards, M., Peeters, T., Schaffers, A.T., Potts, S.G., Kleukers, R., Thomas, C.D., Settele, J., Kunin, W.E., 2006. Parallel declines in pollinators and insect-pollinated plants in Britain and the Netherlands. Science 313, 351–354.

Blüthgen, N., 2010. Why network analysis is often disconnected from community ecology: a critique and an ecologist's guide. Basic Appl. Ecol. 11, 185–195.

Blüthgen, N., Klein, A.M., 2011. Functional complementarity and specialisation: the role of biodiversity in plant–pollinator interactions. Basic Appl. Ecol. 12, 282–291.

Bohan, D.A., Raybould, A., Mulder, C., Woodward, G., Tamaddoni-Nezhad, A., Blüthgen, N., Pocock, M.J.O., Muggleton, S., Evans, D.M., Astegiano, J.,

Massol, F., Loeuille, N., Petit, S., Macfadyen, S., 2013. Networking agroecology: integrating the diversity of agroecosystem interactions. Adv. Ecol. Res. 49, 1–67. Chapter one.

Bommarco, R., Biesmeijer, J.C., Meyer, B., Potts, S.G., Pöyry, J., Roberts, S.P.M., Steffan-Dewenter, I., Öckinger, E., 2010. Dispersal capacity and diet breadth modify the response of wild bees to habitat loss. Proc. R. Soc. Lond. B 277, 2075–2082.

Bond, W.J., 1994. Do mutualisms matter? Assessing the impact of pollinator and disperser disruption on plant extinction. Philos. Trans. R. Soc. Lond. B 344, 83–90.

Breed, M.F., Gardner, M.G., Ottewell, K.M., Navarro, C.M., Lowe, A.J., 2012. Shifts in reproductive assurance strategies and inbreeding costs associated with habitat fragmentation in Central American mahogany. Ecol. Lett. 15, 444–452.

Brualdi, R.A., Ryser, H.J., 1991. Combinatorial Matrix Theory. Cambridge University Press, Cambridge, UK.

Cagnolo, L., Cabido, M., Valladares, G., 2006. Plant species richness in the Chaco Serrano woodland from central Argentina: ecological traits and habitat fragmentation effects. Biol. Conserv. 132, 510–519.

Cagnolo, L., Valladares, G., Salvo, A., Cabido, M., Zak, M., 2009. Habitat fragmentation and species loss across three interacting trophic levels: effects of life-history and food-web traits. Conserv. Biol. 23, 1167–1175.

Calcagno, V., Massol, F., Mouquet, N., Jarne, P., David, P., 2011. Constraints on food chain length arising from regional metacommunity dynamics. Proc. R. Soc. Lond. B 278, 3042–3049.

Carlquist, S., 1966. The biota of long-distance dispersal. II. Loss of dispersibility in Pacific Compositae. Evolution 20, 30–48.

Cheptou, P.-O., 2012. Clarifying Baker's Law. Ann. Bot. 109, 633–641.

Cheptou, P.-O., Massol, F., 2009. Pollination fluctuations drive evolutionary syndromes linking dispersal and mating system. Am. Nat. 174, 46–55.

Cheptou, P.-O., Carrue, O., Rouifed, S., Cantarel, A., 2008. Rapid evolution of seed dispersal in an urban environment in the weed *Crepis sancta*. Proc. Natl. Acad. Sci. U.S.A. 105, 3796–3799.

Chung, F., Lu, L., Vu, V., 2003. Spectra of random graphs with given expected degrees. Proc. Natl. Acad. Sci. 100, 6313–6318.

Cohen, R., Erez, K., ben-Avraham, D., Havlin, S., 2000. Resilience of the internet to random breakdowns. Phys. Rev. Lett. 85, 4626–4628.

Cohen, R., Erez, K., ben-Avraham, D., Havlin, S., 2001. Breakdown of the internet under intentional attack. Phys. Rev. Lett. 86, 3682–3685.

Comins, H.N., Hamilton, W.D., May, R.M., 1980. Evolutionarily stable dispersal strategies. J. Theor. Biol. 82, 205–230.

Cordeiro, N.J., Dangalasi, H.J., McEntee, J.P., Howe, H.F., 2009. Disperser limitation and recruitment of an endemic African tree in a fragmented landscape. Ecology 90, 1030–1041.

Diaz, S., Fargione, J., Chapin III, F.S., Tilman, D., 2006. Biodiversity loss threatens human well-being. PLoS Biol. 4, e277.

Díaz, S., Purvis, A., Cornelissen, J.H.C., Mace, G.M., Donoghue, M.J., Ewers, R.M., Jordano, P., Pearse, W.D., 2013. Functional traits, the phylogeny of function, and ecosystem service vulnerability. Ecol. Evol. 3, 2958–2975.

Dorman, C.F., Fründ, J., Blüthgen, N., Gruber, B., 2009. Indices, graphs and null models: analysing bipartite ecological networks. Open Ecol. J. 2, 7–24.

Ebeling, A., Klein, A.M., Tscharntke, T., 2011. Plant–flower visitor interaction webs: temporal stability and pollinator specialization increases along an experimental plant diversity gradient. Basic Appl. Ecol. 12, 300–309.

Eckert, C.G., Kalisz, S., Geber, M.A., Sargent, R., Elle, E., Cheptou, P.-O., Goodwillie, C., Johnston, M.O., Kelly, J.K., Moeller, D.A., Porcher, E., Ree, R., Vallejo-Marín, M., Winn, A.A., 2010. Plant mating systems in a changing world. Trends Ecol. Evol. 25, 35–43.

Ehrlén, J., Eriksson, O., 2000. Dispersal limitations and patch occupancy in forest herbs. Ecology 81, 1667–1674.

Eilers, E.J., Kremen, C., Greenleaf, S.S., Garber, A.K., Klein, A.M., 2011. Contribution of pollinator-mediated crops to nutrients in the human food supply. PLoS One 6, e21363.

Eklöf, A., Kaneryd, L., Münger, P., 2012. Climate change in metacommunities: dispersal gives double-sided effects on persistence. Proc. R. Soc. Lond. B 367, 2945–2954.

Ekroos, J., Heliölä, J., Kuussaari, M., 2010. Homogenization of lepidopteran communities in intensively cultivated agricultural landscapes. J. Appl. Ecol. 47, 459–467.

Elmqvist, T., Folke, C., Nyström, M., Peterson, G., Bengtsson, J., Walker, B., Norberg, J., 2003. Response diversity, ecosystem change, and resilience. Front. Ecol. Environ. 1, 488–494.

Evju, M., Blumentrath, S., Skarpaas, O., Stabbetorp, O.E., Sverdrup-Thygeson, A., 2015. Plant species occurrence in a fragmented grassland landscape: the importance of species traits. Biodivers. Conserv. 24, 547–561.

Fabian, Y., Sandau, N., Bruggisser, O.T., Aebi, A., Kehrli, P., Rohr, R.P., Naisbit, R.E., Bersier, L.-F., 2013. The importance of landscape and spatial structure for hymenopteran-based food webs in an agro-ecosystem. J. Anim. Ecol. 82, 1203–1214.

Fægri, K., van der Pijl, L., 1979. The Principles of Pollination Ecology, third rev. ed. Pergamon Press, Oxford.

Fahrig, L., 2003. Effects of habitat fragmentation on biodiversity. Annu. Rev. Ecol. Syst. 34, 487–515.

Fahrig, L., 2007. Non-optimal animal movement in human-altered landscapes. Funct. Ecol. 21, 1003–1015.

Fakheran, S., Paul-Victor, C., Heichinger, C., Schmid, B., Grossniklaus, U., Turnbull, L.A., 2010. Adaptation and extinction in experimentally fragmented landscapes. Proc. Natl. Acad. Sci. U.S.A. 107, 19120–19125.

Ferreira, P.A., Boscolo, D., Carvalheiro, L.G., Biesmeijer, J.C., Rocha, P.L., Viana, B.F., 2015. Responses of bees to habitat loss in fragmented landscapes of Brazilian Atlantic Rainforest. Landsc. Ecol. 30, 1–12.

Fischer, J., Lindenmayer, D.B., 2007. Landscape modification and habitat fragmentation: a synthesis. Glob. Ecol. Biogeogr. 16, 265–280.

Fleming, T.H., Sahley, C.T., Holland, J.N., Nason, J.D., Hamrick, J.L., 2001. Sonoran Desert columnar cacti and the evolution of generalized pollination systems. Ecol. Monogr. 71, 511–530.

Fontaine, C., Dajoz, I., Meriguet, J., Loreau, M., 2006. Functional diversity of plant-pollinator interaction webs enhances the persistence of plant communities. PLoS Biol. 4, e1. http://dx.doi.org/10.1371/journal.pbio.0040001.

Fortuna, M.A., Bascompte, J., 2006. Habitat loss and the structure of plant–animal mutualistic networks. Ecol. Lett. 9, 281–286.

Fortuna, M.A., Krishna, A., Bascompte, J., 2013. Habitat loss and the disassembly of mutalistic networks. Oikos 122, 938–942.

Fumero-Cabán, J.J., Meléndez-Ackerman, E.J., 2007. Relative pollination effectiveness of floral visitors of *Pitcairnia angustifolia* (Bromeliaceae). Amer. J. Bot. 94, 419–424.

Galetti, M., Guevara, R., Côrtes, M.C., Fadini, R., Von Matter, S., Leite, A.B., Labecca, F., Ribeiro, T., Carvalho, C.S., Collevatti, R.S., Pires, M.M., Guimarães Jr., P.R., Brancalion, P.H., Ribeiro, M.C., Jordano, P., 2013. Functional extinction of birds drives rapid evolutionary changes in seed size. Science 340, 1086–1089.

Garibaldi, L.A., Steffan-Dewenter, I., Winfree, R., Aizen, M.A., Bommarco, R., Cunningham, S.A., Kremen, C., Carvalheiro, L.G., Harder, L.D., Afik, O., Bartomeus, I., Benjamin, F., Boreux, V., Cariveau, D., Chacoff, N.P., Dudenhöffer, J.H., Freitas, B.M., Ghazoul, J., Greenleaf, S., Hipólito, J., Holzschuh, A., Howlett, B., Isaacs, R., Javorek, S.K., Kennedy, C.M., Krewenka, K.M., Krishnan, S., Mandelik, Y., Mayfield, M.M., Motzke, I., Munyuli, T., Nault, B.A., Otieno, M., Petersen, J., Pisanty, G., Potts, S.G., Rader, R., Ricketts, T.H., Rundlöf, M., Seymour, C.L., Schüepp, C., Szentgyörgyi, H., Taki, H., Tscharntke, T., Vergara, C.H., Viana, B.F., Wanger, T.C., Westphal, C., Williams, N., Klein, A.M., 2013. Wild pollinators enhance fruit set of crops regardless of honey bee abundance. Science 339, 1608–1611.

Garibaldi, L.A., Carvalheiro, L.G., Leonhardt, S.D., Aizen, M.A., Blaauw, B.R., Isaacs, R., Kuhlmann, M., Kleijn, D., Klein, A.M., Kremen, C., Morandin, L., Scheper, J., Winfree, R., 2014. From research to action: enhancing crop yield through wild pollinators. Front. Ecol. Envir. 12, 439–447.

Garibaldi, L.A., Bartomeus, I., Bommarco, R., Klein, A.M., Cunningham, S.A., Aizen, M.A., Boreux, V., Garratt, M.P.D., Carvalheiro, L.G., Kremen, C., Morales, C.L., Schüepp, C., Chacoff, N.P., Freitas, B.M., Gagic, V., Holzschuh, A., Klatt, B.K., Krewenka, K.M., Krishnan, S., Mayfield, M.M., Motzke, I., Otieno, M., Petersen, J., Potts, S.G., Ricketts, T.H., Rundlöf, M., Sciligo, A., Sinu, P.A., Steffan-Dewenter, I., Taki, H., Tscharntke, T., Vergara, C.H., Viana, B.F., Woyciechowski, M., 2015. Trait matching of flower visitors and crops predicts fruit set better than trait diversity. J. App. Ecol. in press.

Gill, R.J., Baldock, K.C.R., Brown, M.J.F., Cresswell, J.E., Dicks, L.V., Fountain, T., Garratt, M.P.D., Gough, L.A., Heard, M.S., Holland, J.M., Ollerton, J., Stone, G.N., Tang, C.Q., Vanbergen, A.J., Vogler, A.P., Woodward, G., Arce, A.N., Boatman, N.D., Brand-Hardy, R., Breeze, T.D., Green, M., Hartfield, C.M., O'Connor, R.S., Osborne, J.L., Phillips, J., Sutton, P.B., Potts, S.G., 2016. Protecting an ecosystem service: Approaches to understanding and mitigating threats to wild insect pollinators. Adv. Ecol. Res. 54, in press.

Girão, L.C., Lopes, A.V., Tabarelli, M., Bruna, E.M., 2007. Changes in tree reproductive traits reduce functional diversity in a fragmented Atlantic forest landscape. PLoS One 2, e908.

Gonzalez, A., Rayfield, B., Lindo, Z., 2011. The disentangled bank: how loss of habitat fragments and disassembles ecological networks. Am. J. Bot. 98, 503–516.

González-Varo, J.P., González-Albaladejo, R., Aparicio-Martínez, A., Arroyo-Marin, J., 2010. Linking genetic diversity, mating patterns and progeny performance in fragmented populations of a mediterranean shrub. J. Appl. Ecol. 47, 1242–1252.

González-Varo, J.P., Biesmeijer, J.C., Bommarco, R., Potts, S.G., Schweiger, O., Smith, H.G., Steffan-Dewenter, I., Szentgyörgyi, H., Woyciechowski, M., Vilà, M., 2013. Combined effects of global change pressures on animal-mediated pollination. Trends Ecol. Evol. 28, 524–530.

Goodwillie, C., Kalisz, S., Eckert, C.G., 2005. The evolutionary enigma of mixed mating systems in plants: occurrence, theoretical explanations, and empirical evidence. Annu. Rev. Ecol. Syst. 36, 47–79.

Guimarães Jr., P.R., Jordano, P., Thompson, J.N., 2011. Evolution and coevolution in mutualistic networks. Ecol. Lett. 14, 877–885.

Haddad, N.M., Brudvig, L.A., Clobert, J., Davies, K.F., Gonzalez, A., Holt, R.D., Lovejoy, T.E., Sexton, J.E., Austin, M.P., Collins, C.D., Cook, W.M., Damschen, E.I., Ewers, R.M., Foster, B.L., Jenkins, C., King, A., Laurance, W.F., Levey, D.J., Margules, C.R., Melbourne, B.A., Nicholls, A.O., Orrock, J.L., Song, D., Townsend, J.R., 2015. Habitat fragmentation and its lasting impact on Earth's ecosystems. Science 1, e1500052.

Hadley, A.S., Betts, M.G., 2012. The effects of landscape fragmentation on pollination dynamics: absence of evidence not evidence of absence. Biol. Rev. 87, 526–544.

Hagen, M., Kissling, D.W., Rasmussen, C., Carstensen, D.W., Dupont, Y.L., Kaiser-Bunbury, C.N., O'Gorman, E.J., Olesen, J.M., De Aguiar, M.A.M., Brown, L.E., Alves-Dos-Santos, I., Guimarães Jr., P.R., Maia, K.P., Marquitti, F.M.D., Vidal, M.M., Edwards, F.K., Genini, J., Jenkins, G.B., Trøjelsgaard, K., Woodward, G., Jordano, P., Ledger, M.E., Mclaughlin, T., Morellato, L.P.C., Tylianakis, J.M., 2012. Biodiversity, species interactions and ecological networks in a fragmented world. Adv. Ecol. Res. 46, 89–120.

Hanski, I., Ovaskainen, O., 2000. The metapopulation capacity of a fragmented landscape. Nature 404, 755–758.

Hegland, S.J., Nielsen, A., Lázaro, A., Bjerknes, A.L., Totland, Ø., 2009. How does climate warming affect plant-pollinator interactions? Ecol. Lett. 12, 184–195.

Hoekstra, J.M., Boucher, T.M., Ricketts, T.H., Roberts, C., 2005. Confronting a biome crisis: global disparities of habitat loss and protection. Ecol. Lett. 8, 23–29.

Hoehn, P., Tscharntke, T., Tylianakis, J.M., Steffan-Dewenter, I., 2008. Functional group diversity of bee pollinators increases crop yield. Proc. R. Soc. Lond. B Biol. Sci. 275, 2283–2291.

Holzschuh, A., Dormann, C.F., Tscharntke, T., Steffan-Dewenter, I., 2011. Expansion of mass-flowering crops leads to transient pollinator dilution and reduced wild plant pollination. Proc. R. Soc. Lond. B 278, 3444–3451.

Holzschuh, A., Steffan-Dewenter, I., Tscharntke, T., 2010. How do landscape composition and configuration, organic farming and fallow strips affect the diversity of bees, wasps and their parasitoids?. J. Anim. Ecol. 79, 491–500.

James, A., Pitchford, J.W., Plank, M.J., 2012. Disentangling nestedness from models of ecological complexity. Nature 487, 227–230.

Jauker, B., Krauss, J., Jauker, F., Steffan-Dewenter, I., 2013. Linking life history traits to pollinator loss in fragmented calcareous grasslands. Landsc. Ecol. 28, 107–120.

Jenkins, D.G., 2011. Ranked species occupancy curves reveal common patterns among diverse metacommunities. Glob. Ecol. Biogeogr. 20, 486–497.

Johnson, S.D., Steiner, K.E., 2000. Generalization versus specialization in plant pollination systems. Trends Ecol. Evol. 15, 140–143.

Jordano, P., 1987. Patterns of mutualistic interactions in pollination and seed dispersal: connectance, dependence asymmetries, and coevolution. Am. Nat. 129, 657–677.

Jordano, P., Bascompte, J., Olesen, J.M., 2003. Invariant properties in coevolutionary networks of plant–animal interactions. Ecol. Lett. 6, 69–81.

Kaiser-Bunbury, C.N., Blüthgen, N., 2015. Integrating network ecology with applied conservation: a synthesis and guide to implementation. AoB Plants 7. plv076.

Kaiser-Bunbury, C.N., Muff, S., Memmott, J., Müller, C.B., Caflisch, A., 2010. The robustness of pollination networks to the loss of species and interactions: a quantitative approach incorporating pollinator behaviour. Ecol. Lett. 13, 442–452.

Keitt, T.H., 2009. Habitat conversion, extinction thresholds, and pollination services in agroecosystems. Ecol. Appl. 19, 1561–1573.

Kennedy, C.M., Lonsdorf, E., Neel, M.C., Williams, N.M., Ricketts, T.H., Winfree, R., Bommarco, R., Brittain, C., Burley, A.L., Cariveau, D., Carvalheiro, L.G., Chacoff, N.P., Cunningham, S.A., Danforth, B.N., Dudenhöffer, J.-H., Elle, E., Gaines, H.R., Garibaldi, L.A., Gratton, C., Holzschuh, A., Isaacs, R., Javorek, S.K., Jha, S., Klein, A.M., Krewenka, K., Mandelik, Y., Mayfield, M.M., Morandin, L., Neame, L.A., Otieno, M., Park, M., Potts, S.G., Rundlöf, M., Saez, A., Steffan-Dewenter, I., Taki, H., Viana, B.F., Westphal, C., Wilson, J.K., Greenleaf, S.S., Kremen, C., 2013. A global quantitative synthesis of local and landscape effects on wild bee pollinators in agroecosystems. Ecol. Lett. 16, 584–599.

Kleijn, D., Winfree, R., Bartomeus, I., Carvalheiro, L.G., Henry, M., Isaacs, R., Klein, A.-M., Kremen, C., M'Gonigle, L.K., Rader, R., Ricketts, T.H., Williams, N.M., Lee, A.N.,

Ascher, J.S., Baldi, A., Batary, P., Benjamin, F., Biesmeijer, J.C., Blitzer, E.J., Bommarco, R., Brand, M.R., Bretagnolle, V., Button, L., Cariveau, D.P., Chifflet, R., Colville, J.F., Danforth, B.N., Elle, E., Garratt, M.P.D., Herzog, F., Holzschuh, A., Howlett, B.G., Jauker, F., Jha, S., Knop, E., Krewenka, K.M., Le Feon, V., Mandelik, Y., May, E.A., Park, M.G., Pisanty, G., Reemer, M., Riedinger, V., Rollin, O., Rundlof, M., Sardinas, H.S., Scheper, J., Sciligo, A.R., Smith, H.G., Steffan-Dewenter, I., Thorp, R., Tscharntke, T., Verhulst, J., Viana, B.F., Vaissiere, B.E., Veldtman, R., Westphal, C., Potts, S.G., 2015. Delivery of crop pollination services is an insufficient argument for wild pollinator conservation. Nat. Commun. 6, 7414. http://dx.doi.org/10.1038/ncomms8414.

Klein, A.M., Vaissiere, B.E., Cane, J.H., Steffan-Dewenter, I., Cunningham, S.A., Kremen, C., Tscharntke, T., 2007. Importance of pollinators in changing landscapes for world crops. Proc. Biol. Sci. 274 (1608), 303–313.

Klein, A.M., Cunningham, S.A., Bos, M., Steffan-Dewenter, I., 2008. Advances in pollination ecology from tropical plantation crops. Ecology 89, 935–943.

Koh, I., Reineking, B., Park, C.R., Lee, D., 2015. Dispersal potential mediates effects of local and landscape factors on plant species richness in maeulsoop forests of Korea. J. Veg. Sci. 26, 631–642.

Kotiaho, J.S., Kaitala, V., Komonen, A., Päivinen, J., 2005. Predicting the risk of extinction from shared ecological characteristics. Proc. Natl. Acad. Sci. U.S.A. 102, 1963–1967.

Krauss, J., Bommarco, R., Guardiola, M., Heikkinen, R.K., Helm, A., Kuussaari, M., Lindborg, R., Öckinger, E., Pärtel, M., Pino, J., Pöyry, J., Raatikainen, K.M., Sang, A., Stefanescu, C., Teder, T., Zobel, M., Steffan-Dewenter, I., 2010. Habitat fragmentation causes immediate and time-delayed biodiversity loss at different trophic levels. Ecol. Lett. 13, 597–605.

Kremen, C., Williams, N.M., Aizen, M.A., Gemmill-Herren, B., LeBuhn, G., Minckley, R., Packer, L., Potts, S.G., Roulston, T., Steffan-Dewenter, I., Vázquez, D.P., Winfree, R., Adams, L., Crone, E.E., Greenleaf, S.S., Keitt, T.H., Klein, A.M., Regetz, J., Ricketts, T.H., 2007. Pollination and other ecosystem services produced by mobile organisms: a conceptual framework for the effects of land-use change.. Ecol. Lett. 10, 299–314.

Kremen, C., Tscharntke, T., 2007. Importance of pollinators in changing landscapes for world crops. Proc. R. Soc. Lond. B 274, 303–313.

Kremen, C., Williams, N.M., Bugg, R.L., Fay, J.P., Thorp, R.W., 2004. The area requirements of an ecosystem service: crop pollination by native bee communities in California. Ecol. Lett. 7, 1109–1119.

Laliberte, E., Wells, J.A., DeClerck, F., Metcalfe, D.J., Catterall, C.P., Queiroz, C., Aubin, I., Bonser, S.P., Ding, Y., Fraterrigo, J.M., McNamara, S., Morgan, J.W., Sánchez Merlos, D., Vesk, D.A., Mayfield, M.M., 2010. Land-use intensification reduces functional redundancy and response diversity in plant communities. Ecol. Lett. 13, 76–86.

Larsen, T.H., Williams, N.M., Kremen, C., 2005. Extinction order and altered community structure rapidly disrupt ecosystem functioning. Ecol. Lett. 8, 538–547.

Laurance, W.F., Nascimento, H.E.M., Laurance, S.G., Andrade, A.C., Fearnside, P.M., Ribeiro, J.E.L., Capretz, R.L., 2006. Rain forest fragmentation and the proliferation of successional trees. Ecology 87, 469–482.

Lavorel, S., Storkey, J., Bardgett, R.D., Bello, F., Berg, M.P., Roux, X., Moretti, M., Mulder, C., Pakeman, R.J., Díaz, S., Harrington, R., 2013. A novel framework for linking functional diversity of plants with other trophic levels for the quantification of ecosystem services. J. Veg. Sci. 24, 942–948.

Leibold, M.A., Holyoak, M., Mouquet, N., Amarasekare, P., Chase, J.M., Hoopes, M.F., Holt, R.D., Shurin, J.B., Law, R., Tilman, D., Loreau, M., Gonzalez, A., 2004.

The metacommunity concept: a framework for multi-scale community ecology. Ecol. Lett. 7, 601–613.

Lever, J.J., Nes, E.H., Scheffer, M., Bascompte, J., 2014. The sudden collapse of pollinator communities. Ecol. Lett. 17, 350–359.

Levins, R., 1969. Some demographic and genetic consequences of environmental heterogeneity for biological control. Bull. Entomol. Soc. Am. 15, 237–240.

Lloyd, D.G., 1992. Self- and cross-fertilization in plants. II. The selection of self-fertilization. Int. J. Plant Sci. 153, 370–380.

Lobo, D., Leao, T., Melo, F.P.L., Santos, A.M.M., Tabarelli, M., 2011. Forest fragmentation drives Atlantic forest of north-eastern Brazil to biotic homogenization. Divers. Distrib. 17, 287–296.

Magrach, A., Laurance, W.F., Larrinaga, A.R., Santamaria, L., 2014. Meta-analysis of the effects of forest fragmentation on interspecific interactions. Conserv. Biol. 28, 1342–1348.

Martén Rodríguez, S.M., Quesada, M., Castro, A.A., Lopezaraiza-Mikel, M., Fenster, C.B., 2015. A comparison of reproductive strategies between island and mainland Caribbean Gesneriaceae. J. Ecol. 103, 1190–1204.

Massol, F., Débarre, F., 2015. Evolution of dispersal in spatially and temporally variable environments: the importance of life cycles. Evolution 69 (7), 1925–1937.

Massol, F., Petit, S., 2013. Interaction networks in agricultural landscape mosaics. Adv. Ecol. Res. 49, 291–338.

Mathiasen, P., Rovere, A.E., Premoli, A.C., 2007. Genetic structure and early effects of inbreeding in fragmented temperate forests of a self-incompatible tree, *Embothrium coccineum*. Conserv. Biol. 21, 232–240.

Matlab, 2011. The MathWorks, Inc., Natick, MA, United States.

McEuen, A.B., Curran, L.M., 2004. Seed dispersal and recruitment limitation across spatial scales in temperate forest fragments. Ecology 85, 507–518.

MEA (Millennium Ecosystem Assessment), 2005. Ecosystems and Human Well-Being: A Framework for Assessment. Island Press, Washington, DC.

Melián, C.J., Bascompte, J., 2002. Food web structure and habitat loss. Ecol. Lett. 5, 37–46.

Melo, F.P.L., Martinez-Salas, E., Benitez-Malvido, J., Ceballos, G., 2010. Forest fragmentation reduces recruitment of large-seeded tree species in a semi-deciduous tropical forest of southern Mexico. J. Trop. Ecol. 26, 35–43.

Memmott, J., Waser, N.M., Price, M.V., 2004. Tolerance of pollination networks to species extinctions. Proc. R. Soc. Lond. B 271, 2605–2611.

Memmott, J., Craze, P.G., Waser, N.M., Price, M.V., 2007. Global warming and the disruption of plant–pollinator interactions. Ecol. Lett. 10, 710–717.

Miklós, I., Podani, J., 2004. Randomization of presence–absence matrices: comments and new algorithms. Ecology 85, 86–92.

Miller-Rushing, A.J., Inouye, D.W., 2009. Variation in the impact of climate change on flowering phenology and abundance: an examination of two pairs of closely related wildflower species. Am. Nat. 96, 1821–1829.

Mitchell, M.G.E., Bennett, E.M., Gonzalez, A., 2015a. Strong and nonlinear effects of fragmentation on ecosystem service provision at multiple scales. Environ. Res. Lett. 10, 094014.

Mitchell, M.G.E., Suarez-Castro, A.F., Martinez-Harms, M., Maron, M., McAlpine, C., Gaston, K.J., Johansen, K., Rhodes, J.R., 2015b. Reframing landscape fragmentation's effects on ecosystem services. Trends Ecol. Evol. 30, 190–198.

Morris, W.F., 2003. Which mutualists are most essential? Buffering of plant reproduction against the extinction of pollinators. In: Kareiva, P., Levin, S.A. (Eds.), The Importance of Species: Perspectives on Expendability and Triage. Princeton University Press, Princeton, pp. 260–280.

Mouquet, N., Matthiessen, B., Miller, T., Gonzalez, A., 2011. Extinction debt in source-sink metacommunities. PLoS One 6, e17567.

Mulder, C., Aldenberg, T., De Zwart, D., Van Wijnen, H.J., Breure, A.M., 2005. Evaluating the impact of pollution on plant-Lepidoptera relationships. Environmetrics 16, 357–373.

Mulder, C., Boit, A., Mori, S., Vonk, J.A., Dyer, S.D., Faggiano, L., Geisen, S., González, A.L., Kaspari, M., Lavorel, S., Marquet, P.A., Rossberg, A.G., Sterner, R.W., Voigt, W., Wall, D.H., 2012. Distributional (in)congruence of biodiversity–ecosystem functioning. Adv. Ecol. Res. 46, 1–88.

Nee, S., May, R.M., 1992. Dynamics of metapopulation: habitat destruction and competitive coexistence. J. Anim. Ecol. 61, 37–40.

Nielsen, A., Bascompte, J., 2007. Ecological networks, nestedness and sampling effort. J. Ecol. 95, 1134–1141.

Öckinger, E., Schweiger, O., Crist, T.O., Debinski, D.M., Krauss, J., Kuussaari, M., Petersen, J.D., Poyry, J., Settele, J., Summerville, K.S., Bommarco, R., 2010. Life--history traits predict species responses to habitat area and isolation: a cross-continental synthesis. Ecol. Lett. 13, 969–979.

Okuyama, T., Holland, J.N., 2008. Network structural properties mediate the stability of mutualistic communities. Ecol. Lett. 11, 208–216.

Olesen, J.M., Bascompte, J., Dupont, Y.L., Jordano, P., 2007. The modularity of pollination networks. Proc. Natl. Acad. Sci. U.S.A. 104, 19891–19896.

Ollerton, J., Winfree, R., Tarrant, S., 2011. How many flowering plants are pollinated by animals? Oikos 120, 321–326.

Podani, J., Schmera, D., 2012. A comparative evaluation of pairwise nestedness measures. Ecography 35, 889–900.

Potts, S.G., Biesmeijer, J.C., Kremen, C., Neumann, P., Schweiger, O., Kunin, W.E., 2010. Global pollinator declines: trends, impacts and drivers. Trends Ecol. Evol. 25, 345–353.

Réjou-Méchain, M., Cheptou, P.-O., 2015. High incidence of dioecy in young successional tropical forests. J. Ecol. 103, 725–732.

Renner, S.S., 1998. Effects of habitat fragmentation on plant-pollination interactions in the tropics. In: Newbery, D.M., Prins, H.H.T., Brown, N.D. (Eds.), Dynamics of Tropical Communities. Blackwell Science, London, pp. 339–360.

Rezende, E.L., Lavabre, J.E., Guimarães, P.R., Jordano, P., Bascompte, J., 2007. Non-random coextinctions in phylogenetically structured mutualistic networks. Nature 448, 925–928.

Richards, A.J., 1997. Plant Breeding Systems, second ed. Chapman & Hall, London.

Ricketts, T.H., Regetz, J., Steffan-Dewenter, I., Cunningham, S.A., Kremen, C., Bogdanski, A., Gemmill-Herren, B., Greenleaf, S.S., Klein, A., Mayfield, M.M., Morandin, L.A., Ochieng, A., Viana, V.F., 2008. Landscape effects on crop pollination services: are there general patterns? Ecol. Lett. 11, 499–515.

Rohr, R.P., Saavedra, S., Bascompte, J., 2014. On the structural stability of mutualistic systems. Science 345, 1253497.

Sabatino, M., Maceira, N., Aizen, M.A., 2010. Direct effects of habitat area on interaction diversity in pollination webs. Ecol. Appl. 20, 1491–1497.

Sala, O.E., Chapin, F.S., Armesto, J.J., Berlow, E., Bloomfield, J., Dirzo, R., et al., 2000. Global biodiversity scenarios for the year 2100. Science 287, 1770–1774.

Scheper, J., Holzschuh, A., Kuussaari, M., Potts, S.G., Rundlöf, M., Smith, H.G., Kleijn, D., 2013. Environmental factors driving the effectiveness of European agri-environmental measures in mitigating pollinator loss—a meta-analysis. Ecol. Lett. 16, 912–920.

Scheper, J., Reemer, M., van Kats, R., Ozinga, W.A., van der Linden, G.T., Schaminée, J.H., Spiepel, H., Kleijn, D., 2014. Museum specimens reveal loss of pollen host plants as key factor driving wild bee decline in The Netherlands. Proc. Natl. Acad. Sci. U.S.A. 111, 17552–17557.

Spiesman, B.J., Inouye, B.D., 2013. Habitat loss alters the architecture of plant-pollinator interaction networks. Ecology 94, 2688–2696.

Steffan-Dewenter, I., Tscharntke, T., 2002. Insect communities and biotic interactions on fragmented calcareous grasslands—a mini review. Biol. Conserv. 104, 275–284.

Steffan-Dewenter, I., Potts, S.G., Packer, L., 2005. Pollinator diversity and crop pollination services are at risk. Trends Ecol. Evol. 20, 651–652.

Suweis, S., Simini, F., Banavar, J.R., Maritan, A., 2013. Emergence of structural and dynamical properties of ecological mutualistic networks. Nature 500, 449–452.

Tabarelli, M., Mantovani, W., Peres, C.A., 1999. Effects of habitat fragmentation on plant guild structure in the montane Atlantic forest of southeastern Brazil. Biol. Conserv. 91, 119–127.

Thébault, E., Fontaine, C., 2010. Stability of ecological communities and the architecture of mutualistic and trophic networks. Science 329, 853–856.

Tilman, D., May, R.M., Lehman, C.L., Nowak, M.A., 1994. Habitat destruction and the extinction debt. Nature 351, 65–66.

Tscharntke, T., Tylianakis, J.M., Rand, T.A., Didham, R.K., Fahrig, L., Batary, P., et al., 2012. Landscape moderation of biodiversity patterns and processes—eight hypotheses. Biol. Rev. 87, 661–685.

Tur, C., Castro-Urgal, R., Traveset, A., 2013. Linking plant specialization to dependence in interactions for seed set in pollination networks. PLoS One 8, e78294.

Tylianakis, J.M., Laliberté, E., Nielsen, A., Bascompte, J., 2010. Conservation of species interaction networks. Biol. Conserv. 143, 2270–2279.

Urban, M.C., Leibold, M.A., Amarasekare, P., De Meester, L., Gomulkiewicz, R., Hochberg, M.E., et al., 2008. The evolutionary ecology of metacommunities. Trends Ecol. Evol. 23, 311–317.

Valiente-Banuet, A., Aizen, M.A., Alcántara, J.M., Arroyo, J., Cocucci, A., Galetti, M., García, M.B., García, D., Gómez, J.M., Jordano, P., Medel, R., Navarro, L., Obeso, J.R., Oviedo, R., Ramírez, N., Rey, P.J., Traveset, A., Verdú, M., Zamora, R., 2015. Beyond species loss: the extinction of ecological interactions in a changing world. Funct. Ecol. 29, 299–307.

van der Niet, T.A., Johnson, S.D., 2012. Phylogenetic evidence for pollinator-driven diversification of angiosperms. Trends Ecol. Evol. 27, 353–361.

Vázquez, D.P., 2005. Degree distribution in plant-animal mutualistic networks: forbidden links or random interactions? Oikos 108, 421–426.

Vázquez, D.P., Aizen, M.A., 2004. Asymmetric specialization: a pervasive feature of plant-pollinator interactions. Ecology 85, 1251–1257.

Vieira, M.C., Almeida-Neto, M., 2015. A simple stochastic model for complex coextinctions in mutualistic networks: robustness decreases with connectance. Ecol. Lett. 18, 144–152.

Vogler, D.W., Kalisz, S., 2001. Sex among the flowers: the distribution of plant mating systems. Evolution 55, 202–204.

Waser, N.M., Chittka, L., Price, M.V., Williams, N.M., Ollerton, J., 1996. Generalization in pollination systems, and why it matters. Ecology 77, 1043–1060.

Wilcock, C., Neiland, R., 2002. Pollination failure in plants: why it happens and when it matters. Trends Plant Sci. 7, 270–277.

Williams, N.M., Crone, E.E., Tai, H.R., Minckley, R.L., Packer, L., Potts, S.G., 2010. Ecological and life-history traits predict bee species responses to environmental disturbances. Biol. Conserv. 143, 2280–2291.

Willson, M., Traveset, A., 2000. The ecology of seed dispersal. In: Fenner, M. (Ed.), The Ecology of Regeneration in Plant Communities. CAB International, Wallingford, UK, pp. 85–110.

Winfree, R., Aguilar, R., Vázquez, D.P., LeBuhn, G., Aizen, M.A., 2009. A meta-analysis of bees' responses to anthropogenic disturbance. Ecology 90, 2068–2076.

CHAPTER SIX

A Network-Based Method to Detect Patterns of Local Crop Biodiversity: Validation at the Species and Infra-Species Levels

Mathieu Thomas[*,†,‡,1], Nicolas Verzelen[§], Pierre Barbillon[¶], Oliver T. Coomes[||], Sophie Caillon[†], Doyle McKey[†,#], Marianne Elias[**], Eric Garine[††], Christine Raimond[‡‡], Edmond Dounias[†], Devra Jarvis[§§,¶¶], Jean Wencélius[††], Christian Leclerc[||||], Vanesse Labeyrie[||||], Pham Hung Cuong[##], Nguyen Thi Ngoc Hue[##], Bhuwon Sthapit[***], Ram Bahadur Rana[†††], Adeline Barnaud[‡‡‡,§§§,¶¶¶], Chloé Violon[††], Luis Manuel Arias Reyes[||||||], Luis Latournerie Moreno[###], Paola De Santis[§§], François Massol[****]

*INRA, UMR 0320/UMR 8120 Génétique Quantitative et Évolution-Le Moulon, Gif-sur-Yvette, France
†CEFE UMR 5175, CNRS-Université de Montpellier, Université Paul-Valéry Montpellier, EPHE, Montpellier, France
‡FRB, CESAB (Centre de synthèse et d'analyse de la biodiversité), Technopôle de l'Environnement Arbois-Méditerranée, Aix-en-Provence, France
§UMR 729 Mathématiques, Informatique et STatistique pour l'Environnement et l'Agronomie, INRA SUPAGRO, Montpellier, France
¶AgroParisTech/UMR INRA MIA, Paris, France
||Department of Geography, McGill University, Montreal, Quebec Canada
#Institut Universitaire de France, Paris, France
**Institut de Systématique, Évolution, Biodiversité ISYEB-UMR 7205-CNRS, MNHN, UPMC, EPHE Muséum national d'Histoire naturelle, Sorbonne Universités, Paris, France
††Université Paris Ouest/CNRS, UMR 7186 LESC, Nanterre, France
‡‡CNRS-UMR 8586 PRODIG, Paris, France
§§Bioversity International, Rome, Italy
¶¶Department of Crop and Soil Sciences, Washington State University, Pullman, Washington, USA
||||CIRAD, UMR AGAP, Montpellier, France
##Plant Resources Center, VAAS, MARD, Hanoi, Viet Nam
***Bioversity International, Office of Nepal, Pokhara City, Nepal
†††Local Initiatives for Biodiversity, Research and Development (LI-BIRD), Pokhara, Kaski, Nepal
‡‡‡IRD, UMR DIADE, Montpellier, France
§§§LMI LAPSE, Dakar, Senegal
¶¶¶ISRA, LNRPV, Centre de Bel Air, Dakar, Senegal
||||||CINVESTAV-IPN Unidad Mérida, Merida, Yucatan, Mexico
###Instituto Tecnológico de Conkal, Division de Estudios de Posgrdo e Investigacion Conkal, Conkal, Yucatan, Mexico
****Unité Evolution, Ecologie et Paléontologie (EEP), CNRS UMR 8198, Université Lille, Lille, France
[1]Corresponding author: e-mail address: thomas@moulon.inra.fr

Advances in Ecological Research, Volume 53
ISSN 0065-2504
http://dx.doi.org/10.1016/bs.aecr.2015.10.002

Contents

1. Introduction 261
2. Description of the Datasets Used in the Meta-Analysis 266
3. Description of the Methodological Framework 273
   3.1 Mathematical Formalism 274
   3.2 Variability of Farms' and Crops' Degrees 275
   3.3 Revealing Data Structure Through LBMs 278
   3.4 Uncovering Outliers Through PCA 281
   3.5 Measuring Originality of Farms' Contributions Through Diversity Measures 286
4. Patterns of Local Crop Diversity: Results of the Meta-Analysis 292
   4.1 Variability of Farms' and Crops' Degrees 292
   4.2 Structure Detection Through Model-Based Clustering (LBM) 297
   4.3 Outlier Detection Through PCA 298
   4.4 Farms' Contributions to Local Diversity 298
5. Discussion 300
   5.1 Contrasted Patterns of Local Crop Diversity at the Species and Infra-Species Levels 300
   5.2 Relevance of Network-Based Methods 304
6. Conclusion 306
Acknowledgements 307
   LBM Representation 308
   Outlier Representation 311
   Statistical Power Study of the Test Measuring the Impact of Crop–Rich and Crop–Poor Farms 312
   Estimation of Nestedness 314
Glossary 316
References 316

Abstract

In this chapter, we develop new indicators and statistical tests to characterize patterns of crop diversity at local scales to better understand interactions between ecological and socio-cultural functions of agroecosystems. Farms, where a large number of crops (species or landraces) is grown, are known to contribute a large part of the locally available diversity of both rare and common crops but the role of farms with low diversity remains little understood: do they grow only common varieties—following a nestedness pattern typical of mutualistic networks in ecology—or do 'crop–poor' farmers also grow rare varieties? This question is pivotal in ongoing efforts to assess the local-scale contribution of small farms to global agrobiodiversity. We develop new network-based approaches to characterize the distribution of local crop diversity (species and infra-species) at the village level and to validate these approaches using meta-datasets from 10 countries. Our results highlight the sources of heterogeneity in crop diversity at the village level. We often identify two or more groups of farms based on their different levels of diversity. In some datasets, 'crop–poor' farms significantly contribute to the local crop diversity. Generally, we find that the distribution of crop diversity is more heterogeneous at the species than at the infra-species level. This analysis reveals the absence of a general pattern of crop diversity distribution, suggesting strong dependence on local agroecological and socio-cultural contexts. These different patterns of crop diversity distribution reflect an heterogeneity in farmers' self-organized action in cultivating and maintaining local crop diversity, which ensures the adaptability of agroecosystems to global change.

1. INTRODUCTION

Agriculture relies on the use of crop plant species to provision human societies with food, clothing, medicinal and narcotic substances, fodder for domestic animals, building materials and more recently with biofuel. Plant crop species were domesticated from wild ancestors, which often display variability in traits adapted to the local environment. During domestication, only a subset of diversity from the wild ancestors was selected, and shaped by the goals of farmers to produce a diversity of landraces, named and managed as distinct entities (Diamond, 2002). Furthermore, different crop species play distinctive, often complementary, roles in agriculture. For instance, including legumes in rotations or in associations with cereals limit the use of external inputs of fertilizer by increasing nutrient inputs through nitrogen fixation (Drinkwater et al, 1998). In many agroecosystems, the end result of these processes of selection among wild diversity, i.e., selection in farmers' fields and adoption of numerous kinds of crops, is a substantial increase in the diversity of cultivated plants, both in terms of the number of species and landrace diversity within species (Jarvis et al., 2008). Modernization of agriculture in industrialized countries during the twentieth century increased agricultural productivity thanks to uniformization, i.e., reduction of the number of crop species and varieties and genetic homogenization of varieties (Bonneuil et al., 2012). This genetic erosion was accompanied by the disruption of interactions among crop and wild species (Macfadyen and Bohan, 2010). This strategy also required an intensive use of fertilizers, pesticides, water and fossil fuels, creating strong environmental perturbations, including habitat fragmentation, soil erosion, water pollution, causing great reduction of wild biodiversity (MEA, 2005). Ecological, economic and social consequences of intensive agriculture are now identified and formalized by the Millennium Ecosystem Assessment. Recommendations for limiting these treats rely on an ecosystemic and transdisciplinary approach to the problem (Mulder et al., 2015). Central to this approach is identifying trade-offs and synergies among ecological functional groups within ecosystems, and estimating their impacts on provisioning, regulating, supporting and cultural services. To maintain ecological synergies, manage trade-offs and sustain ecosystem resilience, values are assigned for these services and stakeholders are influenced through recognition, incentives and rewards based on these values (see Butterfield et al., 2016, for more details). The same conceptual framework is applied to manage agroecosystems (agro-ecology), considering trade-offs between agroecosystem functions (pollination, ecological pest control, etc.) and provision of goods (MEA, 2005). This agro-ecological approach to agriculture is not fully satisfying because it often neglects social and cultural processes directly linked to agriculture, such as local knowledge concerning farming practices, which can make important contribution to ensuring the sustainability of agroecosystems (Jackson et al., 2007; Martin et al, 2010). In this chapter, we consider agricultural systems as socio-agroecosystem in which social and cultural functions need to be examined in addition to ecological functions. More attention must be paid to farmers' practices in highly diversified systems, because these practices play a role in creating and maintaining diversity, which is often of great importance to system functioning. Understanding the interactions of these practices with biological and ecological processes is necessary to improve our understanding and management of synergies and trade-offs occurring in agroecosystems.

A primary requisite for understanding and predicting the sustainability of agroecosystems facing environmental, political, social and economic changes is to assess how these systems manage crop diversity (e.g. Samberg et al., 2013). For instance, in the case of manioc cultivated by the Makushi Amerindians of Guyana, some varieties are specially grown for particular dishes, some grow quickly thereby ensuring early yield, while still others grow slowly and act as an 'ever-present' insurance resource (Elias et al., 2000). Often, diversity is simply valued for its own sake (Boster, 1985), or as a means to foster social relations (Emperaire and Peroni, 2007; Heckler and Zent, 2008). Another example of crop biodiversity maintenance is the great diversity of landraces present in the milpas of Meso-America, which are the end product of several 1000 years of directed selection on maize, beans, squash and chilli peppers by the farmers of the region (Tuxill et al., 2010). Understanding relationships among landraces makes it possible to gain insight into the cultural history. The particular traits exhibited by local varieties grown in milpas today reflect Yucatan farmers' short- and long-term responses to agro-environmental conditions, the ecological demands of crop production and the aesthetic, culinary and religious sensibilities of farmers (Tuxill et al., 2010). Farming practices that maintain crop diversity are of paramount importance in helping crops and farmers adapt to global changes, notably climate change (Vigouroux et al., 2011) and the increasingly rapid emergence of agricultural pests (Diamond, 2002). In addition, cultivating diverse crops and varieties at the landscape level favours ecological and economic sustainability by reducing the need for chemical inputs (Bianchi et al., 2006; Crowder et al., 2010). Crop diversity also provides an insurance value. Although some combinations of species or varieties may be functionally redundant in a agroecosystem, at least at a given time, a subset of species and varieties may confer to the system the capacity to adapt to environmental fluctuations (Di Falco and Perrings, 2005; Jackson et al., 2007; Smale et al., 1998).

The spatial distribution of crop diversity is expected to be partially explained by environmental factors, due to the differential adaptation of crops to local conditions (Mariac et al., 2011). For instance, crops require different physiological adaptations to cope with limiting factors associated with dry and wet climates. Moreover, selective pressures in cultivated environments differ from those in wild environments. However, unless massive inputs liberate crops from environmental constraints, adaptation to local abiotic environments is expected to shape crop diversity—as it shapes the diversity of wild plants—at more or less large spatial scales, over latitudinal or elevational gradients (Vigouroux et al., 2011). At fine spatial scales, local adaptation is also expected to play a role in the distribution of crop diversity, due to the heterogeneity of soil quality of agricultural fields and to variability in local rainfall (Fraser et al., 2012).

In addition to environmental factors, it has been argued that crop diversity can only be understood if social and cultural aspects of the contextual environment are taken into account (Leclerc and Coppens d'Eeckenbrugge, 2012; Rival and McKey, 2008). Agricultural societies have shaped the diversity of their cultivated crops in ways that fit their traditions, habits, myths, social organizations and livelihoods (Delêtre et al., 2011; Labeyrie et al., 2013). Indeed, crops and humans have likely evolved together, as cultural practices may have been shaped by available edible plants and agricultural selection may have answered cultural needs. The study of crop genetic and inter-specific diversity in the context of both environment- and society-driven selective pressures is now taken into account through the $G \times E \times S$ framework (Leclerc and Coppens d'Eeckenbrugge, 2012). Thus, studying the distribution of crop diversity and linking it with both social and environmental factors cannot be based on a solely biological perspective. However, interdisciplinary studies of the distribution of crop diversity must retain quantitative rigour and thus be based on a sound statistical framework. To date, the distribution of crop diversity has been assessed mostly through the use of diversity indices adopted from ecology and economics, indices of richness, evenness, concentration, etc. (e.g. Jarvis et al., 2008). These indices only make use of crop diversity data as an instance of 'type in location' data and this limits the questions that can be addressed. They can help explain why crops are found in the fields they are in, but not why farmers decided to cultivate a given crop, for example. We failed to find any studies that even came near to exploiting the potential of analyses of the network*[1] feature of crop-by-farms datasets which includes social aspects, such as farmer-to-farmer circulation of seeds (and other propagules) of varieties and crop species. These bipartite networks* are composed of two kinds of nodes* representing a farm or a crop (species or landrace); an edge* connects two kinds of nodes and means that a particular crop is grown in a particular farm.

Our main goal in this chapter is to answer the question 'which farms contribute, and how, to the diversity of crops grown in a given village?' by examining inventories of crops species and landraces grown at the farm level. To do this, we offer a novel methodological framework using network-based and null model-based statistical tests. From a methodological perspective, inventory datasets can be construed as bipartite networks, namely crop-by-farm interaction networks*, in the same way as plant–pollinator or host–parasite interaction networks in ecology. In social network analysis, network approaches have been used to assess the properties of network processes linked to social institutions, such as friendship, advice or seed exchange networks ('who interacts with whom' or 'who gives to whom') (Lazega et al., 2012; Reyes-García et al., 2013; Wasserman and Faust, 1994). In ecology, on the other hand, networks have been used to study both contact networks (metapopulations or metacommunities) and structured interaction networks*, such as food webs (herbivore–host plant networks) or mutualistic networks (plant–pollinator networks). When interaction partners can be clearly categorized (plants, pollinators; plants, herbivores and parasitoids), the use of bi- or multi-partite networks is an appropriate approach. In the present study, we develop a framework for the study of crop-by-farm datasets that makes use of the bipartite nature

[1] * indicates that the word or expression is defined in Glossary section.

of the data to reveal potential patterns in the structure of diversity at the scale of the village or the clusters of interacting villages.

Our paper offers an alternative to the nestedness* approach, for several reasons that are detailed below. The study of bipartite networks in ecology is a recent endeavour (Jordano, 1987). Over the past three decades, the topological properties of bipartite networks have been studied to answer a variety of questions, such as whether the networks are stable, robust to species extinctions or additions, functionally redundant, etc. (Astegiano et al., 2015; Gill et al., 2016; Jordano et al., 2003; Thébault and Fontaine, 2010). In particular, the nestedness of mutualistic bipartite networks has often been investigated and studies suggest that nestedness may be the key property explaining the dynamics and structural stability of mutualistic networks (Thébault and Fontaine, 2010). In spatial ecology studies, nested patterns are often explained as resulting from source–sink processes wherein species–rich locations function as sources producing many emigrating individuals which, in turn, contribute to the diversity in species–poor, sink locations (Atmar and Patterson, 1993). In mutualistic interaction networks, nestedness can be understood as arising from feasibility constraints on the existence of specialist–specialist interactions, i.e., nestedness decreases effective inter-specific competition and thus increases the number of species that can coexist (Bascompte and Jordano, 2007; Bastolla et al., 2009). In systems involving social as well as ecological processes, such as in the present case of crop-by-farm interactions, one may ask whether this nestedness pattern holds, as crops present in less diverse (crop–poor) farms could comprise a subset of those cultivated in more diverse farms. Among the Duupa in northern Cameroon, for example, older farmers accumulate crop diversity during their life (sources) and become sources of diversity for young farmers (sinks) (Alvarez et al., 2005). When crops are actively cultivated by farmers, for example, as staple food, copying other farmers' portfolios of crops might result in strong similarities in cultivated diversity among fields, but not necessarily following a nested pattern. Therefore, contrary to the case for ecological systems, certain mechanistic reasons may justify considering crop-by-farm interactions as systematically nested, precluding explanations solely based on source–sink processes.

From a purely methodological perspective, the available indices of network nestedness are inconsistent, both in the value of nestedness metrics and in their associated p-value when confronted with the configuration model*; a null model of partner interactions constrained by degree*, i.e., fixing the degree of rows and columns (Podani and Schmera, 2012). Although nestedness remains a largely verbal concept and its mathematical definition is in need of refinement, being able to detect nested patterns in crop diversity would be useful for characterizing the diversity of strategies used by farmers to cope with different socio-cultural and environmental contexts.

In Section 2, we introduce a meta-dataset of specific and infra-specific crop diversity at the local scale in different agricultural contexts. In Section 3, we describe our methodological framework, graphical representations when appropriate and the tests proposed, illustrated using 'toy', hypothetical examples. Our approach allows us: (i) to test whether the variability in the number of connections per farm and per crop type is different from random expectations under a homogeneous random graph model* (Erdős-Rényi model*); (ii) to reveal structure (modules, cores, etc.) in the dataset using latent-block models* (LBMs); (iii) to uncover 'outliers' (farmers or crop types that do not conform to the general connection pattern) using principal component analyses (PCAs); and (iv) to measure and test the originality of farmers' contributions to overall crop diversity using beta-diversity indices. In Section 4, we perform a meta-analysis applying the methodological framework to our meta-dataset, which allows us to highlight both regularities and particularities among the datasets. Our approach yields graphical representations of the different tests (reordering of interactions in the case of LBMs or principal plane representations for PCAs) and non-parametric tests of our hypotheses, the significance of which is assessed through comparison with a permutation-based null model (the configuration model for graphs with given degrees). These graphical and statistical approaches are designed to be transferable to other similar problems in ecology. In Section 5, we discuss our results and the value and the limits of our approach.

2. DESCRIPTION OF THE DATASETS USED IN THE META-ANALYSIS

Fifty published or unpublished datasets dealing with crop inventories were provided by ethnobiologists, geographers and ecologists for analysis (Tables 1 and 2). These data were collected in 10 different countries (Fig. 1) between 1998 and 2013. For each dataset, a partial set or the full set of farms from the same village was characterized for one of the two classes of operational taxonomic units (OTU) considered: the species or the infra-species (landrace) level. These data were gathered through direct interviews with the plant crop cultivators in the farm, a subset of them or only with the head of the farm. Datasets were retained when the number of characterized farms and the number

Table 1 Description of the 18 Datasets Dealing with Specific Diversity (OTU = Species)

Dataset	Country	Community	Village	Farm Sample Size	Crop Sample Size	Collect Year	Original Article
CL-M01	Kenya	Tharaka	Kamarandi	95	16	2010	Labeyrie et al. (2013)
CV-M01	Cameroon	Tupuri	Gulurgu-Lokoro	15	23	2011	Unpublished data
EG-M05	Cameroon	Duupa farmers	Ninga	14	58	2002	Garine and Raimond (2005)
EG-M08	Cameroon	Duupa farmers	Wante	18	68	2002	Garine and Raimond (2005)
OC-M02	Peru	Corrientes River	Boca del Copal	19	108	2003	Perrault-Archambault and Coomes (2008)
OC-M04	Peru	Corrientes River	San Juan de Trompeteros	35	120	2003	Perrault-Archambault and Coomes (2008)
OC-M05	Peru	Corrientes River	San Juan de Trompetero Nativo	22	108	2003	Perrault-Archambault and Coomes (2008)
OC-M06	Peru	Corrientes River	San Jose de Porvenir	18	84	2003	Perrault-Archambault and Coomes (2008)
OC-M07	Peru	Corrientes River	Nuevo Porvenir	22	83	2003	Perrault-Archambault and Coomes (2008)
OC-M09	Peru	Corrientes River	Nuevo Paraiso	11	83	2003	Perrault-Archambault and Coomes (2008)
OC-M10	Peru	Corrientes River	Nuevo Peruanito	15	88	2003	Perrault-Archambault and Coomes (2008)

Continued

Table 1 Description of the 18 Datasets Dealing with Specific Diversity (OTU = Species)—cont'd

Dataset	Country	Community	Village	Farm Sample Size	Crop Sample Size	Collect Year	Original Article
OC-M11	Peru	Corrientes River	Nuevo Pucacuro	54	161	2003	Perrault-Archambault and Coomes (2008)
OC-M12	Peru	Corrientes River	Santa Rosa	14	124	2003	Perrault-Archambault and Coomes (2008)
OC-M13	Peru	Corrientes River	Santa Elena	30	153	2003	Perrault-Archambault and Coomes (2008)
OC-M14	Peru	Corrientes River	San Jose de Nueva Esperanza	24	139	2003	Perrault-Archambault and Coomes (2008)
OC-M16	Peru	Corrientes River	Valencia	21	147	2003	Perrault-Archambault and Coomes (2008)
SC-M05	Vanuatu	Vanua Lava, Banks group	Eastern coast	15	37	2007–2009	Unpublished data
SC-M06	Ecuador	Huaorani	Guiyero	13	15	2000	Unpublished data

Table 2 Description of the 33 Datasets Dealing with Infra-Specific Diversity (OTU = Landrace)

Dataset	Country	Community	Village	Species	Predominant Propagation Mode	Farms Sample Size	Crop Sample Size	Collect Year	Original Article
AB-M02	Cameroon	Duupa farmers	Wante	Sorghum (*Sorghum bicolor*)	Partially outcrossing	13	23	2003	Unpublished data
CL-M02	Kenya	Tharaka	Kamarandi	Sorghum (*Sorghum bicolor*)	Partially outcrossing	95	20	2010	Labeyrie et al. (2013)
CV-M02	Cameroon	Tupuri	Gulurgu-Lokoro	Sorghum (*Sorghum bicolor*)	Partially outcrossing	15	22	2011	Unpublished data
DJ-M003a	Nepal	Kaski	village9	Rice (*Oryza sativa*)	Inbreeding	33	24	2006	Jarvis et al. (2008)
DJ-M003b	Nepal	Kaski	village10	Rice (*Oryza sativa*)	Inbreeding	52	32	2006	Jarvis et al. (2008)
DJ-M003c	Nepal	Kaski	village11	Rice (*Oryza sativa*)	Inbreeding	24	21	2006	Jarvis et al. (2008)
DJ-M003d	Nepal	Kaski	village14	Rice (*Oryza sativa*)	Inbreeding	25	18	2006	Jarvis et al. (2008)
DJ-M009a	Nepal	Bara	village1	Rice (*Oryza sativa*)	Inbreeding	35	11	2006	Jarvis et al. (2008)
DJ-M009b	Nepal	Bara	village2	Rice (*Oryza sativa*)	Inbreeding	29	12	2006	Jarvis et al. (2008)

Continued

Table 2 Description of the 33 Datasets Dealing with Infra-Specific Diversity (OTU = Landrace)—cont'd

Dataset	Country	Community	Village	Species	Predominant Propagation Mode	Farms Sample Size	Crop Sample Size	Collect Year	Original Article
DJ-M009c	Nepal	Bara	village3	Rice (*Oryza sativa*)	Inbreeding	37	14	2006	Jarvis et al. (2008)
DJ-M009d	Nepal	Bara	village4	Rice (*Oryza sativa*)	Inbreeding	14	8	2006	Jarvis et al. (2008)
DJ-M009e	Nepal	Bara	village5	Rice (*Oryza sativa*)	Inbreeding	31	14	2006	Jarvis et al. (2008)
DJ-M009f	Nepal	Bara	village6	Rice (*Oryza sativa*)	Inbreeding	29	16	2006	Jarvis et al. (2008)
DJ-M012a	Vietnam	Dabac	Cang	Rice (*Oryza sativa*)	Inbreeding	58	42		Jarvis et al. (2008)
DJ-M012b	Vietnam	Dabac	Tat	Rice (*Oryza sativa*)	Inbreeding	57	58		Jarvis et al. (2008)
DJ-M015a	Vietnam	Nghiahung	Dong Lac	Rice (*Oryza sativa*)	Inbreeding	58	42		Jarvis et al. (2008)
DJ-M015b	Vietnam	Nghiahung	Kien Thanh	Rice (*Oryza sativa*)	Inbreeding	57	58		Jarvis et al. (2008)
DJ-M018a	Vietnam	Nhoquan	Quang Mao	Rice (*Oryza sativa*)	Inbreeding	58	42		Jarvis et al. (2008)

DJ-M018b	Vietnam	Nhoquan	Yen Minh	Rice (*Oryza sativa*)	Inbreeding	57	58		Jarvis et al. (2008)
DJ-M030	Mexico	Ichmul	Multi-village	Maize (*Zea mays*)	Outcrossing	101	11		Jarvis et al. (2008)
DJ-M036	Mexico	Yaxcaba	Yaxcaba	Maize (*Zea mays*)	Outcrossing	61	13	1999	Jarvis et al. (2008)
DJ-M039a	Hungary	Dévaványa	village1	Bean (*Phaseolus vulgaris*, *Phaseolus lunatus*, *Vigna unguiculata*)	Inbreeding	13	10		Jarvis et al. (2008)
DJ-M045b	Hungary	Szatmár-Bereg	village2	Bean (*Phaseolus vulgaris*, *Phaseolus lunatus*, *Vigna unguiculata*)	Inbreeding	18	12		Jarvis et al. (2008)
DJ-M045c	Hungary	Szatmár-Bereg	village3	Bean (*Phaseolus vulgaris*, *Phaseolus lunatus*, *Vigna unguiculata*)	Inbreeding	10	12		Jarvis et al. (2008)
DJ-M045d	Hungary	Szatmár-Bereg	village4	Bean (*Phaseolus vulgaris*, *Phaseolus lunatus*, *Vigna unguiculata*)	Inbreeding	12	10		Jarvis et al. (2008)

Continued

Table 2 Description of the 33 Datasets Dealing with Infra-Specific Diversity (OTU = Landrace)—cont'd

Dataset	Country	Community	Village	Species	Predominant Propagation Mode	Farms Sample Size	Crop Sample Size	Collect Year	Original Article
JW-M07	Cameroon		Nulda	Sorghum (*Sorghum bicolor*)	Partially outcrossing	35	22	2008–2009	Unpublished data
JW-M08	Cameroon		Nulda	Sorghum (*Sorghum bicolor*)	Partially outcrossing	45	24	2009–2010	Unpublished data
JW-M09	Cameroon		Nulda	Sorghum (*Sorghum bicolor*)	Partially outcrossing	51	27	2011–2012	Unpublished data
JW-M10	Cameroon		Nulda	Sorghum (*Sorghum bicolor*)	Partially outcrossing	15	21	2012–2013	Unpublished data
ME-M01	Guyana	Makushi Amerindians	Rewa	Cassava (*Manihot esculenta*)	Clonal	24	75	1997–1998	Elias et al. (2000)
SC-M04	Vanuatu	Vanua Lava, Banks group	Eastern coast	Taro (*Colocasia esculenta*)	Clonal	15	34	2007–2009	Unpublished data
SC-M07	Ecuador	Huaorani	Guiyero	Manioc (*Manihot esculenta*)	Clonal	13	29	2000	Unpublished data

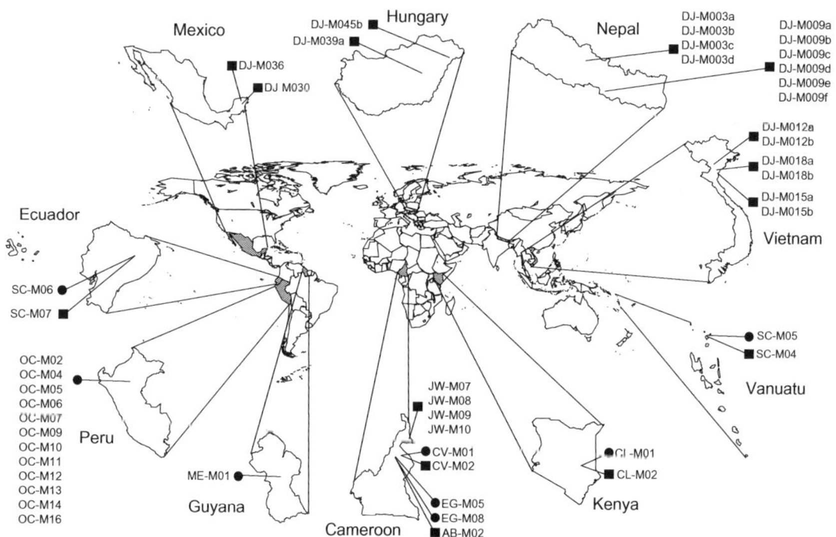

Figure 1 Map showing locations of the different datasets used in the meta-analysis. Filled circles correspond to the datasets collected at the specific level and filled squares correspond to the dataset collected at the infra-specific level.

of crops were both greater than 10. For 18 datasets, information was collected at the species level (Table 1); for 32 datasets, information was collected at the landrace level, which corresponds to the terminal taxon in the farmers' local naming systems, covering seven different species (maize, rice, wheat, bean, manioc, taro and sorghum) which correspond to the major crops of the areas under study (Table 2). These species are characterized by their predominant propagation mode (partially outcrossing, outcrossing, inbreeding and clonal) following the classification proposed by Jarvis et al. (2008). Data were structured following a rectangular incidence matrix* with farms in rows and species or landrace in columns, and represented as a bipartite network. Data collected at the species or infra-species level represent two levels of local crop biodiversity. Underlying processes shaping the distribution of local crop diversity are assumed to be different for these two levels. Therefore, species and infra-species data are analysed and described separately.

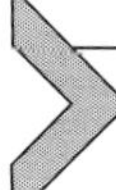

3. DESCRIPTION OF THE METHODOLOGICAL FRAMEWORK

This section introduces the statistical framework for analysing crop-by-farm network data. After defining the main concepts, we detail the four main steps of the analysis. First, the degree distribution of the data is evaluated as a way to test whether a completely random model (Erdős-Rényi model) fits well the data. Second, we use a LBM to investigate more thoroughly the structure of the network. This method pinpoints groups of farms and groups of crops that tend to be highly connected. Third, we test whether this high-level structure (blocks) is different from what can be expected simply by features of the low-level structure such as degree heterogeneity. The methods comprised by these two last steps provide new graphical representations of the network data emphasizing the studied patterns. Finally, complementary analyses based on diversity measures are introduced. In each subsection, toy examples illustrate the purpose, the benefits and the limitations of the proposed methods.

3.1 Mathematical Formalism

In the following, we denote n is the number of farms and m is the number of crops. The incidence matrix (with farms as rows and crops as columns) that summarizes the data is noted $\mathbf{X}$, so that $X_{ij} = 1$ when farm i cultivates crop j. Using this representation (Fig. 2A), we can readily apply statistical methods for binary matrices.

Any incidence matrix can also be treated as the adjacency matrix of some bipartite graph G. More specifically, consider a collection of nodes corresponding to all farms and all crops (species or landraces) and put an edge

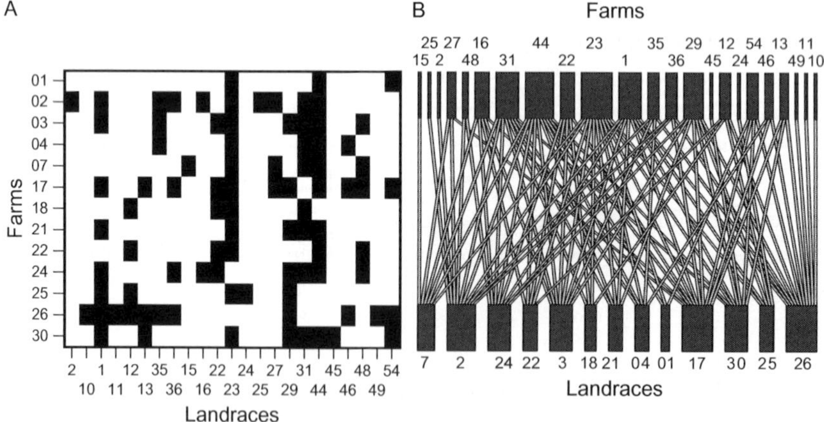

Figure 2 (A) Example of an incidence matrix with farms in lines and crops in columns, and where 0 are black cells and 1 are white cells and (B) example of a crop-by-farm bipartite network between farms and landraces (dataset AB-M02).

between the farm i and the crop j if and only if $X_{ij} = 1$. The obtained network is bipartite (Fig. 2B) as no two farms and no two crops are connected in the network. Building on this equivalence between incidence matrices and bipartite graphs, we can borrow methodologies developed in the field of network analysis (Kolaczyk, 2009).

As these two representations are equivalent, any statistical analysis could be defined either in terms of the incidence matrix or in terms of the bipartite network G. For ease of reading, this chapter makes use of the incidence matrix terminology but we sometimes borrow network notations to emphasize the connection with the literature on network analysis.

Summing over crop, the number of crops cultivated on farm i, C_i, is

$$C_i = \sum_j X_{ij}. \tag{1}$$

Summing over farms, the number of farms where crop j is cultivated, F_j, is

$$F_j = \sum_i X_{ij} \tag{2}$$

Quantities N, C_i, F_j and X_{ij} are finally linked by the following relations:

$$N = \sum_i C_i = \sum_j F_j = \sum_{i,j} X_{ij}. \tag{3}$$

Following network terminology, C_i is also called the farm's degree and F_j the crop's degree.

3.2 Variability of Farms' and Crops' Degrees

3.2.1 Description of the Test on Degree Distributions

First, we evaluate whether all farms in the same village grow a similar number of crop or if there is high heterogeneity between farms' crop richness. Formally, we test whether the degrees F_j follow binomial distributions by considering a statistic T that compares the observed variance of the crops' degree with the one that would have been expected if the degrees C_i were following independent and identically distributed (*iid*) binomial distributions.

$$T_{\text{row}} := \frac{\widehat{\text{Var}}(C)}{n\hat{p}(1-\hat{p})},$$

where $\hat{p} = N/nm$ is the density of the incidence matrix and $\widehat{\text{Var}}(C) = 1/(n-1)\sum_{i=1}^{n}(C_i - m\hat{p})^2$ is the empirical variance of (C_i),

$i=1,\ldots,n$. Large T_{row} values suggest that the farms' crop richness is highly heterogeneous, whereas small T_{row} values suggest more equity. The statistical significance of T is assessed by a parametric bootstrap method working as follows. For $i=1,\ldots,n_{\text{sim}}$, a new incidence matrix $\mathbf{X}^{(i)}$ is generated by sampling independent Bernoulli distributions with parameters $\hat{p}$ in each entry. For all these matrices, the link density $\hat{p}^{(i)}$, the empirical variance of the farms' degrees $\widehat{\text{Var}}^{(i)}(C)$ and the variance ratio $T_{\text{row}}^{(i)}$ are computed. Finally, the left and right p-values are, respectively, $p\,\text{val}_{L,\text{row}} := \frac{\#\{i: T_{\text{row}}^{(i)} < T_{\text{row}}\}}{n}$ and $p\,\text{val}_{R,\text{row}} := \frac{\#\{i: T_{\text{row}}^{(i)} > T_{\text{row}}\}}{n}$.

The crops' degree distribution is evaluated in a similar fashion:

$$T_{\text{col}} := \frac{\widehat{\text{Var}}(F)}{m\hat{p}(1-\hat{p})};$$
$$\widehat{\text{Var}}(F) = \frac{1}{m-1}\sum_{j=1}^{m}\left(F_j - n\hat{p}\right)^2.$$

The corresponding p-values are also evaluated by parametric bootstrap. In our analysis, the parameter n_{sim} is fixed to 10,000.

Under an Erdős-Rényi null model, where all the entries of $\mathbf{X}$ follow independent Bernoulli distributions with identical parameters, and the farms' and crops' degrees follow binomial distributions. Consequently, any small p-value ($p\,\text{val}_{L,\text{row}}, p\,\text{val}_{R,\text{row}}, p\,\text{val}_{L,\text{col}}, p\,\text{val}_{R,\text{col}}$) would indicate that this Erdős-Rényi model is not realistic.

3.2.2 Application of the Test on Degree Distributions to a Toy Example

Figures 3–5 display three examples of incidence matrices. The last two matrices were generated by organizing groups of crops and groups of farms according to a LBM (see presentation in the next subsection). The farms and the crops were sorted by degrees within groups. Note that this structure of groups is generally unknown in real datasets and has to be recovered by statistical inference techniques. In Fig. 3, the incidence matrix was generated from *iid*. Bernoulli random variables. Hence, its row and column degrees follow binomial distributions. This corresponds to the null hypothesis of the test on the variance of degrees. The tests are non-significant for this incidence matrix (Table 3). In Fig. 4, some farms were assumed to grow more crops than others and some crops were assumed to be more common than others. Therefore, as expected, the tests on the variance of degrees show clearly an over-dispersion for farms and crops. In Fig. 5, there exist particular associations between some groups of farms

Figure 3 Incidence matrix with entries generated independently and identically distributed according to a Bernoulli distribution with probability 0.2.

Figure 4 Incidence matrix generated with heterogeneous distribution for different groups of crops and farms (see Fig. 7 in next subsection for details). Some farms grow more crops than other and some crops are more common than others.

and some groups of crops although the degree is quite homogeneous for farms; crops heterogeneity appears because the groups of farms are not of the same size.

As illustrated in these three examples, the tests on the variance of degrees may detect heterogeneity but some particular structure of association may be

Figure 5 Incidence matrix generated with distribution implying particular association between crops and farms (see Fig. 6 in next subsection for details). Two groups of crops are mainly grown by corresponding subgroups of farms.

Table 3 *p*-Values for Tests on the Variability of Degrees for Farms and Crops (Left: Under-Dispersion, Right: Over-Dispersion) Applied on the Three Toy Examples Presented in Figs. 3–5

	Farms		**Crops**	
	Left	**Right**	**Left**	**Right**
Fig. 3	0.8143	0.1857	0.6345	0.3655
Fig. 4	1	<0.001	1	<0.001
Fig. 5	0.1604	0.8396	0.9924	0.0076

missed as in the case of Fig. 5. Indeed, the tests are performed independently on farms and on crops and thus are not able to detect patterns of association.

3.3 Revealing Data Structure Through LBMs

3.3.1 Description of the LBMs

In order to cluster the farms and the crops simultaneously on the basis of the incidence matrix $\mathbf{X}$, we propose to use a probabilistic model called LBM (Govaert and Nadif, 2008; Keribin et al., 2014), which assumes a mixture distribution both on the farms and crops. According to this model, the network is generated relying on latent blocks (also called clusters) of farms and of crops. The probability that a crop j is grown on a farm i is conditioned to these latent blocks and depends only on the block $V(i)$ to which farm i belongs and the block W_j to which crop j belongs. For all $1 \leq i \leq n$, $1 \leq j \leq m$, $1 \leq q \leq Q$ and $1 \leq l \leq L$, the probability that i belongs to block q, that j belongs to block 1 and the conditional probability of X_{ij} given the block V_i and W_j are, respectively, denoted

$$\begin{aligned}
&\mathbb{P}(V_i = q) = \alpha_q, \\
&\mathbb{P}\left(W_j = l\right) = \beta_l, \\
&\mathbb{P}\left(X_{ij} = 1 \mid V_i = q,\ W_j = l\right) = \pi_{ql},
\end{aligned}$$

where $\theta = (\alpha_1, \ldots, \alpha_Q, \beta_1, \ldots, \beta_L, \pi_{11}, \ldots, \pi_{QL})$ is the vector of unknown parameters to be estimated under the obvious constraints $\sum_q \alpha_q = 1$, $\sum_l \beta_l = 1$. This model is quite flexible because it can account not only for situations where there is modularity, i.e., a unique block of crops is associated with each block of farms and these farms tend to grow mainly crops from that block and few from other blocks, but also for situations where there are richer farms (growing significantly more crops than others) and/or more common crops (grown by significantly more farms than others).

The standard procedures to obtain maximum likelihood estimates when dealing with latent variables rely on the expectation–maximization (EM) algorithm (Dempster et al., 1977). However, the computation of the conditional distribution of the latent variables with respect to the observed data is not tractable, which makes the E-step unfeasible. Following Govaert and Nadif (2008), we use a variational approach to cope with this difficulty. The number of blocks of farms Q and the number of blocks of crops L are chosen thanks to the integrated completed likelihood (ICL) criterion as proposed by Keribin et al. (2014). Once the parameters have been estimated, we obtain as a by-product the posterior probabilities $\mathbb{P}(V_i = q|\mathbf{X})$ and $\mathbb{P}(W_i = l|\mathbf{X})$, from which the true blocks are estimated. We can then provide a new representation of the incidence matrix $\mathbf{X}$ where the rows (farms) and the columns (crops) have been re-organized in homogeneous blocks. We used the R package `blockmodels` (Leger, 2015) to perform the estimations and the model selection.

3.3.2 Application of LBM to a Toy Example

Figures 6–8 are illustrations of the block clustering provided by the LBM in three typical cases. The cases of Figs. 6 and 7 are the same as those in Figs. 5 and 4, respectively. The groups were considered as latent/unknown and the

Figure 6 Incidence matrix generated according to a LBM with three blocks of crops, two blocks of farms and $\pi = \begin{pmatrix} 0.5 & 0.1 & 0.6 \\ 0.5 & 0.6 & 0.1 \end{pmatrix}$. (A) Observed incidence matrix and (B) same incidence matrix re-organized and clustered in homogeneous blocks obtained by LBM inference.

Figure 7 LBM clustering when the data are generated with two blocks of farms (rich and poor farms), two blocks of crops (rare and common crops) and $\pi = \begin{pmatrix} 0.7 & 0.3 \\ 0.4 & 0.2 \end{pmatrix}$.

farms and crops were clustered in homogeneous blocks by using the inference procedure described above. This is illustrated in Fig. 6, where the same incidence matrix is plotted before and after re-organization according to the estimated blocks. In Fig. 6, the difference between the two groups of farms comes from the two last groups of crops. The first group of crops is equally

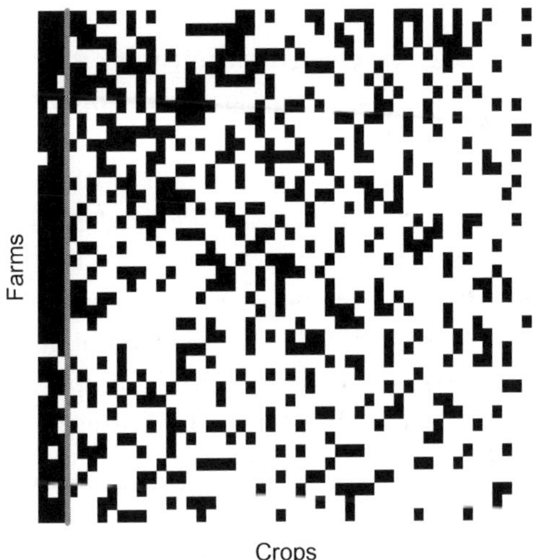

Figure 8 LBM clustering when the data are generated with one block of farms, two blocks of crops (one block with only three crops) and $\pi = (0.9 \ \ 0.3)$.

grown on farms of any group. In contrast, the second group of crops is mainly grown by the second group of farms and the third group of crops is mainly grown by the first group of farms. In Fig. 7, the farms can be separated on the basis of the number of crops that they grow. One group can be said to be 'rich' and the other to be 'poor'. Similarly, two groups are also found for crops, one composed of common crops and the other of rare crops. In Fig. 8, farms are similar and three crops are much more common than the others. Since the difference is quite clear and there are three crops, the ICL criterion for the LBM argues for recognition of a block with only three crops. However, if there are only one or two outlier(s) or if the difference is less clear, this criterion may not separate this (these) outlier(s). This criterion for model selection is not designed for detecting outliers.

3.4 Uncovering Outliers Through PCA

3.4.1 Configuration Model

Fix the degree $(C_i)_{i=1,\ldots,n}$ of each farm and $(F_j)_{j=1,\ldots,m}$ of all crops in $\mathbf{X}$. The (bipartite) configuration model with parameters (C_i) and (F_j) is the uniform distribution over all incidence matrices that leave the degrees C_i and F_j unchanged. In the ecological literature, this model is sometimes referred to as the fixed–fixed null model (Connor and Simberloff, 1979; Ulrich and

Gotelli, 2012; Zaman and Simberloff, 2002). In contrast to the LBM, the configuration model takes as a given that some farms might grow many more crops than others and that some crops are more common than others, but apart from that the incidence matrix is sampled uniformly.

In order to simulate according to the configuration model, we use the tswap sequential algorithm (Miklós and Podani, 2004) implemented in the `permatswap` function of the R package `vegan`. The practitioner has to select burnin and thinning parameters large enough that the algorithm explores well the space of incidence matrices. Although the mixing time of the `tswap` algorithm is unknown, the mixing properties of the sequence can be visually checked using the plot method of `permatswap`.

3.4.2 PCA on Residuals

The expected incidence matrix under the configuration model with degrees (C_i) and (F_j) is denoted $\mathbb{E}_0\left[\mathbf{X}|\left(C_i, F_j\right)\right]$. Alternatively, $\mathbb{E}_0\left[\mathbf{X}|\left(C_i, F_j\right)\right]$ can be seen as the average overall permutations on the entries of $\mathbf{X}$ that keeps the degree sequences for both crops and farms unchanged. Then, the residual matrix $\mathbf{R}$ under the configuration model is the difference between the observed incidence matrix and its expectation under the configuration model

$$R_{ij} = X_{ij} - \mathbb{E}_0\left[X_{ij}|\left(C_i, F_j\right)\right] \tag{4}$$

If the incidence matrix $\mathbf{X}$ was drawn according to the configuration model, then $\mathbf{R}$ would have no particular structure. In order to check the absence of structure, we apply a (non-standardized) PCA on $\mathbf{R}$. As it is customary for a PCA, the projection of the rows (i.e. the farms) along the first principal directions allows (i) discovery of groups of farms that effectively cultivate the same types of crops and (ii) detection of outlier farms whose field crop composition is unusual when the effect of farm richness has been removed. As an example, a farm where a very high diversity is cultivated would not necessarily be an outlier, but this farm will be considered as an outlier if it does not grow some very common crops. The projection of the columns of $\mathbf{R}$ along the first principal directions provides information on outlier crop or groups of crops.

3.4.3 Goodness-of-Fit Test of the Configuration Model

Assessing the statistical significance of the PCA is equivalent to testing whether the network $\mathbf{X}$ has been drawn according to the configuration model. The test rejects the null hypothesis when the largest eigenvalue in the scree plot is unusually large. More precisely, the test is calibrated by permutations $\mathbf{X}^P$ of $\mathbf{X}$ that leaves the degree of each row and column invariant. Denote $\lambda_{\max}$ the largest singular value of $\mathbf{R}$ (i.e. the square-root of the largest eigenvalue of $\mathbf{R}^t\mathbf{R}$), then the p-values are obtained by comparing the singular value $\lambda_{\max}$ to the largest singular values of matrix $\mathbf{R}^P$ arising from permutations $\mathbf{X}^P$.

Under the null hypothesis, the matrix $\mathbf{R}$ is pure noise and all the singular values of $\mathbf{R}$ should be small. Under the presence of outliers or of a few groups of farms that preferentially cultivate some crops, the matrix $\mathbf{R}$ is expected to be the sum of a noisy component and a low-rank component measuring the deviance from the configuration model. As a consequence, the singular value of $\mathbf{R}$ should be higher under the alternative than under the null hypothesis.

Although calibrated differently, the largest singular value statistic has been applied in other problems of community detection (Bickel and Sarkar, 2015).

3.4.4 A New Representation of the Incidence Matrix

Ordering the farms according to the coordinate of their projection along the first principal direction, we denote $\sigma_1(i)$ the farm index associated with the ith smallest coordinate. Similarly, $\sigma_2(j)$ stands for the reordering of the crops according to their projection on the first direction. These permutations (σ_1, σ_2) define a new representation $\mathbf{Y}$ of the incidence matrix:

$$Y_{ij} = X_{\sigma_1(i)\sigma_2(j)} \tag{5}$$

This provides a visualization of the incidence matrix alternative to that offered by the LBM approach.

3.4.5 Toy Examples

Let us describe three typical examples to understand the behaviour of the above statistics. In all these examples, the number n of farms is set to 40 and the number m of crops set to 60.

First, we consider a model with degree heterogeneity. For each farm $i = 1, \ldots, n$ and each crop $j = 1, \ldots, m$, we draw *iid* uniform random variable a_i and b_i in $(0, 1)$. Then, each entry X_{ij} is drawn according to a Bernoulli distribution with parameter $\min(2a_ib_j, 1)$. As a consequence, the incidence matrix $\mathbf{X}$ exhibits large degree heterogeneity among farms (resp. crops) with a low a_i (resp. b_i) value and farms (resp. crops) with a high a_i (resp. b_i). It is therefore not unexpected that the LBM estimation procedure (Fig. 9A)

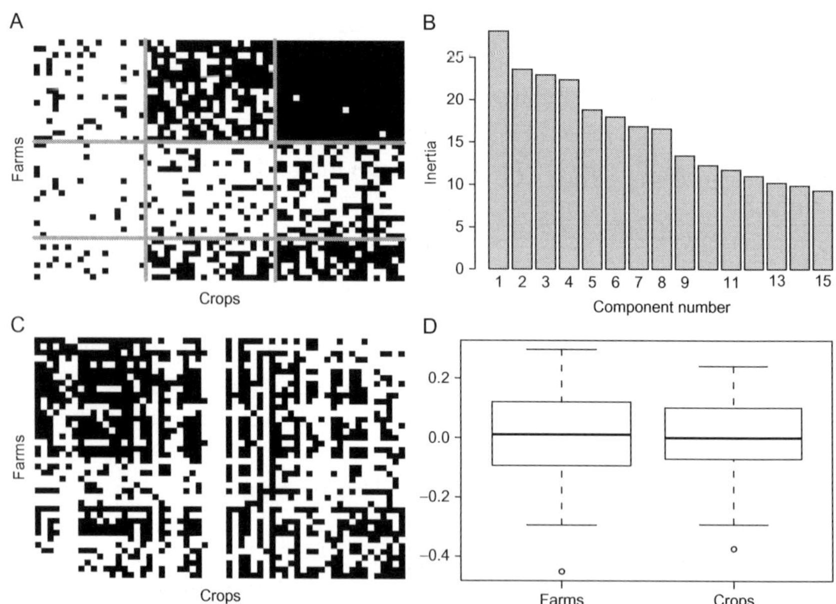

Figure 9 First example to illustrate the method for uncovering outliers through principal component analysis: degree heterogeneity. (A) The LBM representation, (B) the scree plot of the PCA residuals, (C) the representation of the incidence matrix according to the PCA ordering (Eq. 5) and (D) the boxplots of the PCA first coordinates.

recovers several groups of crops and farms. The p-value of configuration model from Section 3.4.3 equals 0.54. Again, this is not surprising, since this incidence matrix has been sampled from a model similar to the configuration model. This implies that the block structure found by the LBM method can be explained by the degree heterogeneity. As the configuration model residuals are completely random here, both the PCA scree plot (Fig. 9B) and the representation (Eq. 5) of the incidence matrix (Fig. 9C) are uninformative. No farms and no crops have outlier PCA coordinates (Fig. 9D).

In the second example, we draw the incidence matrix $\mathbf{X}$ as above. Then, we replace each entry of the first row by independent Bernoulli random variables with parameter 0.5. As a consequence, the first farm is assumed to have a completely different behaviour from all the other farms, as it grows crops regardless of their scarcity (b_j) in the village. The LBM representation (Fig. 10A) is close to that of the first example (Fig. 9A). The p-value of the configuration test is smaller than 10^{-3}, and the scree plot exhibits an unusually large first eigenvalue (Fig. 10B). The first farm is therefore detected as an outlier by the first coordinate representation (Fig. 10D).

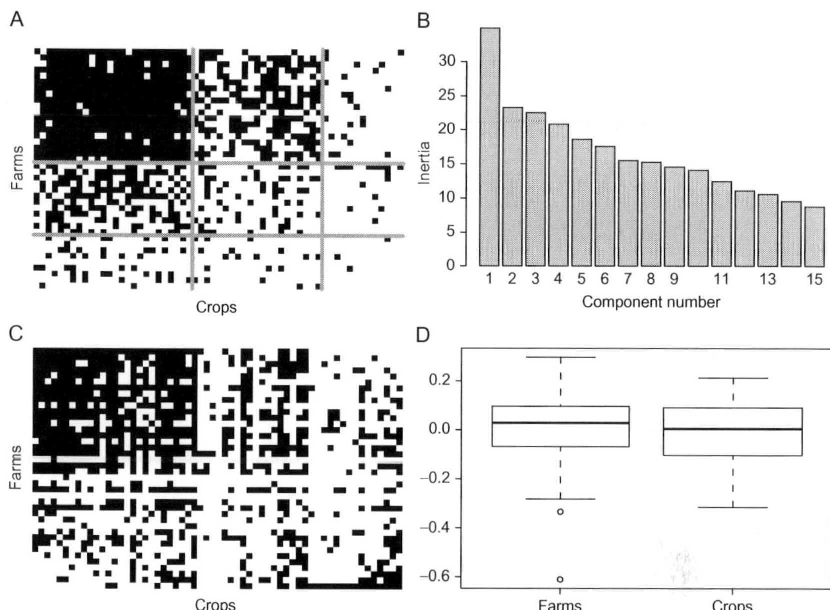

Figure 10 Second example to illustrate the method for uncovering outliers through principal component analysis: outlier. (A) The LBM representation, (B) the scree plot of the PCA residuals, (C) the representation of the incidence matrix according to the PCA ordering (Eq. 5) and (D) the boxplots of the PCA first coordinates.

Finally, the PCA-based representation (Fig. 10C) highlights the unusual behaviour of this farm.

In the last example, we draw random variables a_i and b_j as above. Then, the farms are divided into two groups A_1 and A_2 of size $n/2$ and the crops are divided into two groups B_1 and B_2 of size $m/2$. Then, the entry X_{ij} is drawn according to Bernoulli distribution with parameter $\min(p_{\text{in}}2a_ib_j, 1)$ if $(i,j) \in A_1 \times B_1$ or $(i,j) \in A_2 \times B_2$ and parameter $\min(p_{\text{out}}2a_ib_j, 1)$ if $(i,j) \in A_1 \times B_2$ or $(i,j) \in A_1 \times B_2$ with $p_{\text{in}} = 1.4$ and $p_{\text{out}} = 0.6$. Intuitively, the farms from A_1 (resp. A_2) preferentially grow crops from B_1 (resp. B_2), but the model also allows the degree of the farm and of each crop to be heterogeneous inside the blocks. As a consequence, this model, called degree-corrected, is neither a LBM with 2×2 blocks nor a configuration model but a blend of them. The LBM estimation method recovers too many blocks (Fig. 11A) by grouping farms or crops that are in the same group and have similar degrees. The p-value for the configuration test is found to be smaller than 10^{-3}. This is corroborated by the fact that the scree plot exhibits an unusually large first eigenvalue (Fig. 11B). Contrary to the previous example, this unusually

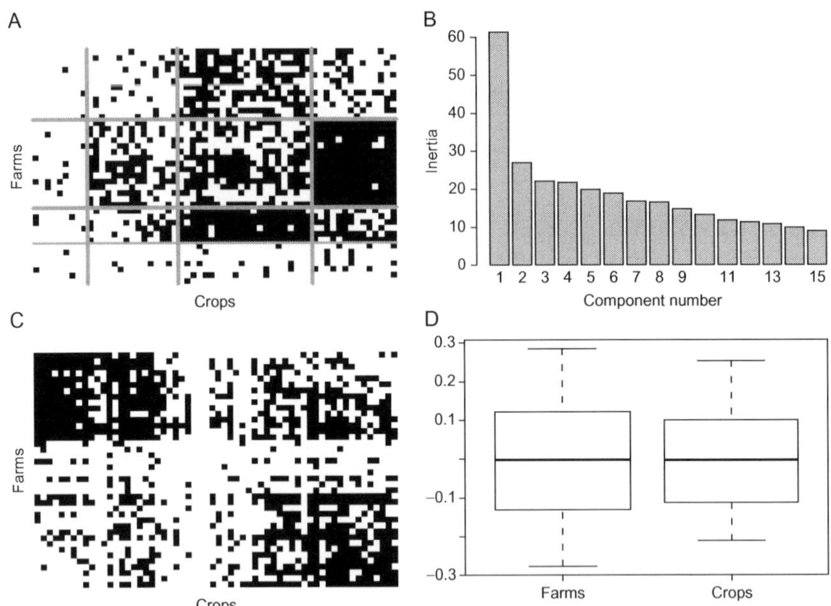

Figure 11 Third example to illustrate the method for uncovering outliers through principal component analysis: blocks. (A) The LBM representation, (B) the scree plot of the PCA residuals, (C) the representation of the incidence matrix according to the PCA ordering (Eq. 5) and (D) the boxplots of the PCA first coordinates.

large singular value is not due to outliers (Fig. 11D) but to the presence of a block structure. The PCA-based matrix representation highlights the presence of these two groups of farms and crops (Fig. 11C).

3.5 Measuring Originality of Farms' Contributions Through Diversity Measures

We will now focus our attention on the distribution of cultivated crop diversity at the level of the sampled location (the village). As mentioned in previous sections, some farms may grow many more crops than others (hence, the high variance in degree among farms in the bipartite network). A question that remains unanswered is whether low-degree farms contribute effectively more or less than high-degree farms to the overall cultivated diversity—'effectively more' being understood as contributing more than expected if crops were chosen randomly from the pool of crops cultivated in the village. In other words, the question is now whether low-degree farms cultivate the most frequent crops in the village only (common crops) or contribute disproportionately to crop diversity by focusing only on crops that are cultivated on very few farms (rare crops).

3.5.1 Theoretical Framework

Further expanding the notations introduced in Section 3.1, we denote p_{ij} the weight associated with the interaction between farm i and crop j among all interactions of farm i:

$$p_{ij} = \frac{X_{ij}}{C_i} \tag{6}$$

The proportion of all the connections in the network that are due to farm i or crop j are, respectively, noted q_i and h_j:

$$q_i = \frac{C_i}{N} \tag{7}$$

$$h_j = \frac{F_j}{N} \tag{8}$$

We note H_i the diversity of crops cultivated on farm i, as measured by Shannon entropy:

$$H_i = -\sum_j p_{ij} \log p_{ij} = \log C_i \tag{9}$$

The average diversity among farms, weighted by the importance q_i of each farm, is denoted H_α:

$$H_\alpha = \sum_i q_i H_i = \frac{1}{N} \sum_i C_i \log C_i \tag{10}$$

The diversity of crops cultivated by all farms, when taken together and weighted by the importance q_i of each farm, is noted H_T and reads as:

$$\begin{aligned} H_T &= -\sum_j \left[\sum_i q_i p_{ij}\right] \log \left[\sum_i q_i p_{ij}\right] = -\sum_j h_j \log h_j \\ &= \log N - \frac{1}{N} \sum_j F_j \log F_j \end{aligned} \tag{11}$$

The difference between H_T and H_α is the turnover in diversity among farms or β diversity, noted H_β:

$$H_\beta = H_T - H_\alpha = \log N - \frac{1}{N}\sum_j F_j \log F_j - \frac{1}{N}\sum_i C_i \log C_i \quad (12)$$

H_β can be further decomposed into individual turnover components, H_{iT}:

$$H_\beta = \sum_i q_i H_{iT} \quad (13)$$

where H_{iT} measures the 'originality' of farm i portfolio of crops when compared to the overall diversity of cultivated crops. An expression for H_{iT} can be found (Lande, 1996):

$$H_{iT} = -\sum_j p_{ij} \log \frac{C_i F_j}{N} \quad (14)$$

3.5.2 Measuring the Diversity Cultivated by Crop–Poor and Crop–Rich Farms

We now focus on measuring the evenness of crops cultivated by a subset I of farms. More specifically, because we are interested in the subset of the most crop–poor or crop–rich farms, we will assume that the set I contains all farms belonging to a certain quantile of the distribution of S_i. The evenness of crops cultivated on farms in set I is noted E_I and reads as

$$E_I = -\frac{\sum_j \left[\sum_{i \in I} q_{i,I} p_{ij}\right] \log \left[\sum_{i \in I} q_{i,I} p_{ij}\right]}{\log(m)}; \quad q_{i,I} = \frac{C_i}{\sum_{i \in I} C_i}. \quad (15)$$

The evenness E_I is the diversity of crops cultivated on all farms in set I divided by the logarithm of the total number m of crops cultivated in the village. It measures the equity of the distribution of crops cultivated on farms in I.

In order to assess whether the evenness is greater in crop–rich farms than crop–poor farms, we compare the value of $E_{\text{Rich}} - E_{\text{Poor}}$ to that of all realizations of the incidence matrix $\mathbf{X}$ under the configuration model (i.e. randomizing connections given degree sequences for both crops and farms) by a permutation test.

3.5.3 Measuring the Impact of Crop–Poor and Crop–Rich Farms

We now focus on measuring the β diversity $H_{\beta,I}$ due to the contribution of a subset I of farms. As previously, the subset I is made up of the most crop–poor or crop–rich farms. We can give an explicit formula for $H_{\beta,I}$ (Lande, 1996):

$$H_{\beta,I} = \sum_{i \in I} q_i H_{iT} = -\sum_i q_i \log q_i + \frac{1}{N}\sum_j \left[\sum\nolimits_{i \in I} X_{ij}\right] \log\left(\frac{1}{F_j}\right) \quad (16)$$

The first term in the right-hand side of Eq. (16) relies on the expression of the α diversity $H_{\alpha,I}$ due to farms in subset I:

$$H_{\alpha,I} = \frac{1}{N}\sum_{i \in I} C_i \log C_i = \frac{\sigma_I \log N}{N} + \sum_{i \in I} q_i \log q_i \quad (17)$$

where σ_I is the 'volume' of interactions due to farms belonging to subset I:

$$\sigma_I = \sum_{i \in I} C_i \quad (18)$$

The second term depends on the correlation between a crop degree F_j and the number of farms within the set I who possess this crop, noted $\varphi_{j,I}$:

$$\varphi_{j,I} = \sum_{i \in I} X_{ij} \quad (19)$$

Plugging Eqs. (17)–(19) into Eq. (16) yields the following expression for $H_{\beta,I}$:

$$H_{\beta,I} = \frac{\sigma_I \log N}{N} - H_{\alpha,I} - \frac{1}{N}\sum_j \varphi_{j,I} \log F_j \quad (20)$$

The quantity $D_I = \frac{1}{N}\sum\nolimits_j \varphi_{j,I} \log F_j$ measures the deficit of originality displayed by the farms in subset I that is due to their cultivation of 'common crops'.

Again, we assess the significance of $H_{\beta,I}$ by a permutation test based on the configuration model. As the set I contains all farms belonging to a certain quantile of the distribution of C_i, all realizations of the incidence matrix $\mathbf{X}$ under the configuration model preserve the set of C_i values to be found in I. As a consequence, the quantity $\frac{\sigma_I \log N}{N} - H_{\alpha,I}$ in the right-hand side of Eq. (16) is invariant with respect to the configuration model. The quantity D_I in the right-hand side of Eq. (16), however, does not satisfy this invariance. Thus, values of $H_{\beta,I}$ that are unusually large for the configuration model mean that farms in subset I contribute more to cultivated biodiversity than expected by the number of types cultivated on farms in I.

3.5.4 *Measuring Originality of Farms' Contributions Through Diversity Measures on Toy Examples*

3.5.4.1 Simulation Model

Two groups of farms are considered: crop–rich (40% of farms) and crop–poor (60% of farms). The crops are divided into two groups with same size: rare and common, consistently with the definition provided at the beginning of Section 3.5. The entries of the incidence matrix are generated as *iid.* Bernoulli random variables with probability p_{ij} (corresponding to farm i and crop j) given by:

$$\text{logit}(p_{ij}) = \mu + \alpha(L_i) + \beta(K_j) + \gamma(L_i : K_j)$$

where logit is the function $x \mapsto \log(x/(1-x))$, L_i indicates the group of farm i, K_j the group of crop j and parameters μ, αs, βs, γs are:

$$\alpha(\text{poor}) = \beta(\text{rare}) = \gamma(\text{poor, rare}) = \gamma(\text{rich, rare}) = \gamma(\text{poor, common}) = 0$$

to ensure identifiability. The interaction term $\gamma(\text{rich, common})$ then drives the respective contributions to diversity of crop–rich and crop–poor farms. Indeed, if the value of this term is zero, the effect of being crop–rich for growing a rare or a common variety will be the same.

3.5.4.2 Three Contrasted Toy Examples

Figures 12–14 correspond, respectively, to the three following settings:

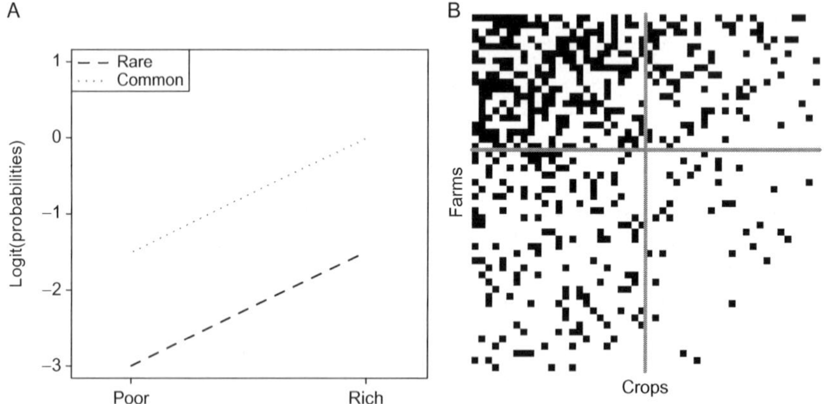

Figure 12 Toy example with equal contributions to diversity of crop–rich and crop–poor farms. $\mu = -3$, $\alpha(\text{rich}) = \beta(\text{common}) = 1.5$, $\gamma(\text{rich, common}) = 0$. (A) Probabilities that a crop is grown on a farm and (B) incidence matrix.

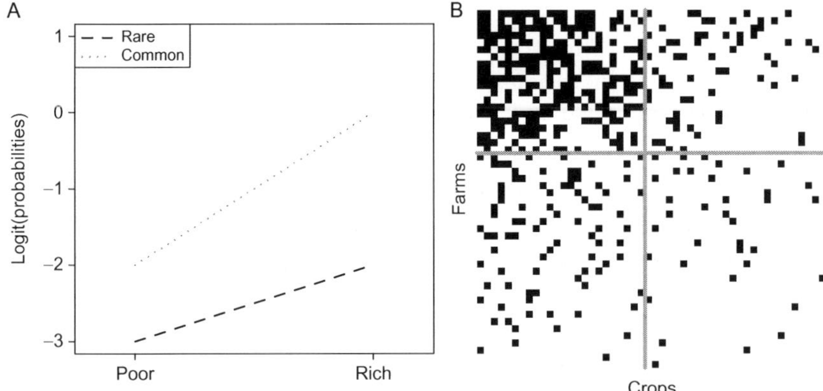

Figure 13 Toy example with greater contribution to diversity of crop–poor farms. $\mu = -3$, $\alpha(\text{rich}) = \beta(\text{common}) = (\text{rich, common}) = 1$. (A) Probabilities that a crop is grown on a farm and (B) incidence matrix.

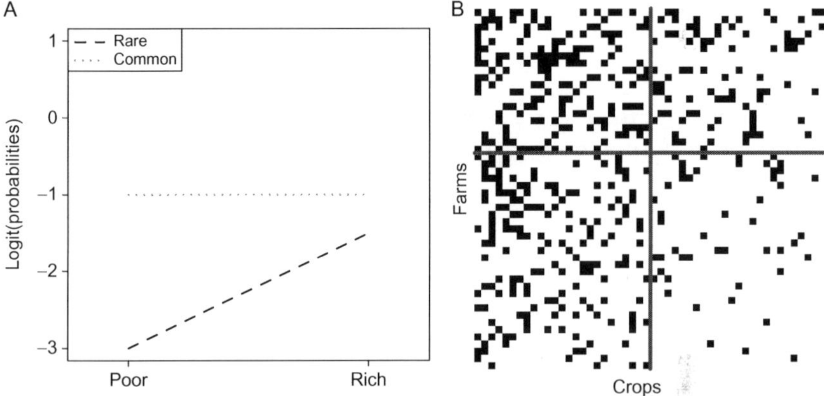

Figure 14 Toy example with greater contribution to diversity of crop–rich farms. $\mu = -3, \alpha(\text{rich}) = -\gamma(\text{rich, common}) = 1.5, \beta(\text{common}) = 2$. (A) Probabilities that a crop is grown on a farm and (B) incidence matrix.

1. Parameters fixed to $\mu = -3$, $\alpha(\text{rich}) = \beta(\text{common}) = 1.5$, $\gamma(\text{rich, common}) = 0$. The crop–rich and crop–poor farms have the same contribution to diversity with respect to their own crop richness. This is ensured by setting the interaction term $\gamma(\text{rich, common})$ to 0.
2. Parameters fixed to $\mu = -3$, $\alpha(\text{rich}) = \beta(\text{common}) = \gamma(\text{rich, common}) = 1$. The crop–poor farms have a greater contribution to diversity, since they grow rare crops and common crop with nearly the same probability, whereas for crop–rich farms, the probability of growing rare crops is clearly smaller than the probability of growing common crops.

3. Parameters fixed to $\mu=-3$, $\alpha(\text{rich})=-\gamma(\text{rich, common})=1.5$, $\beta(\text{common})=2$. The crop–rich farms have a greater contribution to diversity, as they exhibit the same ability of growing rare and common crops.

As shown in Figs. 12–14, these three settings are consistent with what a crop–poor or a crop–rich farm and a rare or a common crop are expected to be.

Results in Table 4 are coherent with the intuitive expectations based on these three models. For the first case, nothing was found to contribute significantly to the distribution of crop diversity. For the two other cases, the tests on evenness and on the contributions to diversity of crop–rich and crop–poor farms agreed. Indeed, in the case of Fig. 13, for instance, the crop–rich farms are found to contribute less than expected to diversity (null hypothesis rejected on the left side), the crop–poor farms are found to contribute more than expected to diversity (null hypothesis rejected on the right side) and the difference of evenness is found to be significantly smaller than expected (null hypothesis rejected on the left side). Based on these three toy examples, a power study was conducted and its results are presented in Appendix.

4. PATTERNS OF LOCAL CROP DIVERSITY: RESULTS OF THE META-ANALYSIS

All the aforementioned methods have been applied to the 50 datasets of the meta-analysis. The results are summarized in Tables 5 and 6 for species and infra-species diversity, respectively.

4.1 Variability of Farms' and Crops' Degrees

The aim of this section is to detect over-dispersion (significant test on the right) or under-dispersion (significant test on the left) of degree distribution for farms and crops, respectively, following the methodology introduced in Section 3.2. Two null hypotheses (H_0) are tested:

1. diversity at species and infra-species level is randomly distributed (i.e. according to a binomial distribution) among farms from the same village (homogeneity of the farm degrees) and
2. crop richness is randomly distributed within the same village (homogeneity of the crop degrees).

4.1.1 Species Diversity

For farms, H_0 was rejected on the right side (16 times over the 18 tested datasets) for the variability of farms' degree (Table 5). There was only

Table 4 *p*-Values for the Contribution Tests Applied to the Three Examples Presented in Figs. 12–14

	Evenness Difference	Significance		Crop–Rich Farms	Significance		Crop–Poor Farms	Significance	
	Rich–Poor Farms	Left	Right	Contribution	Left	Right	Contribution	Left	Right
Fig. 12	0.036	0.433	0.567	0.628	0.587	0.413	0.643	0.413	0.587
Fig. 13	−0.003	<0.001	>0.999	0.638	<0.001	>0.999	0.64	>0.999	<0.001
Fig. 14	0.036	0.989	0.011	0.715	0.996	0.004	0.809	0.004	0.996

Table 5 Statistical Results Obtained for the 18 Datasets Dealing with Specific Diversity

Dataset	Variance of Farms' Degree	Significance Left	Significance Right	Variance of Species' Degree	Significance Left	Significance Right	LBM Farm Cluster Number	LBM Species Cluster Number	Normalized First Singular Value	Significance Right	Evenness Difference Rich–Poor Farms	Significance Left	Significance Right	Crop–Rich Farms Contribution	Significance Left	Significance Right	Crop–Poor Farms Contribution	Significance Left	Significance Right
CL-M01	0.579	<0.001	NS	74.138	NS	<0.001	1	3	0.49	NS	0.167	NS	<0.001	2.187	NS	<0.001	1.796	NS	NS
CV-M01	1.843	NS	<0.05	8.068	NS	<0.001	2	2	2.21	<0.05	0.077	NS	NS	3.054	NS	NS	2.824	NS	NS
EG-M05	7.708	NS	<0.001	6.023	NS	<0.001	2	3	1.07	NS	0.061	NS	NS	3.751	NS	NS	3.538	NS	<0.05
EG-M08	10.207	NS	<0.001	6.3	NS	<0.001	3	3	0.98	NS	0.078	<0.01	NS	3.887	<0.05	NS	3.606	NS	<0.001
OC-M02	6.471	NS	<0.001	4.707	NS	<0.001	2	2	4.7	<0.001	0.08	NS	NS	4.429	NS	NS	4.147	NS	NS
OC-M04	8.353	NS	<0.001	10.016	NS	<0.001	3	3	4.05	<0.05	0.127	NS	<0.001	4.361	NS	<0.001	3.968	<0.001	NS
OC-M05	5.346	NS	<0.001	6.285	NS	<0.001	2	2	2.84	<0.01	0.094	NS	NS	4.323	NS	NS	4.059	NS	NS
OC-M06	5.546	NS	<0.001	5.205	NS	<0.001	2	2	0.75	NS	0.102	NS	NS	4.194	NS	NS	3.859	NS	NS
OC-M07	8.431	NS	<0.001	4.31	NS	<0.001	3	2	2.84	<0.01	0.174	NS	<0.001	4.142	NS	<0.001	3.668	<0.001	NS
OC-M09	16.755	NS	<0.001	2.859	NS	<0.001	2	2	0.17	NS	0.19	NS	NS	4.131	NS	NS	3.734	NS	NS
OC-M10	13.601	NS	<0.001	4.007	NS	<0.001	2	2	1.03	NS	0.144	NS	NS	4.157	NS	NS	3.826	NS	NS
OC-M11	5.801	NS	<0.001	18.977	NS	<0.001	2	4	3.54	<0.01	0.085	NS	<0.001	4.583	NS	<0.001	4.202	<0.001	NS
OC-M12	5.187	NS	<0.001	4.529	NS	<0.001	2	2	1.76	<0.05	0.077	NS	<0.001	4.627	NS	<0.001	4.31	<0.001	NS
OC-M13	10.607	NS	<0.001	9.764	NS	<0.001	3	3	2.28	<0.05	0.093	NS	NS	4.602	NS	NS	4.246	<0.05	NS
OC-M14	9.702	NS	<0.001	6.367	NS	<0.001	3	3	2.56	<0.01	0.081	NS	NS	4.587	NS	<0.05	4.266	NS	NS
OC-M16	9.14	NS	<0.001	6.513	NS	<0.001	3	2	1.61	NS	0.11	NS	NS	4.623	NS	NS	4.219	NS	NS
SC-M05	1.671	NS	NS	9.74	NS	<0.001	1	3	0.35	NS	0.11	NS	NS	3.468	NS	<0.001	3.112	NS	NS
SC-M06	4.595	NS	<0.001	3.662	NS	<0.001	2	2	1.1	NS	0.066	NS	NS	2.678	NS	NS	2.471	NS	NS

NS: non-significant, <0.05: *p*-values ranging from 0.01 to 0.05, <0.01: *p*-values ranging from 0.001 to 0.01 and <0.001: *p*-values lower than 0.001.

Table 6 Statistical Results Obtained for the 33 Datasets Dealing with Infra-Specific Diversity

Dataset	Variance of Farms' Degree	Significance		Variance of Landraces' Degree	Significance		LBM Farm Cluster Number	LBM Landraces Cluster Number	Normalized First Singular Value	Significance	Evenness Difference Rich–Poor Farms	Significance		Crop–Rich Farms' Contribution	Significance		Crop–Poor Farms' Contribution	Significance	
		Left	Right		Left	Right				Right		Left	Right		Left	Right		Left	Right
AB-M02	1.485	NS	NS	3.456	NS	<0.001	1	2	0.82	NS	0.116	NS	NS	2.951	NS	NS	2.667	NS	NS
CL-M02	0.338	<0.001	NS	33.193	NS	<0.001	1	3	2.05	<0.05	0.057	NS	NS	1.978	NS	NS	1.913	NS	NS
CV-M02	0.963	NS	NS	5.731	NS	<0.001	1	2	0.76	NS	0.138	NS	NS	2.943	NS	NS	2.545	<0.05	NS
DJ-M003a	1.191	NS	NS	14.849	NS	<0.001	1	3	0.5	NS	0.224	NS	<0.05	0.449	NS	NS	0.371	NS	NS
DJ-M003b	1.207	NS	NS	17.837	NS	<0.001	1	3	0.69	NS	0.142	NS	NS	0.649	NS	NS	0.561	NS	NS
DJ-M003c	1.164	NS	NS	8.074	NS	<0.001	1	2	0.03	NS	0.194	NS	NS	0.564	NS	NS	0.479	NS	NS
DJ-M003d	0.751	NS	NS	10.915	NS	<0.001	1	2	0.43	NS	0.216	NS	NS	0.497	NS	NS	0.424	NS	NS
DJ-M009a	0.995	NS	NS	13.273	NS	<0.001	1	2	0.68	NS	0.163	NS	NS	0.385	NS	NS	0.463	NS	NS
DJ-M009b	1.033	NS	NS	9.983	NS	<0.001	1	2	0.66	NS	0.378	NS	NS	0.585	NS	NS	0.39	NS	NS
DJ-M009c	0.966	NS	NS	15.084	NS	<0.001	1	2	1.6	NS	0.422	NS	NS	0.648	NS	NS	0.39	NS	NS
DJ-M009d	0.975	NS	NS	5.285	NS	<0.001	1	2	0.57	NS	0.318	NS	NS	0.337	NS	NS	0.362	NS	NS
DJ-M009e	0.848	NS	NS	13.349	NS	<0.001	1	2	0.33	NS	0.271	NS	NS	0.468	NS	NS	0.389	NS	NS
DJ-M009f	1.349	NS	NS	10.697	NS	<0.001	1	2	0.07	NS	0.313	NS	NS	0.596	NS	NS	0.421	NS	NS
DJ-M012a	0.656	<0.05	NS	16.079	NS	<0.001	1	3	1.62	NS	0.124	NS	<0.001	3.206	NS	<0.001	2.794	NS	NS
DJ-M012b	1.18	NS	NS	6.892	NS	<0.001	1	2	0.07	NS	0.086	NS	<0.05	3.576	NS	NS	3.468	NS	NS
DJ-M015a	0.304	<0.001	NS	88.723	NS	<0.001	1	3	0.81	NS	0.097	NS	NS	2.035	NS	NS	1.798	NS	<0.05
DJ-M015b	0.542	<0.05	NS	9.976	NS	<0.001	1	2	1.82	NS	0.099	NS	NS	1.979	NS	NS	1.744	NS	NS
DJ-M018a	0.367	<0.001	NS	9.875	NS	<0.001	1	2	3.95	<0.05	0.246	NS	<0.001	2.354	NS	<0.001	1.782	<0.001	NS
DJ-M018b	0.297	<0.001	NS	45.012	NS	<0.001	1	3	3.82	<0.001	0.153	NS	NS	1.883	NS	NS	1.519	NS	NS
DJ-M030	0.44	<0.001	NS	48.912	NS	<0.001	1	3	3.92	<0.05	0.233	NS	<0.001	1.62	NS	NS	1.21	<0.01	NS

Continued

Table 6 Statistical Results Obtained for the 33 Datasets Dealing with Infra-Specific Diversity—cont'd

Dataset	Variance of Farms' Degree	Significance		Variance of Landraces' Degree	Significance		LBM Farm Cluster Number	LBM Landraces Cluster Number	Normalized First Singular Value	Significance	Evenness Difference Rich–Poor Farms	Significance		Crop–Rich Farms' Contribution	Significance		Crop–Poor Farms' Contribution	Significance	
		Left	Right		Left	Right				Right		Left	Right		Left	Right		Left	Right
DJ-M036	0.637	<0.05	NS	13.24	NS	<0.001	1	2	1.45	NS	0.198	NS	<0.001	2.328	NS	<0.05	1.84	<0.001	NS
DJ-M039a	0.928	NS	NS	0.94	NS	NS	1	1	0.25	NS	0.234	NS	NS	2.236	NS	NS	1.975	NS	NS
DJ-M045b	0.205	<0.001	NS	0.736	NS	NS	1	1	0.03	NS	0.08	NS	NS	2.384	NS	NS	2.321	NS	NS
DJ-M045c	0.746	NS	NS	0.491	NS	NS	1	1	0.45	NS	0.211	NS	NS	2.377	NS	NS	2.221	NS	NS
DJ-M045d	0.357	<0.05	NS	2.295	NS	<0.05	1	1	0.8	NS	0.356	NS	NS	2.099	NS	NS	1.679	NS	NS
JW-M07	0.853	NS	NS	10.789	NS	<0.001	1	3	0.51	NS	0.068	NS	NS	2.644	<0.05	NS	2.465	NS	NS
JW-M08	0.536	<0.01	NS	15.333	NS	<0.001	1	4	0.74	NS	0.115	NS	NS	2.663	NS	NS	2.334	<0.05	NS
JW-M09	0.816	NS	NS	15.148	NS	<0.001	1	4	0.92	NS	0.043	NS	NS	2.873	NS	NS	2.736	NS	NS
JW-M10	0.949	NS	NS	5.73	NS	<0.001	1	2	1.68	NS	0.053	NS	NS	2.76	NS	NS	2.587	NS	NS
ME-M01	3.9	NS	<0.001	7.395	NS	<0.001	2	3	0.84	NS	0.112	NS	NS	3.997	NS	NS	3.649	NS	NS
SC-M04	8.172	NS	<0.001	3.842	NS	<0.001	2	2	0.45	NS	0.099	NS	NS	3.454	NS	NS	3.216	NS	NS
SC-M07	3.335	NS	<0.001	2.998	NS	<0.001	2	2	0.6	NS	0.126	NS	NS	3.188	NS	NS	3.021	NS	<0.001

NS: non-significant, <0.05: *p*-values ranging from 0.01 to 0.05, <0.01: *p*-values ranging from 0.001 to 0.01 and <0.001: *p*-values lower than 0.001.

one case where the test was not significant on both sides (SC-M05) and one case where the test was rejected on the left side (CL-M01). These results indicate that the number of species grown per farm from the same village is generally over-dispersed, with few farms growing more species than expected. For the variability in degree of species, this pattern was even stronger, with a systematically over-dispersed degree distribution.

4.1.2 Infra-Species Diversity

For farms at the infra-specific level, the pattern is completely different as H_0 is rejected on the right side only 3 times over the 32 tested datasets (ME-M01, SC-M04 and SC-M07), and 11 times on the left side (Table 4). These results indicate an under-dispersion of the degree distribution when we consider the distribution of landraces at the village scale. For degree of landraces, H_0 is mostly rejected on the right side with 29 times over the 32 datasets, indicating, as for the species level, an over-dispersion of the degree distribution.

4.2 Structure Detection Through Model-Based Clustering (LBM)

In this section, we seek to detect the existence of patterns within inventory datasets at the village scale using LBM as explained in Section 3.3.

4.2.1 Species Diversity

The clustering method applied to the different datasets detected from one to three clusters for the farms and from two to three clusters for the species (Table 5 and Fig. A1). These results are similar to the toy example illustrated in Fig. 7. Therefore, the clustering seems mostly driven by the heterogeneity in degree of both farms and species. Farms were clustered together because they grow almost the same species. In the case of two clusters for farms, we then define the 'crop–poor' farm cluster as the one with the lower density and the 'crop–rich' farm cluster as the one with the higher density. In the case of two groups for the species, we define the 'rare species' cluster as the one with the lower number of links and the frequent species cluster as the one with the higher number of links.

4.2.2 Infra-Species Diversity

The clustering method detected from one to two clusters for the farms and from one to four clusters for landraces (Table 6 and Figs. A2 and A3). For four datasets (DJ-M039a, DJ-M045b, DJ-M045c and DJ-M045d), only one cluster was detected both for farms and landraces (Table 4). These results of low clustering are consistent with the low variability of the degrees both for the farms and the landraces observed in Section 4.1. Similarly, 26 additional datasets with under-dispersion had only one block for the farms. These findings indicate that for landrace diversity, a lower heterogeneity is generally observed among farms where nearly the same landraces are grown. Only three datasets showed two blocks for the farms (ME-M01, SC-M04 and SC-M07). Nevertheless, it is still possible to distinguish between frequent and rarer landraces.

4.3 Outlier Detection Through PCA

We then applied a PCA to detect farms that are 'outliers' in terms of species and infra-species diversity. See Section 3.4 for methodological details.

4.3.1 *Species Diversity*

Using the test introduced in Section 3.4.3, H_0 was rejected 9 times over the 18 datasets at $\alpha = 0.05$, highlighting the existence of outliers. These outliers are generally two or three farms per dataset (Fig. A4), which can be characterized as farms where a different subset of species is grow compared to other farms with an equivalent degree, belonging to the same cluster.

4.3.2 *Infra-Species Diversity*

H_0 was rejected for 4 datasets over the 32 datasets (CL-M02, DJ-M018a, DJ-M018b and DJ-M030, Fig. A5). These results indicate that in addition to growing almost the same number of landraces, the same portfolio of landraces is grown globally by all farms from the same village. Note that for these four datasets, only one cluster was detected with the LBM (CL-M02, DJ-M018a, DJ-M018b and DJ-M030). Therefore, in this case, we have farms with a particular subset of landraces and having an equivalent degree.

4.4 Farms' Contributions to Local Diversity

In the analyses reported in this section, farms were separated into 'crop–rich' farms and 'crop–poor' farms according to their degree in such a way that arbitrarily 40% of farms were classified as 'crop–rich'. The method described below is not highly sensitive to this threshold value, except for extreme values.

Evenness (E) and contribution (H_β) were computed for each of these two groups as explained in Section 3.5.

4.4.1 Species Diversity

The tests on the difference between E_{rich} and E_{poor} revealed that crop–rich farms had a significantly higher evenness in five cases (CL-M01, OC-M04, OC-M07, OC-M11 and OC-M12). The group of crop–poor farms contributed significantly more than that of crop–rich farms in only one case (EG-M08). H_0 was not rejected in the other cases, indicating that no significant difference in terms of contribution to the global diversity by the crop–rich group of farms compared to the crop–poor group.

Our findings on the difference between E_{rich} and E_{poor} converge with the test of the contributions of crop–rich and crop–poor farms. Indeed, in five cases when the first test was significant on the right side (i.e. a significantly higher contribution to the global diversity by the crop–rich farms than the crop–poor farms), we observed that some crop–rich groups did indeed contribute significantly to the global diversity and that some crop–poor groups contributed significantly less than expected in four of the five cases (Table 5). Two additional datasets showed a significant contribution of the crop–rich farms (OC-M14 and SC-M05) and one additional dataset showed that the crop–poor farms contributed significantly less than expected (OC-M13). The crop–poor farms contributed significantly more than expected in only two cases. In one of these cases (EG-M05), the result is consistent with that of the test on evenness. In the other case (EG-M08), crop–poor farms only showed a significant contribution to global diversity and not to evenness (EG-M08).

4.4.2 Infra-Species Diversity

The tests of the difference between E_{rich} and E_{poor} farms revealed that crop–rich farms had a significantly higher evenness in six cases (Table 6; DJ-M003a, DJ-M012a, DJ-M012b, DJ-M018a, DJ-M030 and DJ-M036). H_0 was not rejected in the other cases, indicating no significant difference in evenness between crop–rich and crop–poor farms. These results were not always convergent with the results of the tests on the contributions of the crop–rich and crop–poor farm groups to diversity at the village level. These latter tests gave convergent results (a significant contribution of a few crop–rich farms to the global diversity) in only two cases (DJ-M018a and DJ-M036) of the six in which the evenness difference was significant. In one additional dataset, few farms from the crop–rich group contributed significantly less than expected (JW-M07). In one additional dataset, the crop–poor farms contributed significantly more than expected (DJ-M012a). In three additional datasets, few farms from the crop–poor group contributed significantly less than expected (CV-M02, DJ-M030 and JW-M08). Finally, in two datasets, the crop–poor farms contributed significantly more than expected (DJ-M0015a and SC-M07).

5. DISCUSSION

5.1 Contrasted Patterns of Local Crop Diversity at the Species and Infra-Species Levels

Applying a set of network-based methods to a meta-dataset of crop diversity reveals distinct sources of heterogeneity in terms of crop distribution at the local scale:

i. crop diversity among farms is generally more heterogeneous at the specific level than at the infra-specific level;
ii. heterogeneity in farms' degrees is one explanation for this heterogeneity, with blocks of low-diversity farms and of high-diversity farms (the same pattern is observed for species and landraces, with blocks of common crops and blocks of rarer crops);
iii. outlier farms with unusual portfolios are another source of heterogeneity and
iv. both low-diversity or high-diversity farms can contribute disproportionately to local diversity by growing rare varieties.

We suggest two main explanations for these general results: heterogeneity in data collection methods and diversity of socio-ecological and environmental contexts. As datasets were collected following different protocols, differences in sampling effort could have an influence on the observed diversity (Perrault-Archambault and Coomes, 2008). An additional source of heterogeneity exists specially at the infra-specific level in the way in which landraces are named and how they are grouped together when they show strong evidence of being the same biocultural object. Nevertheless, a subset of the datasets for landraces were collected in the context of a coordinated global partnership of researchers in order to use a standardized protocol and the same sampling strategy during data collection (Jarvis et al., 2008), and datasets collected in this context also show different patterns (Table 6: DJ-M012a, DJ-M012b, DJ-M015a, DJ-M015b, DJ-M018a, DJ-M018b, DJ-M030, DJ-M036, DJ-M039a, DJ-M045b, DJ-M045c and DJ-M045d). Consequently, variation in the agro-ecological and the socio-cultural contexts, and interactions therein, is likely to have strongly shaped the distribution of local crop diversity.

More specifically, our findings of over-dispersion of the degrees at the specific level and of under-dispersion at the infra-specific level are strengthened by the results of classification using LBM. Indeed, in the cases of over dispersion, two or three blocks of farms are detected whereas for cases of under-dispersion, only one block of farms is detected. Convergence of the results between these two approaches indicates that the variability of the degree distribution is probably the main driver of block structure. It thus makes sense to use as null model a configuration model, controlling for degree, because this would allow assessment of whether other structural drivers, in addition to the degree, act to shape the patterns of diversity. From an ethnobiological or agro-ecological point of view, the block detection means that farms can be distinguished according to the level of diversity they grow. We identify high-diversity and low-diversity farms. Similarly, for crops, we identify common species/landrace (present in fields of most farms) and rare species/landraces (grown on few farms). Such patterns in terms of distribution of local crop diversity are quite common in the literature and consistent with the findings of Jarvis et al. (2008), who found that growing area and landrace diversity are related, and similar to those of Zimmerer (1991) for the distribution of potato biodiversity in Andean Peru.

From an ethnobiological point of view, our findings reflect the differing ways of managing specific (crop species) and infra-specific crop diversity (landraces). Growing numerous species is more complicated than growing numerous landraces, for several reasons. First, each species has its specific needs in terms of soil quality and preparation, sowing date, quantity of labour required and when it is required (Garine and Raimond, 2005). Among landraces of the same species, these needs are not so divergent. Farmers possessing a relatively large land area have more chance to encounter different soil types and quality among their fields. Also, larger farms or those with an extensive social network can expect to have an adequate labour supply (Abizaid et al., 2015) to grow a large portfolio of species (Garine and Raimond, 2005). Thus, farmers with more assets, including social capital and labour, tend to cultivate larger and more numerous fields and have greater crop diversity (Alvarez et al., 2005; Coomes and Ban, 2004; Zimmerer, 1991). Smallholder poverty may limit the diversity of crops that can be raised. Previous studies concluded that certain species are needed to meet basic needs (e.g. food, medicinal, etc.) and other species are more optional, reflected by higher levels of infra-specific diversity for staples compared to other crops (Jarvis et al., 2008), especially under stressful abiotic conditions (Labeyrie et al., 2013). Another possible explanation of the lower heterogeneity for degrees for landraces is that several landraces of the main species may be grown to fill diverse needs driven by cultural and dietary preferences, shifts in market demand and labour availability (Brush and Meng, 1998; Gauchan et al., 2005; Johns et al., 2013), heterogeneity in soil and water resources (Bellon and Taylor, 1993; Bisht et al., 2007), biotic stresses (Finckh and Wolfe, 2006) and the need to enhance pollination levels via outcrossing (Kremen et al., 2002). Much infra-specific diversity is held at the community level rather than within individual farms (Brush et al., 2015; Mulumba et al., 2012). In addition, in agroecosystems where many species are grown, farms maintaining collections of landraces will be few because less varietal diversity of the crop species is available to the farmer due to financial, social or policy constraints. Finally, the reason for a greater heterogeneity of crop diversity at the specific level compared to the infra-specific level may lie in the traits of the crop species considered in the analysis and their reproductive systems. In their broad comparison of nomenclature systems, Jarvis et al. (2008) showed that farmers use more detailed classifications for clonally reproduced crops than for self-fertilizing, partially self-fertilizing or outcrossing crops. This pattern was confirmed in our dataset. The only cases where over-distribution of farm degree was observed at the infra-specific level (ME-M01, SC-M04 and SC-M07) were all villages in which the staple food was provided by clonally propagated species (manioc and taro).

We applied additional tests to detect more detailed patterns in crop diversity within the meta-dataset and the sources of divergence in terms of crop portfolio composition. Our analysis of outliers identified certain farms that held unique portfolios of species or landraces. In most cases, it is the high-diversity farms that mainly contribute to the global diversity. These findings are consistent with the hypothesis of nestedness and of sink–source dynamics described in Alvarez et al. (2005) and Coomes (2010), and frequently postulated to be important, in the dynamics of local diversity, of one or a small number of experts or nodal farmers in a village (Boster, 1983; Kawa et al., 2013; Padoch and Jong, 1991; Peroni and Hanazaki, 2002; Perrault-Archambault and Coomes, 2008; Salick et al., 1997; Subedi et al., 2003; Tapia, 2000).

Nevertheless, it would be incorrect to say that this is a consistent tendency in the meta-dataset. Indeed, we observed the opposite relationship in other datasets whereby low-diversity farms contributed significantly to the local diversity (EG-M05, EG-M08, DJ-M015a and SC-M07). In some cases, one or a few farmers grew rarer species or landraces due to curiosity, for aesthetic reasons, or to maintain a social status of expert at the local level (Elias et al., 2000; Hawkes, 1983; Meilleur, 1998), or to have an object that others do not have (Coomes and Ban, 2004). Possessing a rare species or landrace might, for instance, allow a young farmer to distinguish himself from others to develop niche market (rare vegetables and tobacco) or for other non-economic reasons. Possessing an object that others do not could increase its potential transfer value to other members of the community (Caillon and Lanouguère-Bruneau, 2005). Additional factors influence the distribution of local crop diversity, for instance, the role played by differences associated with gender and generation, access to seed markets, farmers' food preferences and the market value of crops. Patterns of vertical transmission of seeds from mother-in-law to daughter-in-law (Delêtre et al., 2011; Labeyrie et al., 2013) or from father/mother-in-law to son-in-law (Wencélius and Garine, 2015) in patrilinear societies with virilocal rules of residence, i.e., where the son and his wife (wives) stay in the same village of the son's father, generation after generation, may constitute another source of divergence in crop diversity among families from the same village.

Considering now the village unit as a complex system, patterns of crop distribution at both the species and landrace levels are shaped by the self-organized action of the farmers, resulting from the sum of individual choices. This behaviour can be interpreted as a 'collective knowledge' that maintains crop diversity, to cope with multiple environmental and socio-cultural constraints and perturbations, and to maintain cultural cohesion through seed circulation (Emperaire and Peroni, 2007). These self-organized distributions of crop diversity are vulnerable, depending on their pattern and the type of perturbation. For instance, maintenance of crop diversity may be threatened if local crop diversity is concentrated in a few crop–rich farms, should a disaster happen. Local farmer populations can be expected to be more vulnerable to outbreaks and rapid spread of pests or pathogens when crop–rich and crop–poor farms from the same village both grow common species (used as staple food) or common landraces. Therefore, cultivating both common and rare landraces on the same farm increases farms' resilience in case of major pests affecting the most common landraces.

These multiple patterns of crop diversity raise particular concern about the issues around the conservation of crop diversity. By detecting how local diversity is distributed, our methods could help scientists involved in *ex situ* and *in situ* conservation programs to optimize their sampling strategies for plant collection and farmers involvement, respectively. In addition to the statistical methods developed in this chapter, LBM and PCA are visualizations derived from network data, and may serve as useful tools in communicate information about the distribution of crop diversity at the local scale with NGOs, politicians, farmers and all the stakeholders interested by crop management, as suggested by Pocock et al. (2016).

More generally, because these distinct patterns of crop diversity have been detected in different agro-ecological environments and socio-cultural contexts without controlling for other potential factors (and without additional information about each village), it is not yet possible to assess how one particular agro-ecological environment and socio-cultural context shapes the distribution of local crop diversity. Additional studies are needed in this direction to detect the local drivers influencing the observed distribution of crop diversity by collecting data to characterize specific and infra-specific diversity of crop plants and socio-cultural diversity of farmers. Such investigations will help us in understanding trade-offs between ecological and socio-cultural functions within agroecosystems.

5.2 Relevance of Network-Based Methods

The network-based methods introduced in this chapter provide a set of useful tools to analyse the distribution of local diversity in crop species and varieties. Indeed, our framework allowed us to answer four key questions:

1. Are farm and crop degrees more variable than expected under a null model which assumes a homogeneous probability of interaction between potential partners?
2. Are crop-by-farm interactions structured by blocks and, if so, what are the characteristics of these blocks?
3. Are certain crops or certain farms obvious outliers in their pattern of interactions?
4. Do crop–poor (low-degree) and crop–rich (high-degree) farms contribute significantly more or less than expected, based solely on knowledge of their crop–richness (degrees), to the overall diversity of crops cultivated locally?

By combining these different indices, tests and metrics, we provide a realistic and complete picture of the complex structure of crop diversity. This framework readily detected cases, for example, in which crop diversity is different in two different villages (through the LBMs) and identified farms—be they low-degree or high-degree farms—as unique and important providers of crop diversity (through uncovering of outliers in PCA and measures of uniqueness).

One strength of this framework is the use of a hierarchy of null models of increasing complexity. For instance, the simplest model for a bipartite network with variable degrees is the Erdős-Rényi $G(N,p)$ model in which interaction between nodes from the two different categories is restricted (each link has the same probability of occurring). Deviations from this null model allow assessment of degree heterogeneity or the presence of blocks (groups of farms that preferentially cultivate a certain group of species). When looking for more elaborate structures in the network (and not only in degree distributions), we relied on the configuration model, which randomizes interactions while keeping all degrees in the network constant. Consequently, one can disentangle whether the observed patterns, such as the block structure, are simply explained by the degree heterogeneity or are truly emergent properties. Furthermore, the above approaches (LBMs and PCA) provide visualization methods of the network highlighting its different characteristics, e.g., modules or outliers. These graphical representations are complementary to the more usual network representations reviewed in Pocock et al. (2016). It is important to note that our network-based approach can foster transdisciplinarity as it can be extended to datasets from other disciplines, including ecology, to detect particular patterns in bipartite networks (Mulder et al., 2015), especially with the outcome of next-generation sequencing techniques (Vacher et al., 2016). In ecology, the tests could efficiently supplement metrics that are routinely used, such as modularity or nestedness scores (Fortuna et al., 2010). Depending on the size of the dataset, LBMs can be as informative as traditional modularity-computing techniques (or even more informative) in finding underlying structures within bipartite datasets (Leger, 2015). Moreover, LBMs can also elucidate non-modular blocks, such as quasi-partite structures (i.e. when such structures are not exactly bi- or multi-partite but quite close to one of those) within a network. Of course, the power of all such methods depends heavily on the number of nodes in the network, but the application to ecological questions of the set of methods proposed here could readily generate much more informative descriptions of ecological networks than connectance, modularity and nestedness scores alone.

The approach used in this chapter does not rely on a direct estimation of nestedness, because the different methods available to compute nestedness do not converge (Fig. A6). However, the set of methods designed here to uncover the uniqueness of contributions to diversity of crop–rich and crop–poor farms actually provide complementary information on whether specialists interact preferentially with generalists, as assumed under a 'nested' scenario in ecology, or not. We thus suggest that this toolkit could be used as an alternative to the classical methods for detecting nestedness that are usually applied to ecological datasets (Podani and Schmera, 2012). For future use, the code is available at the following URL: http://netseed.cesab.org/.

From a methodological point of view, the configuration model must be accompanied by several caveats. Most prominently, the fact that the degrees of all nodes are constant makes the model highly constrained. Chung and Lu (2002a,b) developed a model that generated graphs with given expected degrees, relaxing the requirement that all samples of the model reproduce exactly the observed degrees. Degrees of networks sampled from this model are allowed to vary slightly around a fixed expected value. Interestingly, the Chung-Lu model has recently been extended into the so-called degree-corrected stochastic block model (Karrer and Newman, 2011) incorporating both degree-heterogeneity parameters as in the Chung-Lu model and a block structure as in the LBM. Such models would allow disentangling the farms' overall crop richness, as well as crop rarity, from the preferences of certain farms for specific groups of crops (block structure). Inference methods for this model have been developed recently (e.g. Lei and Rinaldo, 2014). However, the complexity of these models makes the estimation (and the computation of p-values) unreliable for small networks such as those considered in this study. Nevertheless, the Chung-Lu model and degree-correcting stochastic block models are promising directions for research on larger-scale ecological networks.

6. CONCLUSION

In this chapter, we develop new network-based indicators and statistical tests to characterize patterns of crop diversity at local scales. We applied this methodological framework to a meta-dataset from 10 countries containing inventory data at the specific or infra-specific level. Our results identify different sources of heterogeneity in local crop diversity:

i. diversity at the specific level is generally much more heterogeneous among farms compared to diversity at the infra-specific level;

ii. two or more groups of farms can be identified based on their unique crop richness and

iii. although diversity–rich farms often contribute most to global diversity, in some cases diversity–poor farms contribute equally with rare species and varieties.

This analysis reveals the absence of any general pattern of crop diversity distribution at the village level, indicating a strong dependence on agro-ecological and socio-cultural contexts. These results suggest that local communities adapt self-organized strategies to their growing contexts. Further empirical investigations are needed to disentangle the different drivers shaping crop diversity distribution, more particularly comparing the impacts of biological properties of crops (open-pollinated vs. self-pollinated crop, seed vs. cuttings, annual vs. perennial, etc.), of social organization of farmers (patrilinearity vs. matrilinearity, local community vs. community of practices), of agricultural policy and of diversity of ecological landscapes (open vs. closed systems). Our methodological framework provides a useful approach and an informative overview of patterns in the distribution of diversity. The toolkit developed and applied in this study offers an alternative approach to the classical methods of detecting nestedness, in both ethnographic and ecological datasets. More broadly, this methodological framework—which helps to detect patterns of crop distribution within local social organizations—enables the investigation of trade-offs between ecological and social functions of agroecosystems within a same analytical framework.

ACKNOWLEDGEMENTS

The authors gratefully acknowledge support from France's Foundation pour la Recherche sur la Biodiversité (FRB) that made possible NetSeed, an international collaboration of researchers studying farmer seed networks. Mathieu Thomas was provided with a six-month post-doctoral fellowship within this program. The Centre de Synthèse et d'Analyse sur la Biodiversité (CESAB) provided essential logistical support for regular workshops on the subject, enabling us to develop and mature both the theoretical and methodological aspects of our research. Additional support of the ongoing research collaboration through the MIRES and MADRES networks was provided by the following agencies: the Réseau National des Systèmes Complexes (RNSC), the Institut National de la Recherche Agronomique (INRA) and the Centre National de la Recherche Scientifique (CNRS). We are most grateful to the NETSEED researchers not directly included in the preparation of this chapter for the fruitful discussions in which they participated during the project's different meetings.

APPENDIX. LBM Representation

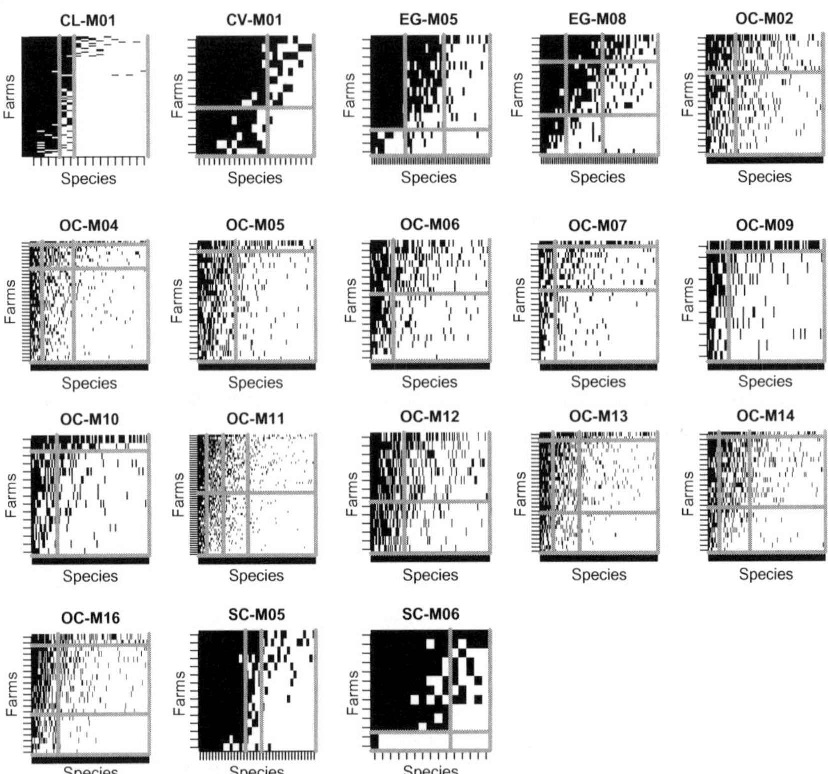

Figure A1 Representation of the incidence matrix for the 18 datasets collected at the specific level. The left panel corresponds to the original matrix without reordering, the right panel corresponds to the reordering based on block detection using the LBM method and density of the graph. The higher density is always on the top left side of the matrix.

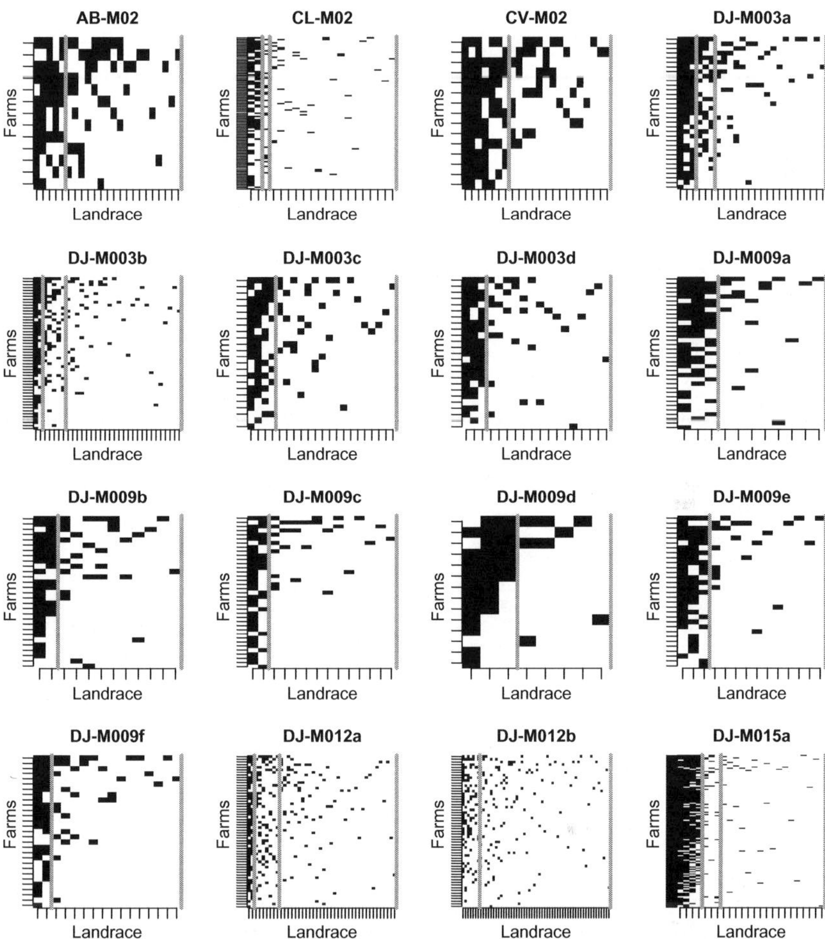

Figure A2 Representation of the incidence matrix for the 32 datasets collected at the infra-specific level. The left panel corresponds to the original matrix without reordering, the right panel corresponds to the reordering based on block detection using the LBM method and density of the graph. The higher density is always on the top left side of the matrix. Part 1.

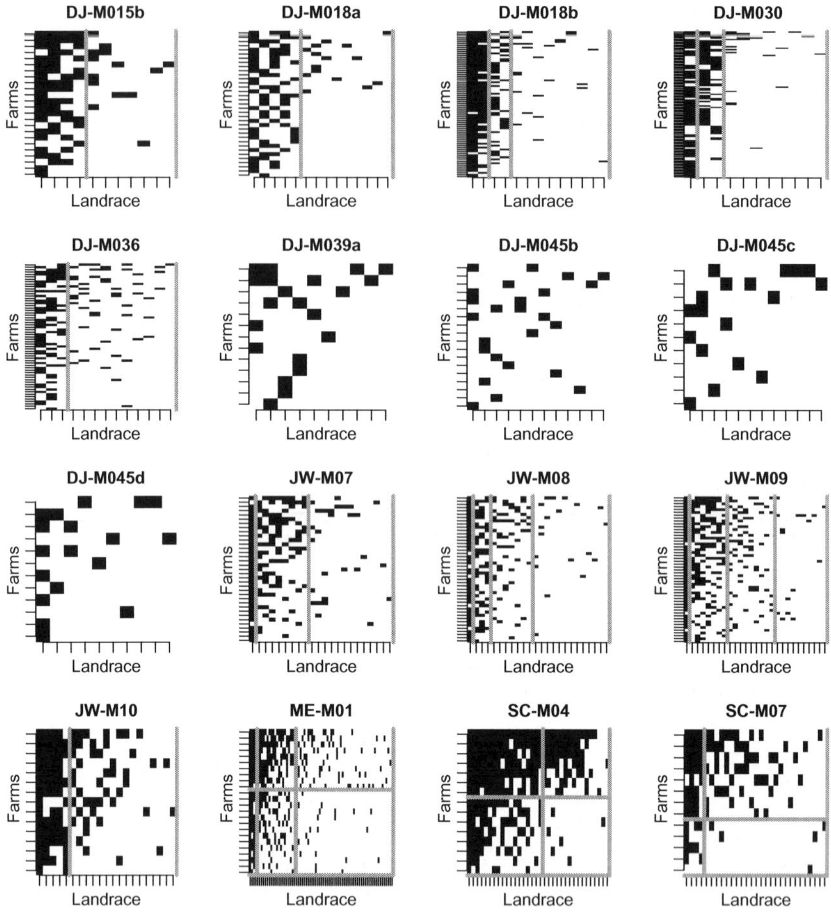

Figure A3 Representation of the incidence matrix for the 32 datasets collected at the infra-specific level. The left panel corresponds to the original matrix without reordering, the right panel corresponds to the reordering based on block detection using the LBM method and density of the graph. The higher density is always on the top left side of the matrix. Part 2.

Outlier Representation

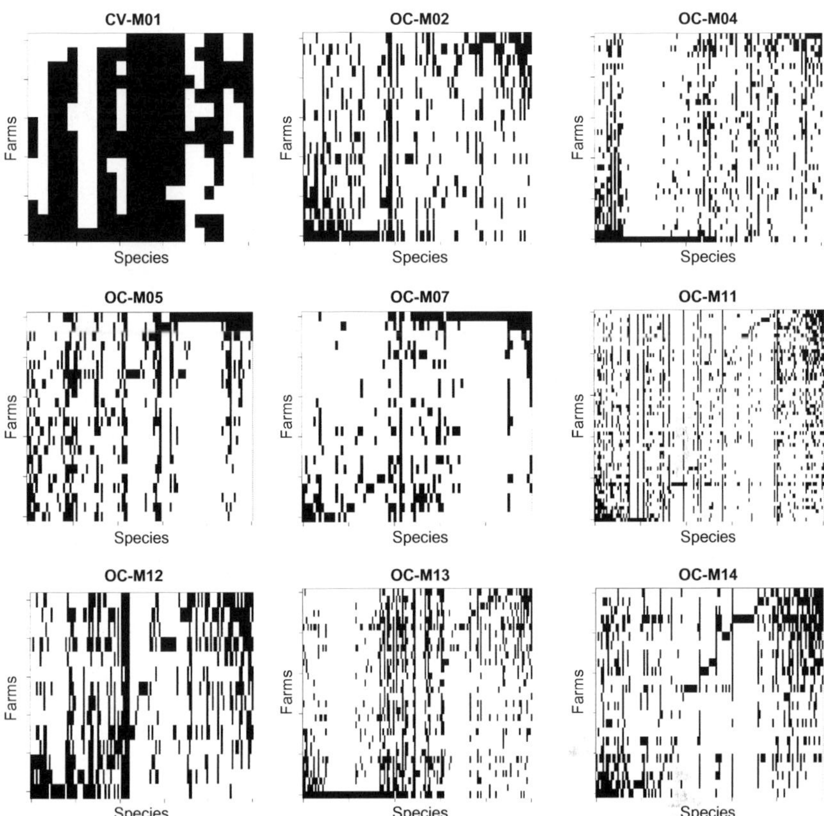

Figure A4 Representation of the PCA residuals on the nine datasets that yielded significant results at the specific level.

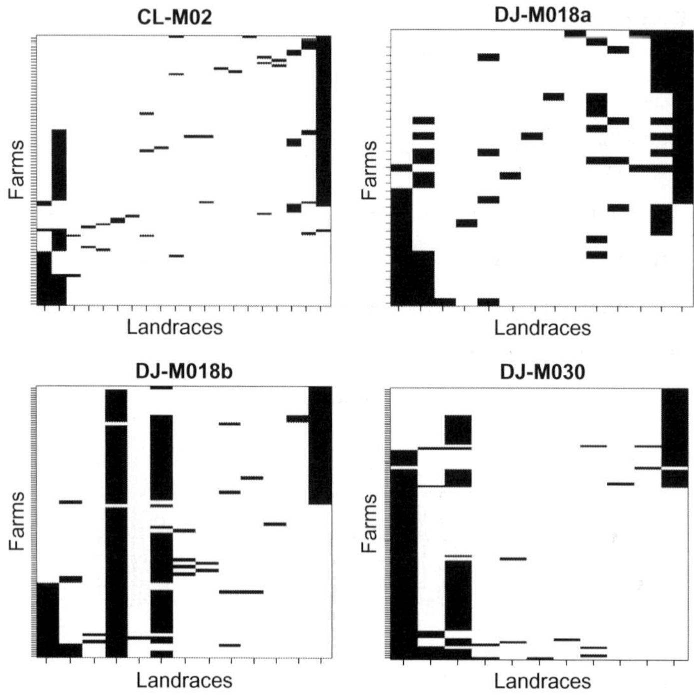

Figure A5 Representation of the PCA residuals on the four datasets that gave significant results at the infra-specific level.

Statistical Power Study of the Test Measuring the Impact of Crop-Rich and Crop-Poor Farms

The same model as in Section 3.5.4 is used for studying the behaviour of this test, introduced in Section 3.5.2. The three different settings of parameters correspond to an edge density of approximately 0.18. Thousand incidence matrices were simulated in each of the three settings with different incidence matrix sizes: $n=20$, 50 and $m=20$, 50.

Table A1 Estimated Probabilities of Rejection of the Null Hypothesis (in %) for the Different Contribution Tests Under the Three Toy Models for Two Alpha Levels, 1% and 5%, When: (a) $n=50, m=50$, (b) $n=50, m=20$, (c) $n=20, m=50$ and (d) $n=20, m=20$

	Fig. 12		Fig. 13		Fig. 14	
Alpha Level	**1.00%**	**5.00%**	**1.00%**	**5.00%**	**1.00%**	**5.00%**
(a)						
E.diff.pvalue.L	0.3	4.1	71	89.5	0.1	0.3
E.diff.pvalue.R	0.4	5.5	0	0	41	64.2
Hbeta.rich.pvalue.L	0.3	4.2	79.2	94.7	0.1	0.3
Hbeta.rich.pvalue.R	0.7	4.9	0	0	39.7	63.6
Hbeta.poor.pvalue.L	0.7	4.9	0	0	39.7	63.6
Hbeta.poor.pvalue.R	0.3	4.2	79.2	94.7	0.1	0.3
(b)						
E.diff.pvalue.L	0.9	4.9	17.2	38.2	0.09	2.6
E.diff.pvalue.R	0.1	5.5	0	0	4.7	14.1
Hbeta.rich.pvalue.L	1.3	4.7	20.3	43.3	0.9	3.1
Hbeta.rich.pvalue.R	0.5	5.3	0	0	5.3	14.8
Hbeta.poor.pvalue.L	0.5	5.3	0	0	5.3	14.8
Hbeta.poor.pvalue.R	1.3	4.7	20.3	43.3	0.9	3.1
(c)						
E.diff.pvalue.L	0.9	6.1	16.8	40	0.3	0.5
E.diff.pvalue.R	1.4	5.7	0	0.2	12.3	29.8
Hbeta.rich.pvalue.L	1.3	6	26.7	52.5	0.3	0.6
Hbeta.rich.pvalue.R	1.1	5.6	0	0	12.3	28.9
Hbeta.poor.pvalue.L	1.1	5.6	0	0	12.3	28.9
Hbeta.poor.pvalue.R	1.3	6	26.7	52.5	0.3	0.6
(d)						
E.diff.pvalue.L	0.7	4.6	5.5	19.1	0.9	3.8
E.diff.pvalue.R	1.1	4.6	0	1.4	2.5	7.7
Hbeta.rich.pvalue.L	0.8	4.6	8.4	22.8	0.7	3.8
Hbeta.rich.pvalue.R	0.8	5.5	0	1.3	2.1	7.4
Hbeta.poor.pvalue.L	0.8	5.3	0	1.3	2.1	7.4
Hbeta.poor.pvalue.R	0.8	4.6	8.5	23.2	0.7	3.8

Table A1 depicts the proportion of rejection (in %) as a function of incidence matrix size when the α-level is set to 1% and 5%. In Settings 2 and 3 (alternative hypothesis), the rejection probability exhibits the same pattern. When there are only $n=20$ farms or $m=20$ crops, the power is quite low, whereas for larger matrices ($n=m=50$), the power is greatly increased. Under the first setting without interaction between richness of the farms and the status of crops, the p-values are nearly uniformly distributed on [0,1]. These simulations confirm that our test is able to detect contrasted contribution to the diversity by 'crop–rich' and 'crop–poor' farms as long as the sample size is large enough.

Estimation of Nestedness

This section describes the nestedness results obtained on the meta-dataset using two methods: the temperature (Rodríguez-Gironés and Santamaría, 2006) and the NODF (Almeida-Neto et al., 2008). Figure A6 represents the p-values computed for each estimator after re-sampling using the configuration model introduced in Section 3.4.1. Our results are consistent with those of Podani and Schmera (2012), because for the same meta-dataset, tests performed with one or the other index were inconsistent.

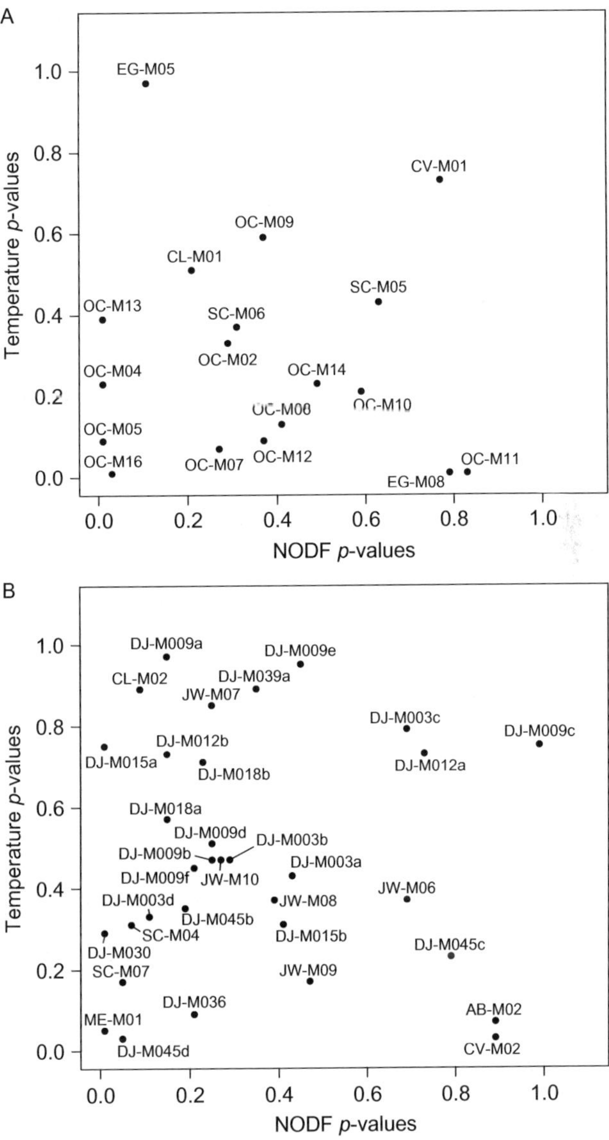

Figure A6 Plot representing on the *x*-axis the NODF *p*-values computed by re-sampling and on the *y*-axis the temperature *p*-values computed by re-sampling. In both cases, re-sampling was performed using the configuration model: (a) for datasets collected at the specific level and (b) for datasets collected at the infra-specific level.

GLOSSARY

Network a set of interconnected actors (human or non-human) and formally modelled by a graph.

Bipartite network a network whose nodes can be partitioned into two disjoint subsets (F to represent the farm and C to represent the crop: species/landraces) such that no edge connects two nodes from F or two nodes from C.

Node synonymous with 'vertex'. A node is the fundamental unit of which graphs are formed.

Edge an edge is a link between two nodes. Every edge has two endpoints in the sets of nodes. In the particular case of bipartite networks, the two endpoints belong to two disjoint subsets of nodes, e.g., farms (F) and crops (C, species or landraces). The presence of an edge indicates that the considered crop is cultivated on the considered farm.

Interaction network a network of nodes that are connected by features. In a crop-by-farm interaction network, crops are cultivated by farmers who are members of the farm.

Nestedness this concept, for which different indices have been devised, aims at quantifying the extent to which nodes of one subset (e.g. F) with low degrees are linked to nodes of the other subset (e.g. C) with high degrees. In the example of crop-by-farm networks, indices of nestedness aim to measure to what extent 'crop–poor' farms grow a subset of the crops cultivated on 'crop–rich' farms.

Degree the number of edges incident to a vertex. A farm's degree is the number of crops cultivated on the considered farm.

Configuration model a random graph model with a prescribed degree sequence. All graphs with this degree sequence obtained by permutation are equiprobable in this model (for details, see Section 3.4.1).

Graph a mathematical concept defined by a finite set of nodes (vertices) connected by edges (links).

Random graph model a generative model of graphs where the set of nodes is deterministic and the edges are drawn according to some probability distribution.

Erdős-Rényi model a random graph model in which all the edges are drawn independently with the same probability p.

Latent-block model a random graph model that assumes that the nodes belong to (unobserved) blocks and that the probability of connection between two nodes depends only on the blocks they belong to. This block structure can be estimated, allowing the clustering of nodes (farms or crops) based on similarities in terms of connectivity properties (see Section 3.3).

Incidence matrix 0/1 matrix **A**. Its rows are indexed by the set of farms F and its columns are indexed by the set of crops C. The entry A_{ij} equals one if and only if crop j is cultivated by farmers on farm i (see Section 3.1).

REFERENCES

Abizaid, C., Coomes, O.T., Takasaki, Y., Brisson, S., 2015. Social network analysis of peasant agriculture: cooperative labor as gendered relational networks. Prof. Geogr. 67, 447–463.

Almeida-Neto, M., Guimarães, P., Guimarães, P.R., Loyola, R.D., Ulrich, W., 2008. A consistent metric for nestedness analysis in ecological systems: reconciling concept and measurement. Oikos 117, 1227–1239.

Alvarez, N., Garine, E., Khasah, C., Dounias, E., Hossaert-McKey, M., McKey, D., 2005. Farmers' practices, metapopulation dynamics, and conservation of agricultural biodiversity on-farm: a case study of sorghum among the Duupa in sub-Sahelian Cameroon. Biol. Conserv. 121, 533–543.

Astegiano, J., Guimarães Jr., P.R., Cheptou, P.-O., Vidal, M.M., Mandai, C.Y., Ashworth, L., Massol, F., 2015. Persistence of plants and pollinators in the face of habitat loss: insights from trait-based metacommunity models. Adv. Ecol. Res. 53, 201–257.

Atmar, W., Patterson, B.D., 1993. The measure of order and disorder in the distribution of species in fragmented habitat. Oecologia 96, 373–382.

Bascompte, J., Jordano, P., 2007. Plant–animal mutualistic networks: the architecture of biodiversity. Annu. Rev. Ecol. Evol. Syst. 38, 567–593.

Bastolla, U., Fortuna, M.A., Pascual-García, A., Ferrera, A., Luque, B., Bascompte, J., 2009. The architecture of mutualistic networks minimizes competition and increases biodiversity. Nature 458, 1018–1020.

Bellon, M.R., Taylor, J.E., 1993. 'Folk' soil taxonomy and the partial adoption of new seed varieties. Econ. Dev. Cult. Chang. 41, 763–786.

Bianchi, F., Booij, C.J.H., Tscharntke, T., 2006. Sustainable pest regulation in agricultural landscapes: a review on landscape composition, biodiversity and natural pest control. Proc. R. Soc. B Biol. Sci. 273, 1715–1727.

Bickel, P.J., Sarkar, P., 2015. Hypothesis testing for automated community detection in networks. J. Roy. Stat. Soc. B Stat. Meth. (early view).

Bisht, I., Mehta, P., Bhandari, D., 2007. Traditional crop diversity and its conservation on-farm for sustainable agricultural production in Kumaon Himalaya of Uttaranchal State: a case study. Genet. Resour. Crop. Evol. 54, 345–357.

Bonneuil, C., Goffaux, R., Bonnin, I., Montalent, P., Hamon, C., Balfourier, F., Goldinger, I., 2012. A new integrative indicator to assess crop genetic diversity. Ecol. Indic. 23, 280–289.

Boster, J., 1983. A comparison of the diversity of Jivaroan gardens with that of the tropical forest. Hum. Ecol. 11, 47–68.

Boster, J.S., 1985. Selection for perceptual distinctiveness: evidence from Aguaruna cultivars of *Manihot esculenta*. Econ. Bot. 39, 310–325.

Brush, S.B., Meng, E., 1998. Farmers' valuation and conservation of crop genetic resources. Genet. Resour. Crop. Evol. 45, 139–150.

Brush, S.B., Bellon, M.R., Hijmans, R.J., Ramirez, Q.O., Perales, H.R., Etten, J.v., 2015. Assessing maize genetic erosion. Proc. Natl. Acad. Sci. U.S.A. 112, E1.

Butterfield, B.J., et al., 2016. Tradeoffs and compatibilities among ecosystem services: biological, physical and economic drivers of multifunctionality. Adv. Ecol. Res. 54, in press.

Caillon, S., Lanouguère-Bruneau, V., 2005. Gestion de l'agrobiodiversité dans un village de Vanua Lava (Vanuatu): stratégies de sélection et enjeux sociaux. J. Soc. Océanistes 120–121, 129–148.

Chung, F., Lu, L., 2002a. The average distances in random graphs with given expected degrees. Proc. Natl. Acad. Sci. U.S.A. 99, 15879–15882.

Chung, F., Lu, L., 2002b. Connected components in random graphs with given expected degree sequences. Ann. Comb. 6, 125–145.

Connor, E.F., Simberloff, D., 1979. The assembly of species communities: chance or competition? Ecology 60, 1132–1140.

Coomes, O.T., 2010. Of stakes, stems, and cuttings: the importance of local seed systems in traditional Amazonian societies. Prof. Geogr. 62, 323–334.

Coomes, O.T., Ban, N., 2004. Cultivated plant species diversity in home gardens of an Amazonian peasant village in Northeastern Peru. Econ. Bot. 58, 420–434.

Crowder, D.W., Northfield, T.D., Strand, M.R., Snyder, W.E., 2010. Organic agriculture promotes evenness and natural pest control. Nature 466, 109–112.

Delêtre, M., McKey, D.B., Hodkinson, T., 2011. Marriage exchanges, seed exchanges, and the dynamics of manioc diversity. Proc. Natl. Acad. Sci. U.S.A. 108, 18249–18254.

Dempster, A.P., Laird, N.M., Rubin, D.B., 1977. Maximum likelihood from incomplete data via the EM algorithm. J. R. Stat. Soc. Ser. B 39, 1–38.

Di Falco, S., Perrings, C., 2005. Crop biodiversity, risk management and the implications of agricultural assistance. Ecol. Econ. 55, 459–466.

Diamond, J., 2002. Evolution, consequences and future of plant and animal domestication. Nature 418, 700–707.

Drinkwater, L.E., Wagoner, P., Sarrantonio, M., 1998. Legume-based cropping systems have reduced carbon and nitrogen losses. Nature 396, 262–265.

Elias, M., Rival, L., Mckey, D., 2000. Perception and management of cassava (*Manihot esculenta* Crantz) diversity among Makushi Amerindians of Guyana (South America). J. Ethnobiol. 20, 239–265.

Emperaire, L., Peroni, N., 2007. Traditional management of agrobiodiversity in Brazil: a case study of manioc. Hum. Ecol. 35, 761–768.

Finckh, M.R., Wolfe, M.S., 2006. Diversification strategies. In: Cooke, B.M., Jones, D., Gareth, B.K. (Eds.), The Epidemiology of Plant Diseases. Springer, Netherlands, pp. 269–307.

Fortuna, M.A., Stouffer, D.B., Olesen, J.M., Jordano, P., Mouillot, D., Krasnov, B.R., Poulin, R., Bascompte, J., 2010. Nestedness versus modularity in ecological networks: two sides of the same coin? J. Anim. Ecol. 79, 811–817.

Fraser, J., Alves-Pereira, A., Junqueira, A., Peroni, N., Clement, C., 2012. Convergent adaptations: bitter manioc cultivation systems in fertile anthropogenic dark earths and floodplain soils in Central Amazonia. PLoS One 7, e43636.

Garine, E., Raimond, C., 2005. La culture intensive fait-elle disparaître la biodiversité? In: Dynamique de la Biodiversité et Modalité d'Accès aux Milieux et aux Ressources. Institut Français de la Biodiversité, Paris, France, pp. 25–28.

Gauchan, D., Smale, M., Chaudhary, P., 2005. Market-based incentives for conserving diversity on farms: the case of rice landraces in Central Tarai, Nepal. Genet. Resour. Crop. Evol. 52, 293–303.

Gill, R.J., et al., 2016. Protecting an ecosystem service: approaches to understanding and mitigating threats to wild insect pollinators. Adv. Ecol. Res. 54, in press.

Govaert, G., Nadif, M., 2008. Block clustering with Bernoulli mixture models: comparison of different approaches. Comput. Stat. Data Anal. 52, 3233–3245.

Hawkes, J.G., 1983. The Diversity of Crop Plants. Harvard University Press, London, UK.

Heckler, S., Zent, S., 2008. Piaroa manioc varietals: hyperdiversity or social currency. Hum. Ecol. 36, 679–697.

Jackson, L.E., Pascual, U., Hodgkin, T., 2007. Utilizing and conserving agrobiodiversity in agricultural landscapes. Agric. Ecosyst. Environ. 121, 196–210.

Jarvis, D.I., Brown, A.H., Cuong, P.H., Collado-Panduro, L., Latournerie-Moreno, L., Gyawali, S., Tanto, T., Sawadogo, M., Mar, I., Sadiki, M., 2008. A global perspective of the richness and evenness of traditional crop-variety diversity maintained by farming communities. Proc. Natl. Acad. Sci. U.S.A. 105, 5326–5331.

Johns, T., Powell, B., Maundu, P., Eyzaguirre, P.B., 2013. Agricultural biodiversity as a link between traditional food systems and contemporary development, social integrity and ecological health. J. Sci. Food Agric. 93, 3433–3442.

Jordano, P., 1987. Patterns of mutualistic interactions in pollination and seed dispersal: connectance, dependence asymmetries, and coevolution. Am. Nat. 129, 657–677.

Jordano, P., Bascompte, J., Olesen, J.M., 2003. Invariant properties in coevolutionary networks of plant–animal interactions. Ecol. Lett. 6, 69–81.

Karrer, B., Newman, M.E., 2011. Stochastic blockmodels and community structure in networks. Phys. Rev. E 83, 016107.

Kawa, N.C., McCarty, C., Clement, C.R., 2013. Manioc varietal diversity, social networks, and distribution constraints in rural Amazonia. Curr. Anthropol. 54, 764–770.

Keribin, C., Brault, V., Celeux, G., Govaert, G., 2014. Estimation and selection for the latent block model on categorical data. Stat. Comput. 25, 1–16.

Kolaczyk, E.D., 2009. Statistical Analysis of Network Data: Methods and Models. Springer Science & Business Media, New York, USA.

Kremen, C., Williams, N.M., Thorp, R.W., 2002. Crop pollination from native bees at risk from agricultural intensification. Proc. Natl. Acad. Sci. U.S.A. 99, 16812–16816.

Labeyrie, V., Rono, B., Leclerc, C., 2013. How social organization shapes crop diversity: an ecological anthropology approach among Tharaka farmers of Mount Kenya. Agric. Hum. Values 31, 97–107.

Lande, R., 1996. Statistics and partitioning of species diversity, and similarity among multiple communities. Oikos 76, 5–13.

Lazega, E., Mounier, L., Snijders, T., Tubaro, P., 2012. Norms, status and the dynamics of advice networks: a case study. Soc. Networks 34, 323–332.

Leclerc, C., Coppens d'Eeckenbrugge, G., 2012. Social organization of crop genetic diversity. The G × E × S interaction model. Diversity 4, 1–32.

Leger, J.-B., 2015. Blockmodels: Latent and Stochastic Block Model Estimation by a 'V-EM' Algorithm. In: R package version 1.1.1.

Lei, J., Rinaldo, A., 2014. Consistency of spectral clustering in stochastic block models. Ann. Stat. 43, 215–237.

Macfadyen, S., Bohan, D.A., 2010. Crop domestication and the disruption of species interactions. Basic Appl. Ecol. 11, 116–125.

Mariac, C., Jehin, L., Saïdou, A.-A.A., Vigouroux, Y., 2011. Genetic basis of pearl millet adaptation along an environmental gradient investigated by a combination of genome scan and association mapping. Mol. Ecol. 20, 80–91.

Martin, J.F., Roy, E.E.D., Diemont, S., Ferguson, B.G., 2010. Traditional ecological knowledge (TEK): ideas, inspiration, and designs for ecological engineering. Ecol. Eng. 36, 839–849.

Meilleur, B.A., 1998. Clones within clones: cosmology and esthetics and Polynesian crop selection. Anthropologica 40, 71–82.

Miklós, I., Podani, J., 2004. Randomization of presence–absence matrices: comments and new algorithms. Ecology 85, 86–92.

Millennium Ecosystem Assessment (MEA), 2005. Ecosystems and Human Well-Being: Biodiversity Synthesis. World Resources Institute, Washington, DC, USA.

Mulder, C., et al., 2015. 10 years later: revisiting priorities for science and society a decade after the millennium ecosystem assessment. Adv. Ecol. Res. 53, 1–53.

Mulumba, J.W., Nankya, R., Adokorach, J., Kiwuka, C., Fadda, C., De Santis, P., Jarvis, D.I., 2012. A risk-minimizing argument for traditional crop varietal diversity use to reduce pest and disease damage in agricultural ecosystems of Uganda. Agric. Ecosyst. Environ. 157, 70–86.

Padoch, C., Jong, W.d., 1991. The house gardens of Santa Rosa: diversity and variability in an Amazonian agricultural system. Econ. Bot. 45, 166–175.

Peroni, N., Hanazaki, N., 2002. Current and lost diversity of cultivated varieties, especially cassava, under swidden cultivation systems in the Brazilian Atlantic Forest. Agric. Ecosyst. Environ. 92, 171–183.

Perrault-Archambault, M., Coomes, O.T., 2008. Distribution of agrobiodiversity in home gardens along the Corrientes River, Peruvian Amazon. Econ. Bot. 62, 109–126.

Pocock, M.J.O., et al., 2016. The visualisation of ecological networks, and their use as a tool for engagement, advocacy and management. Adv. Ecol. Res. 54, in press.

Podani, J., Schmera, D., 2012. A comparative evaluation of pairwise nestedness measures. Ecography 35, 889–900.

Reyes-García, V., Molina, J.L., Calvet-Mir, L., Aceituno-Mata, L., Lastra, J.J., Ontillera, R., Parada, M., Pardo-de Santayana, M., Rigat, M., Vallès, J., 2013. 'Tertius gaudens': germplasm exchange networks and agroecological knowledge among home gardeners in the Iberian Peninsula. J. Ethnobiol. Ethnomed. 9, 1–11.

Rival, L., McKey, D., 2008. Domestication and diversity in manioc (*Manihot esculenta* Crantz ssp. *esculenta*, Euphorbiaceae). Curr. Anthropol. 49, 1119–1128.

Rodríguez-Gironés, M.A., Santamaría, L., 2006. A new algorithm to calculate the nestedness temperature of presence–absence matrices. J. Biogeogr. 33, 924–935.

Salick, J., Cellinese, N., Knapp, S., 1997. Indigenous diversity of cassava: generation, maintenance, use and loss among the Amuesha, Peruvian upper Amazon. Econ. Bot. 51, 6–19.

Samberg, L., Shennan, C., Zavaleta, E., 2013. Farmer seed exchange and crop diversity in a changing agricultural landscape in the southern highlands of Ethiopia. Hum. Ecol. 41, 477–485.

Smale, M., Hartell, J., Heisey, P.W., Senauer, B., 1998. The contribution of genetic resources and diversity to wheat production in the Punjab of Pakistan. Am. J. Agric. Econ. 80, 482–493.

Subedi, A., Chaudhary, P., Baniya, B.K., Rana, R.B., Tiwari, R.K., Rijal, D.K., Sthapit, B.R., Jarvis, D., 2003. Who maintains crop genetic diversity and how? Implications for on-farm conservation and utilization. Cult. Agric. 25, 41–50.

Tapia, M.E., 2000. Mountain agrobiodiversity in Peru. Mt. Res. Dev. 20, 220–225.

Thébault, E., Fontaine, C., 2010. Stability of ecological communities and the architecture of mutualistic and trophic networks. Science 329, 853–856.

Tuxill, J., Reyes, L.A., Moreno, L.L., Uicab, V.C., Jarvis, D.I., 2010. All maize is not equal: maize variety choices and Mayan foodways in rural Yucatan, Mexico. In: Staller, J., Carrasco, M. (Eds.), Pre-Columbian Foodways: Interdisciplinary Approaches to Food, Culture, and Markets in Ancient Mesoamerica. Springer, New York, USA.

Ulrich, W., Gotelli, N.J., 2012. A null model algorithm for presence–absence matrices based on proportional resampling. Ecol. Model. 244, 20–27.

Vacher, C., et al., 2016. Learning ecological networks from next-generation sequencing data. Adv. Ecol. Res. 54, in press.

Vigouroux, Y., Barnaud, A., Scarcelli, N., Thuillet, A.-C., 2011. Biodiversity, evolution and adaptation of cultivated crops. C. R. Biol. 334, 450–457.

Wasserman, S., Faust, K., 1994. Social Network Analysis: Methods and Applications. Cambridge University Press, Cambridge, UK.

Wencélius, J., Garine, É., 2015. Dans les sillons de l'alliance. Ethnographie de la circulation des semences de sorgho dans l'Extrême-Nord du cameroun. Les Cahiers d'Outre Mer 265, 93–116.

Zaman, A., Simberloff, D., 2002. Random binary matrices in biogeographical ecology—instituting a good neighbor policy. Environ. Ecol. Stat. 9, 405–421.

Zimmerer, K.S., 1991. Labor shortages and crop diversity in the Southern Peruvian Sierra. Geogr. Rev. 81, 414–432.

INDEX

Note: Page numbers followed by "*f*" indicate figures, "*t*" indicate tables, "*b*" indicate boxes and "*ge*" indicate glossary.

A

Abiotic factors
ecosystem function, 60–62
Agriculture
agro-ecological approach to, 261–262
in industrialized countries, 261–262
Agroecosystems, 261–262
species diversity, 301–302
Anthropogenic land-use change, 202–203
Anthropogenic stressors, 63*f*, 69–70
Aquatic DFWs, trophic levels in, 101*t*
Aquatic ecosystems, 99–100
ARtificial Intelligence for ESs (ARIES), 36–37

B

Baker's law, 207
BEF. *See* Biodiversity and ecosystem functioning (BEF)
Biodiversity
as driver of ecosystem functioning, 62–70
on ecosystem functioning, 182–183
and effects of species interactions, 65–68
vs. single ecosystem function, 166–168
Biodiversity and ecosystem functioning (BEF), 30*b*, 162–163
biodiversity *vs.* single EFs, 166–168
complementarity effects, 164–165
early intuition and establishing hypotheses, 164–166
and FWT
principles for integration, 181–184
research for management of multiple ESs, 184–187, 185*f*
limitations, 173–174
linking multiple functions and scaling of mechanisms, 168–173
linking multiple interactions with, 177–179
model, 164*f*, 181
multi-trophic, 165–166
principles, 170*t*
sampling effects, 164–165
selection effects, 164–165
topologies of, 181–182
Biodiversity–ecosystem functioning (B-EF), 56–57, 59–60, 65–66, 68
Biodiversity–ecosystem service (B-ES), 56–57, 60, 68
Bipartite incidence matrices, with determined degree sequences, 243–246
Bipartite networks, 264–265, 275
simplest model for, 305
topological properties of, 265
Breeding system
dispersal ability and, 212
plant, 206–208

C

Clean Water Act, 6–7
Complex system theory, 4–5
Configuration model
local crop biodiversity, 305–306
principal component analyses, 281–284
Consumers role, in ecosystem functions, 173–174
Crop diversity, 262–263. *See also* Local crop biodiversity
genetic study, 263–264
spatial distribution, 263–264

D

Detrital food webs (DFWs), 99–100
food-web topology metrics of, 137*t*
model, 103*f*
terrestrial, 99–100, 119*f*
trophic levels in aquatic and terrestrial, 101*t*
visualization of case study of, 140*f*
Dispersal ability
and breeding systems, 212
matters for metacommunity persistence, 231–233
variation in, 231–232

Diversity. *See also* Crop diversity; Species diversity
  and ecosystem functioning, 62–65
  phylogenetic, 23–24
Diversity effect, definition, 58*ge*
Donor-control model, 99–100

E

Eco-evolutionary dynamics, 19
Ecological network, 18, 120–121, 123–124, 178–179, 182–183
Ecological resilience, definition, 58*ge*, 73–74
Ecological stoichiometry, 104–105, 115–120, 124
Ecosystem
  aquatic, 99–100
  and societies, functional attributes and networks, 11–15, 12–13*f*
  terrestrial, 99–100, 105–106
    productivity in, 30*b*
Ecosystem function (EF) /ecosystem functioning, 98, 162–163
  abiotic factors, 60–62
  biodiversity and, 30*b*
    effects of species interactions, 65–68
  biodiversity as driver of, 62–70
  biodiversity on, 182–183
  biodiversity *vs.* single, 166–168
  definition, 58*ge*
  diversity and, 62–65
  predicting functional redundancy and outcomes for, 77–82
  role of consumers in, 173–174
  role of food webs in, 179–180
  species interactions affect, 183–184
Ecosystem multifunctionality, 169
Ecosystem process
  definition, 58*ge*
  description, 60–61, 182, 185–186
Ecosystem service (ES), 162–163
  BEF–FWT research for management of multiple, 184–187
  categories, 57*f*
  conceptual framework of, 6*f*
  coupling models to data, 37–38
  definition, 56–57, 58*ge*
  delivery, global connections role, 39
  management of, 82–83
  multitrophic interactions, 31*f*
  network approaches to, 15–22, 16*f*
  portfolio theory for management of, 76–77
  predominance of, 3–4
  and process, trophic levels, 101*t*
  species interactions and, 78*b*
  sustainable delivery of, 38–39
  'value' of services, prioritize, 35–37
Ecosystem service research
  data analysis, 131*t*
    data set, characteristics, 105–112, 107*t*, 109–111*f*
    data sources and selection criteria, 127–128
    futures, 124–127
    meta-analytic procedures, 129–133
    publication and experimental biases, 134–148
    quantifying predator effects, 128–129
  description, 121–124
  effects of categorical variables, 112–121, 114–115*t*, 116–117*f*
  potential biasing factors, 135*t*
EF. *See* Ecosystem function (EF)
Effect traits, 58*ge*, 62–64
Engineering resilience, definition, 58*ge*
Erdős-Rényi model, 266, 276, 305
ES. *See* Ecosystem service (ES)
Expectation–maximization (EM) algorithm, latent-block models, 279

F

Feeding behaviour, 179–180
Fishing down effect, 39–40
Food web. *See also* Detrital food webs (DFWs)
  energy and material fluxes through, 182–183
  network of empirical aquatic, 20*f*
  role in ecosystem functioning, 179–180
  structure, 176–177
    by single interaction types, 176–177
  topology metrics, of DFWs, 137*t*
Food web theory (FWT), 69–70, 163, 164*f*
  and BEF
    principles for integration, 181–184

research for management of multiple ES, 184–187, 185*f*
compartmentalization effects, 174–175
early intuition and establishing hypotheses, 174–176
food web structure, 176–177
interaction effects, 174–175
limitations, 179–180
linking multiple interactions, 177–179
models of, 181
principles, 170*t*
trophic structure effects, 174–175
Freshwater ecosystem, 99–100, 105–106
case study, 21*b*
FWT. *See* Food web theory (FWT)

G

General linear mixed models (GLMM), 129–130

H

Habitat fragmentation, 202–203
on local species persistence, 210–211
on plant–pollinator network, 210
traits modulating wild plant response to, 204–206
Habitat loss, 202–203
for higher dispersal and autonomous self-pollination, 234–236
plant-pollinator metacommunity persistence of, 224*f*, 226*f*
species persistence in, 210–212
Harbouring plant species, 231–233
Heterogeneity, temporal, 207–208
Human-dominated landscapes, pollination services in, 236–238, 241–242

I

Incidence matrix, 266–273
farms and crops columns, 274*f*, 275–276, 277*f*
latent-block models, 278–281, 280*f*
principal component analyses, 281–285, 284–285*f*
Infra-species diversity
local crop biodiversity, 297, 299–304
latent-block models, 295*t*, 297
principal component analyses, 298
Insurance effect hypothesis, 74–75
Integrated completed likelihood (ICL), latent-block models, 279
Inter-specific diversity, genetic study, 263–264

L

"Landscape-moderated" filtering, 235
Latent-block models (LBMs), 266, 273–274, 276–279
application to toy example, 279–281
block clustering, 280–281*f*
EM algorithm, 279
ICL criterion, 279
incidence matrix, 278–281, 280*f*
infra-species diversity, 295*t*, 297
network-based methods, 305–306
species diversity, 297
LBMs. *See* Latent-block models (LBMs)
Local crop biodiversity
Chung-Lu model, 306
collective knowledge, 303
during domestication, 261–262
farms contributions to, 298–300
genetic study, 263–264
infra-species diversity, 297, 299–304, 309–310*f*
latent-block models, 278–281
methodological framework, 273–274
crops/farms degree distribution, 275–278, 277–278*f*
incidence matrix, 274*f*
mathematical formalism, 274–275
in mutualistic interaction networks, 265
nestedness estimation, 314, 315*f*
network-based methods, 304–306
operational taxonomic units, 266–273, 267*t*, 269*t*
principal component analyses, 281–286
in social network analysis, 264–265
spatial distribution, 263–264
species diversity, 292–297, 294*t*, 299–304, 308*f*
statistical power study, 312, 313*t*

M

Makushi Amerindians of Guyana, 262–263
MEA. *See* Millennium Ecosystem Assessment (MEA)

Metacommunity, 70–72
mass-effect paradigms in, 71*f*
plant dispersal within, 228*f*
plant pollination generalization within, 230*f*
Metacommunity model, 211
trait-based, 210–212
caveats of, 239–241
construction of, 213–214
predictions of, 231
Metacommunity persistence
dispersal ability matters for, 231–233
plant–pollinator (*see* Plant–pollinator metacommunity persistence)
Metacommunity theory, 210–211
Meta-ecosystem, 58*ge*, 70–72
Microbial biomass, dynamical responses of, 103–104
Millennium Ecosystem Assessment (MEA), 56–57, 261–262
impact of, 7–11, 9–10*f*
implementation, 15–22
provisioning services, 28–29
regulating services, 26–28
research priorities one decade after, 22–39
supporting services, 29–32
underpinning knowledge: from functioning to services, 23–26
Multitrophic BEF, 165–166
Multitrophic interaction, ecosystem service, 31*f*

N

Nestedness
on distribution of degrees, 247–248
measurement, 246*b*
Nested structure, 208–209, 231–232, 236
Network approach
to ecosystem service, 15–22, 16*f*
Network connectance, 227, 236–237
Network theory, 17–22
Non-fragmented landscapes, 222*f*
Nutrient cycling, 99–100, 101*t*, 105

O

Operational taxonomic units (OTUs)
local crop biodiversity, 266–273, 267*t*, 269*t*
Outcrossing–disperser syndrome, 207–208

P

PCAs. *See* Principal component analyses (PCAs)
Phylogenetic diversity, 23–24
Plant breeding system, 206–208
Plant crop species, 261–262
Plant dispersal
coefficient of variation of, 219, 228*f*
within metacommunities, 228*f*
Plant dispersal ability
description, 207, 214
evolution of, 207–208
neutrality in, assuming, 223–227
Plant pollination generalization, within metacommunities, 230*f*
Plant–pollinator interactions, 208–210, 218
Plant–pollinator metacommunity persistence, 220–227, 221*f*
of habitat loss, 224*f*, 226*f*
in non-fragmented landscapes, 222*f*
Plant–pollinator network
constructing theoretical, 212–213
habitat fragmentation on, 210
organization, 208–210, 227
Plant species, harbouring, 231–233
Plant traits, 227–231
theoretical scenarios linking, 214–219, 216*f*
Pollination
generalization, dispersal ability and, 206–208
Pollination service
in human-dominated landscapes, 236–238, 241–242
Pollination syndrome, 30
Pollinator, 208–209
contrasting effects of lower dependence on, 238–239
density, 231–232
diversity, habitat loss and fragmentation effects on, 202–204
on species persistence, 238–239
Portfolio theory, ecosystem service management, 76–77
Predator–prey interactions, 174–175
Principal component analyses (PCAs), 266
configuration model, 281–283
crop rich–poor farms, 288–289
goodness-of-fit test, 282–283

incidence matrix, 281–285, 284–285*f*
infra-species diversity, 298
on residuals, 282, 311–312*f*
species diversity, 298
theoretical framework, 287–288
toy examples, 283–286, 284–286*f*
simulation model, 290
three contrasted, 290–292, 290–291*f*

R

Random graph model, 266
Resilience
aspects into portfolio theory, 76–77
cross-scale model of, 74–75
definition, 73–74
and stability of functioning, 72–76
Response traits, 58*ge*, 62–64, 63*f*
Restricted maximum likelihood (REML), 129–130

S

Self-pollination
habitat loss, 234–236
rates, 227, 229*f*, 231–232
SEM. *See* Structural equation modelling (SEM)
Soil
microbial-driven process in, 165–166
Spatiotemporal dimensions, metacommunities and meta-ecosystems, 70–72
Spatiotemporal scales, 72–76
Species diversity
agroecosystems, 301–302
local crop biodiversity, 292–297, 294*t*, 299–304
latent-block models, 297
principal component analyses, 298
Species interactions, 162–163, 164*f*
affect ecosystem functions, 183–184
BEF and, 164–174
biodiversity and effects of, 65–68
and ecosystem services, 78*b*
FWT and, 174–180
Species persistence
in fragmented landscapes, 208–210
habitat fragmentation on local, 210–211
in habitat loss, 210–212
pollinators on, 238–239
Species traits, 77
diversity and ecosystem functioning, 62–65, 63*f*
Structural equation modelling (SEM), 15–17
Sustainable delivery, of ecosystem service, 38–39

T

Temporal heterogeneity, 207–208
Terrestrial DFWs, 100–103, 119*f*
trophic levels in, 101*t*
Terrestrial ecosystems, 99–100, 105–106
productivity in, 30*b*
Trait-based metacommunity model, 210–212
caveats of, 239–241
construction, 213–214
predictions of, 231
Traits
modulation, wild plant response to habitat fragmentation, 204–206
plant, 227–231
response, 62–64, 63*f*
species (*see* Species traits)
theoretical scenarios linking plant, constructing, 214–219, 216*f*
Trophic interactions, 179–180
multi, 173–174

V

Van't Hoff's rule, 60–61

W

Water-soluble organic matter, 98
Wilderness Act, 6–7

ADVANCES IN ECOLOGICAL RESEARCH VOLUME 1–53

CUMULATIVE LIST OF TITLES

Aerial heavy metal pollution and terrestrial ecosystems, **11**, 218

Age determination and growth of Baikal seals (*Phoca sibirica*), **31**, 449

Age-related decline in forest productivity: pattern and process, **27**, 213

Allometry of body size and abundance in 166 food webs, **41**, 1

Analysis and interpretation of long-term studies investigating responses to climate change, **35**, 111

Analysis of processes involved in the natural control of insects, **2**, 1

Ancient Lake Pennon and its endemic molluscan faun (Central Europe; Mio-Pliocene), **31**, 463

Ant-plant-homopteran interactions, **16**, 53

Anthropogenic impacts on litter decomposition and soil organic matter, **38**, 263

Arctic climate and climate change with a focus on Greenland, **40**, 13

Arrival and departure dates, **35**, 1

Assessing the contribution of micro-organisms and macrofauna to biodiversity-ecosystem functioning relationships in freshwater microcosms, **43**, 151

A belowground perspective on Dutch agroecosystems: how soil organisms interact to support ecosystem services, **44**, 277

The benthic invertebrates of Lake Khubsugul, Mongolia, **31**, 97

Big data and ecosystem research programmes, **51**, 41

Biodiversity, species interactions and ecological networks in a fragmented world **46**, 89

Biogeography and species diversity of diatoms in the northern basin of Lake Tanganyika, **31**, 115

Biological strategies of nutrient cycling in soil systems, **13**, 1

Biomanipulation as a restoration tool to combat eutrophication: recent advances and future challenges, **47**, 411

Biomonitoring of human impacts in freshwater ecosystems: the good, the bad and the ugly, **44**, 1

Bray-Curtis ordination: an effective strategy for analysis of multivariate ecological data, **14**, 1

Body size, life history and the structure of host-parasitoid networks, **45**, 135
Breeding dates and reproductive performance, **35**, 69
Can a general hypothesis explain population cycles of forest Lepidoptera? **18**, 179
Carbon allocation in trees; a review of concepts for modeling, **25**, 60
Catchment properties and the transport of major elements to estuaries, **29**, 1
A century of evolution in *Spartina anglica*, **21**, 1
Changes in substrate composition and rate-regulating factors during decomposition, **38**, 101
The challenge of future research on climate change and avian biology, **35**, 237
Climate change and eco-evolutionary dynamics in food webs, **47**, 1
Climate change impacts on community resilience: evidence from a drought disturbance experiment **46**, 211
Climate change influences on species interrelationships and distributions in high-Arctic Greenland, **40**, 81
Climate influences on avian population dynamics, **35**, 185
Climatic and geographic patterns in decomposition, **38**, 227
Climatic background to past and future floods in Australia, **39**, 13
The climatic response to greenhouse gases, **22**, 1
Coevolution of mycorrhizal symbionts and their hosts to metal-contaminated environment, **30**, 69
Community genetic and competition effects in a model pea aphid system, **50**, 239
Communities of parasitoids associated with leafhoppers and planthoppers in Europe, **17**, 282
Community structure and interaction webs in shallow marine hardbottom communities: tests of an environmental stress model, **19**, 189
A complete analytic theory for structure and dynamics of populations and communities spanning wide ranges in body size, **46**, 427
Complexity, evolution, and persistence in host-parasitoid experimental systems with *Callosobruchus* beetles as the host, **37**, 37
Connecting the green and brown worlds: Allometric and stoichiometric predictability of above- and below-ground networks, **49**, 69
Conservation of the endemic cichlid fishes of Lake Tanganyika; implications from population-level studies based on mitochondrial DNA, **31**, 539
Constructing nature: laboratory models as necessary tools for investigating complex ecological communities, **37**, 333

Construction and validation of food webs using logic-based machine learning and text mining, **49**, 225
The contribution of laboratory experiments on protists to understanding population and metapopulation dynamics, **37**, 245
The cost of living: field metabolic rates of small mammals, **30**, 177
Decomposers: soil microorganisms and animals, **38**, 73
The decomposition of emergent macrophytes in fresh water, **14**, 115
Delays, demography and cycles; a forensic study, **28**, 127
Dendroecology; a tool for evaluating variations in past and present forest environments, **19**, 111
Determinants of density-body size scaling within food webs and tools for their detection, **45**, 1
Detrital dynamics and cascading effects on supporting ecosystem services, **53**, 97
The development of regional climate scenarios and the ecological impact of green-house gas warming, **22**, 33
Developments in ecophysiological research on soil invertebrates, **16**, 175
The direct effects of increase in the global atmospheric CO_2 concentration on natural and commercial temperate trees and forests, **19**, 2; **34**, 1
Distributional (In)congruence of biodiversity—ecosystem functioning, **46**, 1
The distribution and abundance of lake dwelling Triclads-towards a hypothesis, **3**, 1
DNA metabarcoding meets experimental ecotoxicology: Advancing knowledge on the ecological effects of Copper in freshwater ecosystems, **51**, 79
Do eco-evo feedbacks help us understand nature? Answers from studies of the trinidadian guppy, **50**, 1
The dynamics of aquatic ecosystems, **6**, 1
The dynamics of endemic diversification: molecular phylogeny suggests an explosive origin of the Thiarid Gastropods of Lake Tanganyika, **31**, 331
The dynamics of field population of the pine looper, *Bupalis piniarius* L. (Lep, Geom.), **3**, 207
Earthworm biotechnology and global biogeochemistry, **15**, 369
Ecological aspects of fishery research, **7**, 114
Eco-evolutionary dynamics of agricultural networks: implications for sustainable management, **49**, 339
Eco-evolutionary dynamics: experiments in a model system, **50**, 167

Eco-evolutionary dynamics of individual-based food webs, **45**, 225
Eco-evolutionary dynamics in a three-species food web with intraguild predation: intriguingly complex, **50**, 41
Eco-evolutionary dynamics of plant–insect communities facing disturbances: implications for community maintenance and agricultural management, **52**, 91
Eco-evolutionary interactions as a consequence of selection on a secondary sexual trait, **50**, 143
Eco-evolutionary spatial dynamics: rapid evolution and isolation explain food web persistence, **50**, 75
Ecological conditions affecting the production of wild herbivorous mammals on grasslands, **6**, 137
Ecological networks in a changing climate, **42**, 71
Ecological and evolutionary dynamics of experimental plankton communities, **37**, 221
Ecological implications of dividing plants into groups with distinct photosynthetic production capabilities, **7**, 87
Ecological implications of specificity between plants and rhizosphere microorganisms, **31**, 122
Ecological interactions among an Orestiid (Pisces: Cyprinodontidae) species flock in the littoral zone of Lake Titicaca, **31**, 399
Ecological studies at Lough Ine, **4**, 198
Ecological studies at Lough Hyne, **17**, 115
Ecology of mushroom-feeding Drosophilidae, **20**, 225
The ecology of the Cinnabar moth, **12**, 1
Ecology of coarse woody debris in temperate ecosystems, **15**, 133; **34**, 59
Ecology of estuarine macrobenthos, **29**, 195
Ecology, evolution and energetics: a study in metabolic adaptation, **10**, 1
Ecology of fire in grasslands, **5**, 209
The ecology of pierid butterflies: dynamics and interactions, **15**, 51
The ecology of root lifespan, **27**, 1
The ecology of serpentine soils, **9**, 225
Ecology, systematics and evolution of Australian frogs, **5**, 37
Ecophysiology of trees of seasonally dry Tropics: comparison among phonologies, **32**, 113
Ecosystems and their services in a changing world: an ecological perspective, **48**, 1
Effect of flooding on the occurrence of infectious disease, **39**, 107

Effects of food availability, snow, and predation on breeding performance of waders at Zackenberg, **40**, 325
Effect of hydrological cycles on planktonic primary production in Lake Malawi Niassa, **31**, 421
Effects of climatic change on the population dynamics of crop pests, **22**, 117
Effects of floods on distribution and reproduction of aquatic birds, **39**, 63
The effects of modern agriculture nest predation and game management on the population ecology of partridges (*Perdix perdix* and *Alectoris rufa*), **11**, 2
El Niño effects on Southern California kelp forest communities, **17**, 243
Empirically characterising trophic networks: What emerging DNA-based methods, stable isotope and fatty acid analyses can offer, **49**, 177
Empirical evidences of density-dependence in populations of large herbivores, **41**, 313
Endemism in the Ponto-Caspian fauna, with special emphasis on the Oncychopoda (Crustacea), **31**, 179
Energetics, terrestrial field studies and animal productivity, **3**, 73
Energy in animal ecology, **1**, 69
Environmental warming in shallow lakes: a review of potential changes in community structure as evidenced from space-for-time substitution approaches, **46**, 259
Environmental warming and biodiversity-ecosystem functioning in freshwater microcosms: partitioning the effects of species identity, richness and metabolism, **43**, 177
Estimates of the annual net carbon and water exchange of forests: the EUROFLUX methodology, **30**, 113
Estimating forest growth and efficiency in relation to canopy leaf area, **13**, 327
Estimating relative energy fluxes using the food web, species abundance, and body size, **36**, 137
Evolution and endemism in Lake Biwa, with special reference to its gastropod mollusc fauna, **31**, 149
Evolutionary and ecophysiological responses of mountain plants to the growing season environment, **20**, 60
The evolutionary ecology of carnivorous plants, **33**, 1
Evolutionary inferences from the scale morphology of Malawian Cichlid fishes, **31**, 377

Explosive speciation rates and unusual species richness in haplochromine cichlid fishes: effects of sexual selection, **31**, 235
Extreme climatic events alter aquatic food webs: a synthesis of evidence from a mesocosm drought experiment, **48**, 343
The evolutionary consequences of interspecific competition, **12**, 127
The exchange of ammonia between the atmosphere and plant communities, **26**, 302
Faster, higher and stronger? The Pros and Cons of molecular faunal data for assessing ecosystem condition, **51**, 1
Faunal activities and processes: adaptive strategies that determine ecosystem function, **27**, 92
Fire frequency models, methods and interpretations, **25**, 239
Floods down rivers: from damaging to replenishing forces, **39**, 41
Food webs, body size, and species abundance in ecological community description, **36**, 1
Food webs: theory and reality, **26**, 187
Food web structure and stability in 20 streams across a wide pH gradient, **42**, 267
Forty years of genecology, **2**, 159
Foraging in plants: the role of morphological plasticity in resource acquisitions, **25**, 160
Fossil pollen analysis and the reconstruction of plant invasions, **26**, 67
Fractal properties of habitat and patch structure in benthic ecosystems, **30**, 339
Free air carbon dioxide enrichment (FACE) in global change research: a review, **28**, 1
From Broadstone to Zackenberg: space, time and hierarchies in ecological networks, **42**, 1
From natural to degraded rivers and back again: a test of restoration ecology theory and practice, **44**, 119
Functional traits and trait-mediated interactions: connecting community-level interactions with ecosystem functioning, **52**, 319
The general biology and thermal balance of penguins, **4**, 131
General ecological principles which are illustrated by population studies of Uropodid mites, **19**, 304
Generalist predators, interactions strength and food web stability, **28**, 93
Genetic correlations in multi-species plant/herbivore interactions at multiple genetic scales: implications for eco-evolutionary dynamics, **50**, 263
Genetic and phenotypic aspects of life-history evolution in animals, **21**, 63

Geochemical monitoring of atmospheric heavy metal pollution: theory and applications, **18**, 65
Global climate change leads to mistimed avian reproduction, **35**, 89
Global persistence despite local extinction in acarine predator-prey systems: lessons from experimental and mathematical exercises, **37**, 183
Habitat isolation reduces the temporal stability of island ecosystems in the face of flood disturbance, **48**, 225
Heavy metal tolerance in plants, **7**, 2
Herbivores and plant tannins, **19**, 263
High-Arctic plant–herbivore interactions under climate influence, **40**, 275
High-Arctic soil CO_2 and CH_4 production controlled by temperature, water, freezing, and snow, **40**, 441
Historical changes in environment of Lake Titicaca: evidence from Ostracod ecology and evolution, **31**, 497
How well known is the ichthyodiversity of the large East African lakes? **31**, 17
Human and environmental factors influence soil faunal abundance-mass allometry and structure, **41**, 45
Human ecology is an interdisciplinary concept: a critical inquiry, **8**, 2
Hutchinson reversed, or why there need to be so many species, **43**, 1
Hydrology and transport of sediment and solutes at Zackenberg, **40**, 197
The Ichthyofauna of Lake Baikal, with special reference to its zoogeographical relations, **31**, 81
Impact of climate change on fishes in complex Antarctic ecosystems, **46**, 351
Impacts of warming on the structure and functioning of aquatic communities: individual- to ecosystem-level responses, **47**, 81
Implications of phylogeny reconstruction for Ostracod speciation modes in Lake Tanganyika, **31**, 301
Importance of climate change for the ranges, communities and conservation of birds, **35**, 211
Increased stream productivity with warming supports higher trophic levels, **48**, 285
Individual-based food webs: species identity, body size and sampling effects, **43**, 211
Individual variability: the missing component to our understanding of predator–prey interactions, **52**, 19
Individual variation decreases interference competition but increases species persistence, **52**, 45

Industrial melanism and the urban environment, **11**, 373
Individual trait variation and diversity in food webs, **50**, 203
Inherent variation in growth rate between higher plants: a search for physiological causes and ecological consequences, **23**, 188; **34**, 283
Insect herbivory below ground, **20**, 1
Insights into the mechanism of speciation in Gammarid crustaceans of Lake Baikal using a population-genetic approach, **31**, 219
Interaction networks in agricultural landscape mosaics, **49**, 291
Integrated coastal management: sustaining estuarine natural resources, **29**, 241
Integration, identity and stability in the plant association, **6**, 84
Intrinsic and extrinsic factors driving match–mismatch dynamics during the early life history of marine fishes, **47**, 177
Inter-annual variability and controls of plant phenology and productivity at Zackenberg, **40**, 249
Introduction, **38**, 1
Introduction, **39**, 1
Introduction, **40**, 1
Isopods and their terrestrial environment, **17**, 188
Lake Biwa as a topical ancient lake, **31**, 571
Lake flora and fauna in relation to ice-melt, water temperature, and chemistry at Zackenberg, **40**, 371
The landscape context of flooding in the Murray–Darling basin, **39**, 85
Landscape ecology as an emerging branch of human ecosystem science, **12**, 189
Late quaternary environmental and cultural changes in the Wollaston Forland region, Northeast Greenland, **40**, 45
Linking biodiversity, ecosystem functioning and services, and ecological resilience: Towards an integrative framework for improved management, **53**, 55
Linking spatial and temporal change in the diversity structure of ancient lakes: examples from the ecology and palaeoecology of the Tanganyikan Ostracods, **31**, 521
Litter fall, **38**, 19
Litter production in forests of the world, **2**, 101
Long-term changes in Lake Balaton and its fish populations, **31**, 601
Long-term dynamics of a well-characterised food web: four decades of acidification and recovery in the broadstone stream model system, **44**, 69

Macrodistribution, swarming behaviour and production estimates of the lakefly *Chaoborus edulis* (Diptera: Chaoboridae) in Lake Malawi, **31**, 431
Making waves: the repeated colonization of fresh water by Copepod crustaceans, **31**, 61
Manipulating interaction strengths and the consequences for trivariate patterns in a marine food web, **42**, 303
Manipulative field experiments in animal ecology: do they promise more than they can deliver? **30**, 299
Marine ecosystem regime shifts induced by climate and overfishing: a review for the Northern Hemisphere, **47**, 303
Mathematical model building with an application to determine the distribution of Durshan® insecticide added to a simulated ecosystem, **9**, 133
Mechanisms of microthropod-microbial interactions in soil, **23**, 1
Mechanisms of primary succession: insights resulting from the eruption of Mount St Helens, **26**, 1
Mesocosm experiments as a tool for ecological climate-change research, **48**, 71
Methods in studies of organic matter decay, **38**, 291
The method of successive approximation in descriptive ecology, **1**, 35
Meta-analysis in ecology, **32**, 199
Microbial experimental systems in ecology, **37**, 273
Microevolutionary response to climatic change, **35**, 151
Migratory fuelling and global climate change, **35**, 33
The mineral nutrition of wild plants revisited: a re-evaluation of processes and patterns, **30**, 1
Modelling interaction networks for enhanced ecosystem services in agroecosystems, **49**, 437
Modelling terrestrial carbon exchange and storage: evidence and implications of functional convergence in light-use efficiency, **28**, 57
Modelling the potential response of vegetation to global climate change, **22**, 93
Module and metamer dynamics and virtual plants, **25**, 105
Modeling individual animal histories with multistate capture–recapture models, **41**, 87
Mutualistic interactions in freshwater modular systems with molluscan components, **20**, 126
Mycorrhizal links between plants: their functioning and ecological significances, **18**, 243
Mycorrhizas in natural ecosystems, **21**, 171

The nature of species in ancient lakes: perspectives from the fishes of Lake Malawi, **31**, 39
Networking agroecology: Integrating the diversity of agroecosystem interactions, **49**, 1
A network-based method to detect patterns of local crop biodiversity: Validation at the species and infra-species levels, **53**, 259
Nitrogen dynamics in decomposing litter, **38**, 157
Nocturnal insect migration: effects of local winds, **27**, 61
Nonlinear stochastic population dynamics: the flour beetle *Tribolium* as an effective tool of discovery, **37**, 101
Nutrient cycles and H^+ budgets of forest ecosystems, **16**, 1
Nutrients in estuaries, **29**, 43
On the evolutionary pathways resulting in C_4 photosynthesis and crassulacean acid metabolism (CAM), **19**, 58
Origin and structure of secondary organic matter and sequestration of C and N, **38**, 185
Oxygen availability as an ecological limit to plant distribution, **23**, 93
Parasitism between co-infecting bacteriophages, **37**, 309
Persistence of plants and pollinators in the face of habitat loss: Insights from trait-based metacommunity models, **53**, 201
Scaling-up trait variation from individuals to ecosystems, **52**, 1
Temporal variability in predator–prey relationships of a forest floor food web, **42**, 173
The past as a key to the future: the use of palaeoenvironmental understanding to predict the effects of man on the biosphere, **22**, 257
Towards an integration of biodiversity–ecosystem functioning and food web theory to evaluate relationships between multiple ecosystem services, **53**, 161
Pattern and process of competition, **4**, 11
Permafrost and periglacial geomorphology at Zackenberg, **40**, 151
Perturbing a marine food web: consequences for food web structure and trivariate patterns, **47**, 349
Phenetic analysis, tropic specialization and habitat partitioning in the Baikal Amphipod genus *Eulimnogammarus* (Crustacea), **31**, 355
Photoperiodic response and the adaptability of avian life cycles to environmental change, **35**, 131
Phylogeny of a gastropod species flock: exploring speciation in Lake Tanganyika in a molecular framework, **31**, 273

Phenology of high-Arctic arthropods: effects of climate on spatial, seasonal, and inter-annual variation, **40**, 299
Phytophages of xylem and phloem: a comparison of animal and plant sapfeeders, **13**, 135
Population and community body size structure across a complex environmental gradient, **52**, 115
The population biology and Turbellaria with special reference to the freshwater triclads of the British Isles, **13**, 235
Population cycles in birds of the Grouse family (Tetraonidae), **32**, 53
Population cycles in small mammals, **8**, 268
Population dynamical responses to climate change, **40**, 391
Population dynamics, life history, and demography: lessons from *Drosophila*, **37**, 77
Population dynamics in a noisy world: lessons from a mite experimental system, **37**, 143
Population regulation in animals with complex life-histories: formulation and analysis of damselfly model, **17**, 1
Positive-feedback switches in plant communities, **23**, 264
The potential effect of climatic changes on agriculture and land use, **22**, 63
Predation and population stability, **9**, 1
Predicted effects of behavioural movement and passive transport on individual growth and community size structure in marine ecosystems, **45**, 41
Predicting the responses of the coastal zone to global change, **22**, 212
Predictors of individual variation in movement in a natural population of threespine stickleback (*Gasterosteus aculeatus*), **52**, 65
Present-day climate at Zackenberg, **40**, 111
The pressure chamber as an instrument for ecological research, **9**, 165
Primary production by phytoplankton and microphytobenthos in estuaries, **29**, 93
Principles of predator-prey interaction in theoretical experimental and natural population systems, **16**, 249
The production of marine plankton, **3**, 117
Production, turnover, and nutrient dynamics of above and below ground detritus of world forests, **15**, 303
Quantification and resolution of a complex, size-structured food web, **36**, 85
Quantifying the biodiversity value of repeatedly logged rainforests: gradient and comparative approaches from borneo, **48**, 183
Quantitative ecology and the woodland ecosystem concept, **1**, 103

Realistic models in population ecology, **8**, 200
References, **38**, 377
The relationship between animal abundance and body size: a review of the mechanisms, **28**, 181
Relative risks of microbial rot for fleshy fruits: significance with respect to dispersal and selection for secondary defence, **23**, 35
Renewable energy from plants: bypassing fossilization, **14**, 57
Responses of soils to climate change, **22**, 163
Rodent long distance orientation ("homing"), **10**, 63
The role of body size in complex food webs: a cold case, **45**, 181
The role of body size variation in community assembly, **52**, 201
Scale effects and extrapolation in ecological experiments, **33**, 161
Scale dependence of predator-prey mass ratio: determinants and applications, **45**, 269
Scaling of food-web properties with diversity and complexity across ecosystems, **42**, 141
Scaling from traits to ecosystems: developing a general trait driver theory via integrating trait-based and metabolic scaling theories, **52**, 249
Secondary production in inland waters, **10**, 91
Seeing double: size-based and taxonomic views of food web structure, **45**, 67
The self-thinning rule, **14**, 167
Shifts in the Diversity and Composition of Consumer Traits Constrain the Effects of Land Use on Stream Ecosystem Functioning, **52**, 169
A simulation model of animal movement patterns, **6**, 185
Snow and snow-cover in central Northeast Greenland, **40**, 175
Soil and plant community characteristics and dynamics at Zackenberg, **40**, 223
Soil arthropod sampling, **1**, 1
Soil diversity in the Tropics, **21**, 316
Soil fertility and nature conservation in Europe: theoretical considerations and practical management solutions, **26**, 242
Solar ultraviolet-b radiation at Zackenberg: the impact on higher plants and soil microbial communities, **40**, 421
Some economics of floods, **39**, 125
Spatial and inter-annual variability of trace gas fluxes in a heterogeneous high-Arctic landscape, **40**, 473
Spatial root segregation: are plants territorials? **28**, 145

Species abundance patterns and community structure, **26**, 112
Stochastic demography and conservation of an endangered perennial plant (*Lomatium bradshawii*) in a dynamic fire regime, **32**, 1
Stomatal control of transpiration: scaling up from leaf to regions, **15**, 1
Stream ecosystem functioning in an agricultural landscape: the importance of terrestrial–aquatic linkages, **44**, 211
Structure and function of microphytic soil crusts in wildland ecosystems of arid to semiarid regions, **20**, 180
Studies on the cereal ecosystems, **8**, 108
Studies on grassland leafhoppers (Auchenorrhbyncha, Homoptera) and their natural enemies, **11**, 82
Studies on the insect fauna on Scotch Broom *Sarothamnus scoparius* (L.) Wimmer, **5**, 88
Sustained research on stream communities: a model system and the comparative approach, **41**, 175
Systems biology for ecology: from molecules to ecosystems, **43**, 87
The study area at Zackenberg, **40**, 101
Sunflecks and their importance to forest understorey plants, **18**, 1
A synopsis of the pesticide problem, **4**, 75
The temperature dependence of the carbon cycle in aquatic ecosystems, **43**, 267
Temperature and organism size – a biological law for ecotherms? **25**, 1
Terrestrial plant ecology and ^{15}N natural abundance: the present limits to interpretation for uncultivated systems with original data from a Scottish old field, **27**, 133
Theories dealing with the ecology of landbirds on islands, **11**, 329
A theory of gradient analysis, **18**, 271; **34**, 235
Throughfall and stemflow in the forest nutrient cycle, **13**, 57
Tiddalik's travels: the making and remaking of an aboriginal flood myth, **39**, 139
Towards understanding ecosystems, **5**, 1
Trends in the evolution of Baikal amphipods and evolutionary parallels with some marine Malacostracan faunas, **31**, 195
Trophic interactions in population cycles of voles and lemmings: a model-based synthesis **33**, 75
The use of perturbation as a natural experiment: effects of predator introduction on the community structure of zooplanktivorous fish in Lake Victoria, **31**, 553

The use of statistics in phytosociology, **2**, 59
Unanticipated diversity: the discovery and biological exploration of Africa's ancient lakes, **31**, 1
Understanding ecological concepts: the role of laboratory systems, **37**, 1
Understanding the social impacts of floods in Southeastern Australia, **39**, 159
Using fish taphonomy to reconstruct the environment of ancient Lake Shanwang, **31**, 483
Using large-scale data from ringed birds for the investigation of effects of climate change on migrating birds: pitfalls and prospects, **35**, 49
Vegetation, fire and herbivore interactions in heathland, **16**, 87
Vegetational distribution, tree growth and crop success in relation to recent climate change, **7**, 177
Vertebrate predator–prey interactions in a seasonal environment, **40**, 345
Water flow, sediment dynamics and benthic biology, **29**, 155
When ranges collide: evolutionary history, phylogenetic community interactions, global change factors, and range size differentially affect plant productivity, **50**, 293
When microscopic organisms inform general ecological theory, **43**, 45
10 years later: Revisiting priorities for science and society a decade after the millennium ecosystem assessment, **53**, 1
Zackenberg in a circumpolar context, **40**, 499
The zonation of plants in freshwater lakes, **12**, 37.

CPI Antony Rowe
Eastbourne, UK
November 27, 2015